This book is to be returned on or before
the last stamp below

The
Technique
of
Lighting for
Television
and Film

Third Edition

The Technique of Lighting for Television and Film

Third Edition

Gerald Millerson

Focal Press

Focal Press
An imprint of Butterworth-Heinemann
Linacre House, Jordan Hill, Oxford OX2 8DP
A division of Reed Educational and Professional Publishing Ltd

R A member of the Reed Elsevier plc group

OXFORD   JOHANNESBURG   BOSTON
MELBOURNE   NEW DEHLI   SINGAPORE

First published 1972
Reprinted 1974, 1977, 1978
Second edition 1982
Reprinted 1983, 1985, 1986, 1988, 1989
Third edition 1991
Reprinted 1992, 1993, 1994, 1995, 1996

© **Gerald Millerson 1991**

British Library Cataloguing in Publication Data
Millerson, Gerald
    The technique of lighting for television and film – 3rd ed.
    1. Television programmes. Lighting   2. Cinematography. Lighting
    I. Title
    778.59

ISBN 0 240 51299 5

Library of Congress Cataloguing in Publication Data
Millerson, Gerald
    The technique of lighting for television and film/Gerald
    Millerson – 3rd ed.
    p. cm.
    Includes bibliographical references and index
    ISBN 0 240 51299 5
    1. Cinematography. Lighting   2. Television. Lighting
    I. Title   II. Series
    TR899.M48   1991
    778.5′2343–dc20   90–20266

Printed and bound in Great Britain by
The Bath Press, Avon

Contents

Preface to the Third Edition

This is a practical study of the art and craft of lighting for the screen. Whether you are lighting for a single-camera unit on location or a multi-camera studio production, you will find here detailed discussions of the underlying principles and techniques involved.

In its new format, this international sourcebook has been extensively restyled to reflect current developments. It doesn't assume that you have previous knowledge or experience, and has been specially designed to put the exact information you need at your fingertips.

The new text is sectionalized so that it is more adaptable to lighting courses, and personal study. It takes the guesswork out of lighting, and covers all aspects of the craft, from the basic physics of illumination to everyday practicalities.

Here you will find explanations of the mechanics, techniques and aesthetics of lighting treatment. You will become familiar with everyday problems and practical solutions, and the text will remind you of alternative effective methods.

As you will see, lighting techniques need to be flexible, for they must satisfy a wide range of situations, and this book will help you to *understand* rather than merely *imitate* routines.

Skillful lighting involves a subtle blend of systematic mechanics and a sensitive visual imagination. It requires anticipation, perceptiveness, patience and know-how. But learning through practice alone can take a deal of time. This book is a distillation of many years of experience, with advice and guidance that will bring you successful results right from the start.

Your own skills will develop with practice. Scrutinize the play of light and shade in the world about you. Critically assess the lighting treatment in the motion pictures and TV productions you watch. If you have a videotape recorder available, use it to analyze scenes. Critiques of other people's work will guide you towards a personal style and self-confidence.

Above all, be curious. Experiment, and you will derive endless pleasure from the persuasive magic of lighting. This book was written to encourage you to do exactly that!

G.M.

Acknowledgements

It is not possible to thank individually the many colleagues and students with whom I have had the opportunity to freely exchange opinions and ideas over the years. This book, which is a summary of many facets of the art and craft of lighting, reflects the work of many specialists: lighting crews, cameramen, scenic designers, scenic artists, video operators, video engineers, etc. Successful lighting relies on close teamwork and an appreciation of each others' problems.

I would like, in particular, to thank the scenic designers of the imaginative studio settings used for the demonstrations included in this book:

Sally Hulke – *Sinister Street,*
Peter Kindred – *Christ Recrucified,*
Colin Shaw – *Ryan International.*

In addition to my own photographs throughout this book are the fine stills from *Great Expectations* and *Oliver Twist* kindly provided by The Rank Organisation Ltd.

I am also indebted to the British Broadcasting Corporation for the opportunities to use studio facilities, and to colleagues for their invaluable help and encouragement.

G.M.

1 An introduction

Why bother?

Of course we take *light* for granted. It is part of our everyday world; a phenomenon so familiar that we cease to think about it.

When shooting on location there is usually light of some sort around. It may not be ideal, but at least it allows the camera to see what is going on. If whatever you are looking at happens to be in shadow, move it into the light; if the surroundings are insufficiently bright, add illumination of some kind. In the studio, surely all you need is overall lighting from a large bank of lamps. So why all this mystique about lighting techniques? Why does one need to study the obvious?

Most of us start with questions like these. It is certainly puzzling when you find that a straightforward interview has needed a dozen lamps, and then see a particularly impressive lighting effect which was achieved with just a *single lamp*!

Why do we need lighting techniques?

There are a number of reasons why carefully arranged lighting is desirable. Some are self-evident, others are less obvious:

- *Light intensity* To produce high-quality pictures the light level (intensity) needs to be appropriate for the sensitivity of the medium you're using, and the lens aperture (*f*-stop). As we shall see later, if there is too little light, focusing can be very difficult (limited depth of field) and various picture defects become prominent (e.g. picture noise).
- *Lighting contrast* Bright areas and deep shadows can be an embarrassment, for the camera can only handle a limited contrast range. Suitable lighting can often overcome this problem.
- *Inconsistent pictorial effect* A subject may be strongly lit from one camera position and silhouetted from another. It may look less attractive as the angle changes. Careful lighting can correct these differences.
- *Light must suit the production techniques* If, for example, a shadow falls on the subject as the camera moves closer, this will not only be a distraction but may prevent the audience from seeing the subject clearly.
- *Inappropriate illumination* Poorly arranged, light can produce flat, two-dimensional pictures. Effective lighting creates an illusion of depth and solidity; subjects stand out from the background.

15

- *Unsuitable illumination* Badly chosen lighting treatment can make a picture appear artificial and unconvincing. Suitably arranged lighting builds an impression of realism, a completely convincing environment.
- *Light should develop a suitable ambience* The environmental atmosphere and mood should be appropriate for the occasion.
- *Lighting affects a subject's appearance* Depending on the way light falls on a subject, it may appear beautiful, ugly, mysterious, distorted, stark, crude . . .
- *Uncontrolled illumination can cause distracting effects* It can result in shadows, hotspots, reflections, that take the audience's attention. Careful lighting can avoid or suppress these problems.

What can lighting do?

From what we have already seen it is clear that lighting in television and film is about much more than *just making things visible*. Skillful lighting allows you to adjust and manipulate the impression conveyed on the screen. You do this by carefully controlling the strength and quality of the light, by arranging its angle and coverage.

A magical property of light is its flexibility – the ease with which it can be controlled. You can totally transform a situation in an instant at the touch of a switch. Or you can alter it gradually and imperceptibly, without your audience realizing how subtly it is changing. Simply by adjusting the relative intensities of the same group of lamps you can transform a scene's appearance or the prevailing atmosphere.

Lighting potentials

Let's look at typical ways in which lighting allows us to control the picture:

- *Light can reveal form, texture, detail.* We can strongly emphasize features, or light them so that they are barely discernible by adjusting the coverage, quality and direction of the light.
- *Light can conceal.* Carefully arranged shadows can restrict what is visible.
- *Light gradations can suggest surface contouring.* Even where none exists, shading can make a flat surface appear curved or sloping.
- *Light can imply features* that are not really there Light patterns and cast shadows can convincingly suggest features as diverse as distant buildings, water, falling snow, windows, foliage . . .
- *Light can modify our impressions of distance and size in a picture.*
- *Light can adjust the color of subjects.* It can add color to neutral surfaces, and alter color values within the scene.
- *Light can suppress all surface contouring and detail, and reveal only the subject's outline.* Silhouette style.

- *Light can suppress all surface contouring and concentrate on subject detail.* Notan style.
- *Light can emphasize solidity and form.* Chiaroscuro style.
- *Light influences how the audience responds to a picture.* Light can intrigue, mystify, excite . . .
- *Lighting can guide the audience's interest.* It can concentrate their attention and move it from one area to another, gradually or abruptly.
- *Lighting can create compositional relationships for the camera.* By forming and adjusting tonal masses.
- *Light can develop an atmosphere or mood.* Lighting can build up a particular ambience.
- *Light can imply time of day and weather.* It can suggest that a scene is lit by strong sunlight or moonlight, stormy conditions, etc.
- *Light can establish environmental associations.* Light patterns, shadows, can suggest that action is taking place in a prison, forest, church, etc.
- *Light can isolate a subject.* For example, within a spotlight.
- *Light can create visual continuity.* It can join and unify a series of separate subjects.
- *Light can provide visual movement.* Patterns of moving light and shadow, flashing lights, color changes, etc. can create exciting effects.

What is bad lighting?

It could be argued that there is no such thing as *bad* lighting; simply lighting that is *inappropriate* for that particular occasion. If you a lit a newsreader with an overhead light, the skull-like result would be quite unacceptable. So would ghoulish underlighting from a low lamp. However, both treatments could create an excellent effect in a dramatic situation, where you want to shock your audience.

In practice, there are all kinds of 'bad lighting'. It may, for example:

- Distract our attention – e.g. a wrongly cued lighting change that makes us over-aware of production mechanics.
- Produce an ugly or unattractive effect – e.g. the unflattering effect of two nose shadows, black eye sockets, hot tops to heads.
- Prevent us from seeing the subject properly – e.g. one person's shadow falls across the face of another.
- Create an entirely inappropriate ambiance – badly designed lighting can cause an attractive studio setting to look phoney or unattractive. Even a real locale can appear unconvincing if ineptly lit!

Learning to light

There are really three broad approaches you can follow when lighting for the camera.

The first is *trial and error*. You move lamps around until you get acceptable results. This takes a lot of time and patience – and luck. It is possible to learn about lighting this way, and achieve great pictures. But more often they are a disappointing miss.

In the second method one learns *routines*. Lamps are located to illuminate the subject in a regular pattern. Results can be very satisfactory, but there is a sameness about all the pictures. Atmosphere is sacrificed for technique. This mechanical approach teaches you little about the artistic potentials of lighting.

The third method, which we shall follow here, is *creative analysis*. You build up lighting treatment methodically from a real appreciation of how light behaves. Here we shall consider how light modifies and enhances; how it alters the appearance of subjects; how it can develop a mood. You will discover how to manipulate light; how to use light creatively to achieve exactly the effects you are seeking. When problems arise, *creative analysis* enables you to discover and correct them systematically.

From this foretaste of the potentials of *lighting* let us begin our journey into this fascinating field by exploring the nature of light itself.

2 The nature of light

In this chapter we shall be discussing the fundamentals of *light* itself. We shall examine the nature of light, how it behaves, the ways in which it varies, the principles of color.

What is light?

Radiant energy is emitted from many natural and man-made sources, including the sun, radioactive ores, radio transmitters, electric lamps . . . It travels out through space in the form of *electromagnetic radiation* (waves of energy with associated electric and magnetic fields) at the incredible speed of over 186 000 miles a second (3×10^{10} cm/s).

Although we can classify this radiated energy by measuring the *frequency* at which it vibrates (in hertz or cycles per second), with light it is generally more convenient to measure its *wavelength* instead – i.e. the distance between one wave-crest and the next – in *nanometers*. (In a given material the energy's wavelength is inversely proportional to its frequency.)

The characteristic nature of this electromagnetic radiation changes with its rate of vibration or wavelength, and as you can see in Figure 2.1, the complete electromagnetic spectrum covers a vast range of

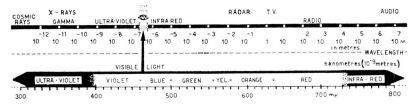

Fig. 2.1 The electromagnetic spectrum
Electromagnetic energy is propagated over an extremely wide frequency range, from almost zero to over 1×10^{22} hertz (cosmic rays). Visible light occupies only a minute segment of this spectrum; the visible spectrum comprising gradual transitions from one hue to the next. Hues are identified by wavelength measurement. Various units are used: nanometers or millimicrons (10^{-9} m), or Angstrom units (10^{-10} m).

diverse phenomena. We are going to take a close look at the small spectral band that our eyes perceive as light – the *visible spectrum*.

White light

When a light source emits energy over most of the visible spectrum we normally see the overall effect as 'white light'. Throughout this

19

Table 2.1   Spectral hues

Each spectral hue occupies a region of the spectrum, e.g.:

Red	Orange	Yellow	Green	Blue	Indigo	Violet
610–720	590–610	570–590	510–550	450–480	430–450	380–430

Wavelength in 10^{-9} meters (nm)
1 nanometer (nm) = 1 millimicron = 0.000001 mm = m^{-9}

The eye can distinguish about 150 separate color differences in the spectrum.

spectrum there is a gradual transition from one color to the next, and it is a matter of opinion as to where each region we call 'red – orange – yellow – green – blue – indigo – violet' actually begins and ends. Those shown in Table 2.1 are typical. The primary colors for NTSC television systems are specified as red = 610 nm; green = 535 nm; blue = 470 nm. (The respective EBU/PAL primaries are 610, 540, 465 nm.)

If only *part* of the spectrum is visible, as in colored light, we judge the resulting color mixture from its most prominent hues. Thus at sunset, we may assess the light as 'orange' although it actually contains a certain amount of red, yellow and green light – but very little blue.

To reveal which colors of the spectrum are present in a light beam we have to pass it through a prism, a spectroscope, or a diffraction grating; which will break the light up into its component *spectral colors/spectral hues*.

Unlike certain animals and insects, our eyes cannot see radiation beyond the visible spectrum, so anything illuminated by infra-red or ultra-violet light remains dark to us; but visible to them. To detect light in these regions we need to use special film, video equipment or measuring instruments.

Light and shade

The screened image

In daily life, stereoscopic vision enables us to assess depth and distance with remarkable accuracy. In addition, there are various secondary visual clues we have learned to associate with distance and dimension. For example –

- *Perspective:* the way size seems to diminish with distance, and parallel lines converge.
- *Parallactic movement:* the relative rates at which planes are displaced as they move across our field of vision.
- *Overlap:* surfaces appear to overlap others that are more distant.
- *Diminishing detail:* detail becomes less visible with distance.

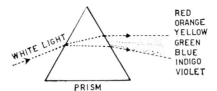

Fig. 2.2 The prism
The speed of light changes with the medium through which it travels. Passing from one substance to another, e.g. air to glass, its path is deflected (refraction). When a narrow light beam is interrupted by a prism, its component colours become refracted to differing degrees, and spread in a spectral band. Regions of this displayed spectrum are strong or weak, according to hue proportions in the analyzed light.

However, there are important differences between our direct experience and the *flat, two-dimensional images* we see on a screen. Watching a picture, we can only see what the cameraman or the director has chosen to show us. We have no free selection, and the shot's restrictions limit our visual information.

We lack direct stereoscopic clues. Instead we have to rely on the secondary clues, and interpret size, distance, depth, by comparing tone, line and color. As a result, we quite often make inaccurate assessments about what we are seeing.

Tonal values within a picture play an important part in the way we interpret it. They not only influence the general mood but modify our impression of dimension and distance. We find, for instance, that in a picture, lighter-toned areas seem larger and more distant than darker ones. The distribution of tone, and the way subjects are arranged within the frame, create compositional effects that we never experience in daily life.

Lighting techniques take advantage of such illusions to manipulate space and distance.

In order to use light creatively we need to look afresh at the visual world, to reconsider many fundamentals that we have been taking for granted all our lives. We shall find that things are not what they seem!

Light intensity

When working with light we are continually reminded of the considerable differences there can be between

● What we *think* we are seeing – i.e. our *subjective* impressions, and
● What is *actually there* – i.e. *objective* reality.

We make comparative judgments. For example, a dark room can be filled with the light from a single candle. But take that candle out into the sunlight, and even its flame looks insignificant.

Judging light levels

Our vision usually needs to adapt to local lighting conditions (high or low intensity) before we can judge tonal variations. At low intensities we are able to detect even slight tonal differences. But at higher and higher light levels it requires progressively greater variations for us to be able to see any increases.

That is because the eye does not perceive differences in light intensities as they really are; its response is inherently *logarithmic*. We cannot really on our unaided eyes to assess light levels when deciding whether there is too much or too little for the camera system. Instead we have to use a light meter of some sort to measure:

- *Incident light:* the intensity of light falling on the subject.
- *Surface brightness:* the amount of light reflected from specific parts of the scene.
- *Integrated reflected light:* the average amount of reflected light reaching the camera.

We shall look at these different methods in detail on page 379.

Lighting terms

Some of the terms used to classify light intensities are quite general, while others are specific measurements. When we speak of *brightness* or *luminosity*, for example, we are describing our *impressions* of the amount of light emitted by a source, transmitted through a medium, or reflected from a surface, not its true value.

As you will see later, these impressions can be very unreliable. One subject can *look* brighter than another simply because it is placed in front of a background of a different tone.

Lightness describes the relative amount of light actually reflected from a surface (particularly colored). The proportion of the incident to reflected light gives us the surface's *reflection coefficient* or *reflectivity*.

When making accurate measurements of the quantity of light emitted from a surface we use the term *luminance*. Doubling the intensity of the illumination produces double the surface luminance. Snow has a high luminance, while black velvet has an extremely low luminance. (Both are color-free neutrals, so have zero saturation.)

Surface brightness

There can be considerable differences between a surface's actual 'tone' and its effective brightness on camera. It is worth our looking at the reasons for this in some detail, for they are the basis on which we make many visual judgments. They directly affect how we interpret pictures.

The nature of the surface

■ *The surface's reflectance*   i.e. the proportion of the incident light it reflects.

As Table 2.3 shows, there are considerable differences in the reflectance values of materials. When light falls on fresh snow, up to 93–97% is reflected. Quite low intensity illumination makes it clearly visible. Black velvet, on the other hand, may only reflect as little as

Table 2.2   Color TV paints reflectance range

	Munsell value	Reflectance (%)	Typical colors
Maximum picture white	8	58	White
			Cream
Light values		51.5	Ochre
	7.5	50	Eau de nil
		48	Golden yellow
		47	French blue
		45	Lime
		45	Pink
		45	Powder blue
	7	43	Light gray
(light skin)		43	Gray-green
		43	Rose
		42	Mustard
		42	Cement
		42	Blue-gray
		41	Light stone
		37	Wedgewood blue
		37	Buff
	6.5	35	Orange
		33	Sea green
Medium values		32	Pastel blue
		31	Jade
	6	30	Cobalt
(dark skin)		29	Lavender
		27	Dove grey
		26	Beige
	5.5	25	Mushroom
		24	Tan
		23	Copper
		22	Vermilion
		22	Lilac
	5	20	Sage
		17	Scarlet
	4.5	15	Dark coffee
Dark values	4	12	Indian red
		11	Maroon
		10	Ultramarine
	3.5	9	Burnt sienna
		8	Prussian blue
		7	Purple
		7	Dark gray
	2	3	Black

Table 2.3   Typical subject reflectance values

Light reflected (%)	Subject
100	Ideal reflector
97–93	Magnesium carbonate, fresh snow
92–87	Polished silver
90	White plaster
88–75	White gloss paint
85	Aluminium foil
80–60	White porcelain, china, white paper
65	Chromium
60–30	White cloth
40–30	Light skin tones
30–20	Cement
30–15	Green leaves
20	Bronzed skin tones
15–10	Brickwork
10–5	Black paper
5–1	Black paint
1	Black cloth
1–0.3	Black velvet

1/100 or 3/1000 of the incident light, so that modeling or surface detail may not be obvious, even under intense lighting.

■ *Its surface finish*   Whether it is rough, smooth, matte, shiny. As you can see in Figure 2.4, a *rough* surface appears equally bright, whatever angle we view it from, for the light is diffusely reflected. But when we look at a *glossy* surface its apparent brightness can vary considerably; depending on our position relative to the incident light. If light is coming from behind us the surface may look fairly dark. On the other hand, if the light is coming *towards* us it will bounce up from the shiny surface and appear extremely bright.

Fig. 2.3   Surface brightness
Apparent surface brightness depends on both surface tone and incident light intensity. Each of these examples would appear equally bright. The surface texture and finish affect how brightness changes with light direction and the viewing angle.

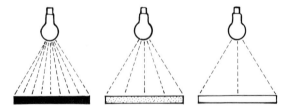

In practical lighting, such glossy surfaces can pose problems, particularly when we see light sources reflected as localized hotspots. It is frustrating to find when you are shooting polished or metallic surfaces that they will often appear quite *dark* on camera; except from certain angles, where the full intensity of the lighting is reflected towards the camera. These strong specular reflections can not only be distracting but can create technical problems for video cameras (burn-in, streaking, comet-tailing).

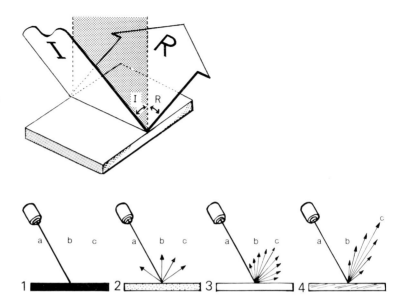

Fig. 2.4 part 1   The basic laws of reflection
The light's reflected angle (R) is always equal to that of the incident light (I). The imaginary 'normal' plane is at right angles to the reflecting surface, where the rays meet. The (I), (R), and 'normal' always lie in the same plane. The reflected image appears laterally reversed, and is as far behind the reflecting surface as the subject is in front of it.

Fig. 2.4 part 2   Surface reflection
1. Complete absorption (e.g. black velvet). Little or no reflection. Surface appears dark from all viewpoints.
2. Diffuse reflection (rough irregular surface). Light scatters in all directions, fairly bright from all viewpoints.
3. Spread reflection (glossy surface). Fairly dark at a; fairly bright at b; bright at c.
4. Specular reflection (shiny surface). Fairly dark at a, b. Very bright at c.

Some wall coverings have a dark background decorated with a pattern of metallic foil. To the eye this can provide an attractive rich-looking appearance. On camera, you are likely to find that the surface looks dull and featureless from most angles. But from certain directions, where the pattern catches the light, it shines out brightly. Consequently, as the camera angle changes the pattern comes and goes.

■ *The angle of the surface*   Unless a surface is *perfectly matte*, its brightness will vary to some extent with the camera's position. When shooting straight onto a surface (i.e. a plane at right angles across the lens axis) it will appear brightest when the light is coming from around the camera position. But if we angle the surface or the light, its brightness gradually falls. The change becomes more rapid as the angle increases, for it follows a *cosine law* (Figure 2.5). As you would expect, this effect is most obvious with smoother surfaces.

■ *The form of the surface*   Although we do not normally give such effects much thought, from our earliest years we soon learn to recognize that this is how light behaves. We subconsciously use such visual clues to interpret form and shape in the flat picture. So where, for example, we see the brightness of a surface gradually falling off towards its edges we 'know' that the surface must be curved. Similarly, where a surface has an even overall tone we interpret it as flat. We accept variations in surface brightness or *shading* as indications of surface contours. Darker areas are depressions, and lighter areas are raised – or at least, that is what we assume as we look at the picture.

In fact, if the Scenic Artist has painted a flat surface that way, or the Lighting Director has deliberately arranged light to give those

Fig. 2.5 part 1   Surface angle – cosine law

When a surface is angled to a light beam, the same amount of light spreads over a larger area, so that it becomes less brightly iluminated.

$$(I) = \frac{Lumens}{(Distance)^2} \times \cos \theta$$

This means that for practical purposes up to 45° illumination is reasonably constant; 45–60° light falls to ½ (1 stop open); 60–70° light falls to ¼ (2 stops open); 70–80° light falls to ⅛ (3 stops open); 80–90° light falls to zero.

Fig. 2.5 part 2   Surface angle – light direction

Surface brightness varies with the angle of the surface and the lighting angle, relative to your viewpoint. Vertical planes look brightest when lit from your viewpoint; horizontal planes, when light shines straight towards you.

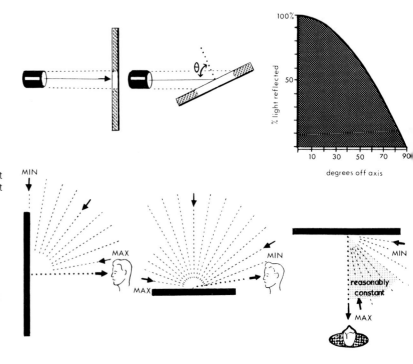

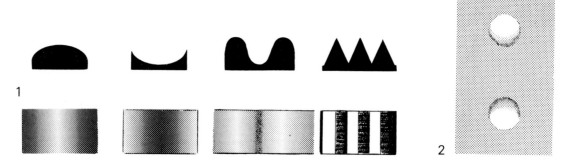

Fig. 2.6   Tonal gradation

1. In a flat picture tonal gradation is usually interpreted as surface contour.
2. Direction of shading, too, can influence interpretation. This shading implies that the surface has a bump and a depression; but invert the page, and the illusion is reversed

impressions, we'll not be able to judge from the screen image exactly what *is* there in front of the camera!

The incident light

■ *The light intensity*   The amount of light falling onto a surface will depend not only on the light output of the source but also on how close that source is to the surface. The intensity of some lighting fixtures falls off more rapidly with distance than others.

Most surfaces around us are lit from several directions. Even where there is apparently a single source such as the sun, there will also be sky light, and light reflected from the ground and other nearby surfaces.

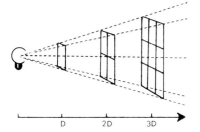

Fig. 2.7   Inverse square law
With increased distance, the light emitted from a given part of a point source will fall rapidly, as it spreads over a progressively larger area. This fall-off in light level is inversely proportional to the distance squared, i.e $1/d^2$. Thus, doubling the lamp distance would reduce the light to $\frac{1}{4}$. A lens at *f*/6.3 would need to open to *f*/3.2 to compensate. This law is only strictly true for the uniformly diverging rays from a point source, unmodified by a lens or reflector. Nor does it apply to parallel light rays, which ideally do not decrease in intensity with distance.

It is easy to overlook how often *shadowing* cuts off illumination from one direction or another, altering the apparent surface tones, and increasing local tonal contrast.

■ *The clarity of the surface*   Atmospheric haze, smoke, mist can considerably alter tonal values in the scene and reduce contrast. We're all familiar with the way distant landscape seems nearer or further away under different atmospheric conditions.

When a surface is not sharply focused, its apparent surface brightness changes. With smaller areas in particular, tonal gradation is reduced or even lost. Patterned tones tend to average out to an overall value. (See *Detail*.)

■ *The color of the incident light*   As we shall see shortly (Figure 2.15), the color of the incident light can modify the apparent brightness of a surface.

■ *Subjective effects*   The tone and color of surrounding areas can have a considerable effect on how bright a surface appears to be. Where a surface area is small in proportion to surrounding areas (or seems to be, because it is distant), this subject effect can be very strong.

■ *Polarizing effects*   Reflected light can alter our impressions of surface brightness; particularly when shooting shiny, translucent or transparent surfaces. (See *Polarization*.)

Gray scale (achromatic values)

If a surface absorbs all or most of the light falling onto it we judge it to be black. The less light absorbed, the greater the proportion of the illumination reflected, and the lighter that surface appears to be. So we can build up a progressive scale of brightnesses – from black, through dark grays to mid- and light grays, up to white.

Under suitable conditions our eyes can distinguish anything from fifty to a hundred different *tones* between black and white. But our judgment is influenced by the area of these tones, and the prevailing light levels. A mid-gray surface can appear to be white when strongly lit; dark gray or even black when left unlit.

There are a number of technical processes in which it is extremely useful to have a *continuous tonal scale* showing progressive changes from black to white. A *tonal wedge* of this kind can be used to check the tonal reproduction of a film emulsion. Placed in front of a TV camera or a video film scanner (telecine), it will show at a glance the system's response over the entire tonal range – its *transfer characteristic*. (The resultant video signal has a 'sawtooth' shape.)

Because one tone merges imperceptibly into the next on this type of tonal scale it is not easy to refer to specific values. When we want to do so, it is often more convenient to use a *step wedge* or *gray scale*. Here tones are selected at regular intervals throughout the range, so

Fig. 2.8   The gray scale

The standard gray scale takes the maximum-to minimum tonal range readily reproduced by the TV system (overall 20:1 contrast), and divides it into 10 steps. (Relative white reflectance 60.3%.) Each step is $\sqrt{2}$ times brightness of the next; the logarithmic scale looking linear to the eye. (What appears mid-gray is only 17½% reflectance; not the 50% you might expect.)

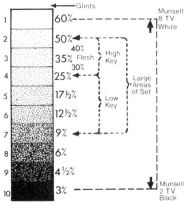

Fig. 2.9   Achromatic values

A continuous tonal wedge displays a progressive series of tonal values between black and white. A step wedge takes regular tonal intervals, and forms tonal blocks, each proportionally lighter than the one below.

(1) This tonal scale can be calibrated for reference purposes. Too many steps (2) would make individual selection and matching difficult.

Too few steps (3) would provide only a coarse indication of a system's reproduction of tonal gradation. Where a system cannot reproduce the entire range (4) tonal extremes become merged.

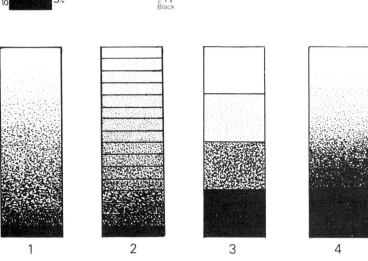

1   2   3   4

that each step looks proportionally lighter than the previous one. Slight steps can be difficult to distinguish, for a tone needs to be some 2% brighter than its neighbor (in daylight) for us to detect any difference at all.

The actual number of steps chosen for a gray scale will depend on its purpose. For many applications, 10 steps will provide a good indication of a system's performance. Five may suffice. For greater precision, 20 or more steps may be needed.

In television, a 10-step gray scale shown in Figure 2.8 has been widely used for many years. (The earlier limits of Munsell 2.5 to 8.5 were modified to 2 and 8 with the advent of color TV.) It can be used to check the performance of all video equipment. (The video signal from a gray scale appears as a series of well-defined steps in a 'staircase waveform'.)

Tonal contrast and tonal gradation

Picture quality is very dependent on how effectively we can reproduce the various tones in the original scene. There are two aspects of tonal reproduction we are interested in here – tonal contrast and tonal gradation.

■ *Tonal contrast (subject brightness range)*   Here we are comparing the values of tones. We may be judging the difference between any two tones within the scene; e.g. the contrast between the tones of a person's face, and their background. Or the difference between tonal extremes; e.g. the lightest and darkest tones in the scene or the picture.

Where the subject itself has highly contrasting tones the picture will have a bold, clear-cut, well-defined look. There may be few subtle half-tones.

If tonal contrasts are slight, there will be little differentiation between planes, and they may appear to merge. The picture is likely to look flat, and lack dynamism.

Fig. 2.10   Tonal contrast
The contrast range (brightness range) of this tonal scale is shown (*left*). In any tonal comparison (*right*) we assess relative contrast ratios. One may be three times the lightness of another; a 3:1 contrast ratio.

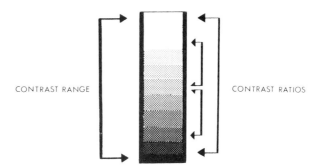

CONTRAST RANGE   CONTRAST RATIOS

■ *Tonal gradation*   The clarity with which the system reproduces different tonal values. In a picture with a wide range of half-tones we can discern even slight modeling, surface form, texture. But if a picture with subtle half-tones is reproduced through a high-contrast system (e.g. a TV receiver with exaggerated contrast setting), these slight tonal variations merge, and modeling is lost.

Contrast range/subject brightness range

All picture-reproducing systems have their inherent limitations. Some can only reproduce tones accurately over a very restricted contrast range, while others offer a much wider tonal coverage.

If subject tones are comparatively limited there is no problem. But if a scene contains a considerable tonal range we can only expect to reproduce a restricted part of it reasonably accurately. We can't, for example, hope to reproduce tonal gradation in both the snow on the mountain *and* the deepest shadows beneath the forest trees. The contrast range is too great for the system. Instead, we must select the tones we are most interested in, and adjust the *exposure* to suit them, and/or illuminate the shadows to reduce the overall contrast.

The *contrast range* that can be reproduced varies at different stages of any process. For instance, whereas a *film negative* can register contrast ratios of up to 200:1 (i.e. where the brightest areas are 200 times as bright as the darkest), a glossy bromide print may reach only about 60:1. Reproduced in a book, the contrast ratio of the same shot may be as little as 8:1 or less. A projected transparency, on the other hand, can reproduce a range of 160:1. While a black and white motion picture can provide a reproducible contrast range of 100:1 it is safer when working with color film to assume limits of around 30:1, with maximum contrast between adjacent areas of 35:1 to 40:1. For the TV screen maximum contrasts are typically around 10:1 to 20:1 for adjacent tones and 30:1 on widely spaced areas.

The contrast range from black to white tells us only part of the story. It does not reveal how well half-tones are reproduced between these extremes.

A number of factors affect the system's performance, and reduce the usable contrast range. At the camera, lens flares (internal reflections from light shining into the lens), dust on the lens, strong ultra-violet light, haze, can all reduce image contrast.

If you adjust the controls of a TV receiver to obtain bright, strongly contrasted pictures, tonal reproduction becomes coarsened, gradation is lost in the lightest tones and the shadows, defocusing and other visual defects develop (e.g. 'doming' on picture tubes produces spurious color patches).

Tonal range and pictorial effect

Whether a reduced contrast range and tonal restrictions really matter depends on the nature of your picture, and its purpose. If it is important to show a wide range of tones clearly and accurately (e.g. for a medical photograph) any tonal distortions may be an embarrassment. If a picture relies for its effect on subtle half-tones, the shot may lose much of its appeal when tonal values are degraded.

Where, for example, we are mainly concerned with tonal gradation in the lightest areas of the scene, and care little about darker or intermediate tonal values, the fact that the overall tonal range is limited may be quite unimportant.

Paradoxically, there are many occasions when we deliberately falsify tones; exaggerating or reducing tonal contrast, lightening or darkening tones to improve the dramatic effect of the picture. Even where the original scene is not particularly contrasty, we can create this effect by lighting treatment, processing, or video adjustments. But more of that later, when we discuss picture control (Chapter 13).

Tonal values and picture impact

The impact a picture makes can be strongly influenced by the proportions and range of its tones. In general, we find that darker tones tend to give a shot visual weight, and have a considerable effect on pictorial balance. Light tones create emphasis, adding vitality and crispness. Graded tone blends areas together, guiding the eye from darker to lighter parts.

Bold, well-defined, high-contrast effects can be arresting, vigorous, powerful. They can convey a strong illusion of depth. But inappropriately used, high contrast can appear harsh, unsympathetic, even crude.

Where the predominating tones are mid to light gray with little or no shadow detail, the *high-key* effect can be light, delicate, beautiful. At worst, it may appear dull, lacking in vitality, flat.

Where mid to lowest tones predominate, with few highlight details, the *low-key* effect can be dramatic and mysterious. Overdone, though, the result can become depressing and difficult to see.

Detail

What has *detail* to do with lighting? Well, as you will see, if detail is unclear due to poor system performance or bad focusing ('soft pictures'), picture clarity suffers. Subtle modeling is lost in face tones, subjects tend to merge with their backgrounds, pattern tones deteriorate. And these issues directly affect our lighting techniques.

Wherever a picture contains tiny adjacent areas of differing tone, we interpret this as *detail*; whether it results from small features close to the camera or larger ones some distance away.

Even where a subject has quite clear-cut borders or edges (boundary transitions), it may be quite difficult to distinguish it from the surroundings, unless we can see strong tonal contrasts. For example, a pattern consisting of a series of light gray stripes against a white background, or dark gray on a black background, can contain as much actual detail as a black-on-white design, but they will be far less visible. Low contrast not only makes it harder to focus accurately on detail, but such detail soon merges with distance, or as focus softens.

Sharpness of detail can affect picture quality in two ways:

● When fine patterns are defocused they average out to a mid tone. Narrow black and white stripes appear gray; close color patterns mix to a new hue.
● Tonal gradation is diminished, and subtle contouring is lost. You can prove this by shooting a tonal wedge and defocusing the camera while watching a picture monitor, and a video waveform monitor.

These effects are particularly noticeable when the depth of field is restricted (e.g. at large lens apertures – $f/1.9$). Because backgrounds appear sharply focused in some shorts and defocused in others, there are subjective changes in brightness and color as shots are intercut.

The finest detail a process can resolve – the maximum *resolution* or *definition* – varies considerably between systems. In photographic processes the resolution of detail is measured in lines or cycles per millimeter. The system's overall resolution will depend on the performance of the optics, camera mechanics and the film emulsion's properties, and may typically be around 20–70 lines per millimeter. A color slide can resolve about 30 lines per millimeter.

In television the smallest detail that can be resolved is determined by the system's line standards (525 horizontal lines, 625 lines), for this decides how rapidly the system can change from one tonal level to another during the television scanning process (maximum video bandwidth).

TV resolution is usually quoted relative to the number of lines that can be discerned *across* the picture. In a 525-line system the maximum practical resolution is around 334/340 lines (4.18 MHz bandwidth); although a studio monitor may be able to display up to 640 lines (8 MHz). A 625-line system can resolve up to 430 lines horizontally (5.5 MHz), e.g. 25–40 lines/mm. Typical TV receivers may only display around 300 lines (3.75 MHz), although newer high-definition receivers can resolve over 560 lines. In practice, the maximum detail visible will also be influenced by the screen size, viewing distance, and receiver adjustment.

Color

Light is *invisible*. We don't see light; we see the *effect* of light as it falls onto material surfaces. (The rays of light in a smoke-laden or dusty atmosphere are in fact reflections from the particles there.) Whether

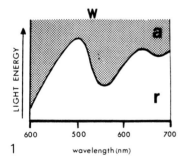

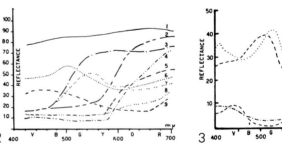

Fig. 2.11   Colored surfaces
1. (W) White light falls upon a surface. (a) Light is absorbed by the surface. (r) Remaining light is reflected and recognized as surface color.
2. Surfaces are seldom colored in pure spectral hues. Most colors spread over a wide portion of the spectrum. According to the relative proportions of hues, we interpret subjects as having a particular color: (1) White card; (2) Orange; (3) Canary yellow; (4) Red rose petal; (5) Signal red; (6) Pale blue; (7) Light green; (8) Maroon; (9) Sky blue.
3. Sometimes colors that match under one light (e.g. daylight) may appear dissimilar under another (*metameric*) due to marked differences in extreme red and blue content. Thus in (A) two green plastic materials differ and in (B) two blue-green paints differ in spectrophotometric curves.

the light itself appears to be 'white' or has a distinct hue will affect the apparent color of the illuminated surfaces.

When illumination falls onto a surface a certain amount of the overall light spectrum is *selectively absorbed* (depending on its molecular structure), and is lost within it in the form of heat. The rest is reflected, and it is this remaining light we interpret as the color of that surface.

Where light passes through transparent or translucent materials (e.g. color medium) part of the spectrum is absorbed, and the remaining *transmitted light* we recognize as its color.★

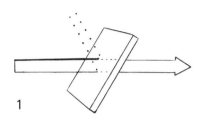 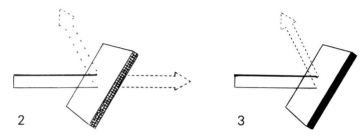

Fig. 2.12   Transmission of materials
1. Transparent medium (specular transmission). As light meets a transparent material, some is reflected, a little is absorbed, most passes through. A colored medium transmits its own color and absorbs all others. The proportion passed is its transmission factor (transmittance). Transmission of 100% equals a transparency of 1. A glass–air surface transmits about 96–90% of the incident light.
2. Translucent medium (diffuse transmission). Translucent material permits some light to pass, but prevents clear visibility through the medium (e.g. frosted glass). Opacity denotes the proportion of light passed by a material. If ⅓ passes, its opacity is 3; transparency = 1/opacity; density = log opacity.
3. Opaque medium. No light passes through an opaque medium. The amount reflected depends upon surface absorption, color, texture, lighting angle, etc.

Figure 2.11 shows how the light reflected from various materials spreads over much of the spectrum. This graph can tell us a lot about the surface color. If the reflectance curve is relatively flat it includes similar proportions of all hues, and so is comparatively *neutral*. Where the curve peaks, that hue is prominent. Where it dips, there

★Further color sources:

Interference   Light reflected from inner and outer surfaces of thin films mutually adds/subtracts, creating spectral colors (oil films, bubbles, 'Newton's rings').
Diffraction   Spectral colors arising from fine, closely scribed grooves (iridescence on beetles, diffraction gratings).
Dispersion   Spectral light arising from 'prismatic dispersion' due to differences in the refraction of transparent media to light of different wavelengths (rainbows, prisms).

Certain natural sources do not produce effective results on camera:

Fluorescence   Materials absorbing energy at one wavelength, and re-radiating it at another while energized.
Photo-luminescence   Materials excited into luminescence by electric fields.
Phosphorescence   Materials that continue to glow after their excitation source is removed. (Similarly, Electroluminescence, Cathodoluminescence, and Catho-dophosphorescence are generally successful.)

Fig. 2.13 Saturation
As a pure hue becomes admixed with white light it displays a less-defined chromatic character. A 'white' surface reflects all spectral colors; a 'pink' surface predominantly reflects red light, together with a proportion of all others; a 'red' surface reflects only a limited spectral range.

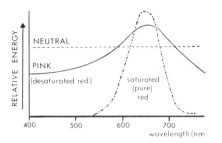

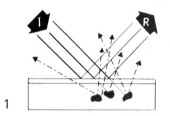

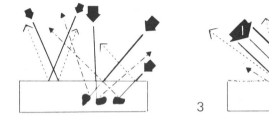

Fig. 2.14 Saturation and surface
A glossy material comprises a smooth transparent surface, beneath which lie randomly shaped colored pigments.
1. Under strongly directional lighting, reflection from the smooth surface layer is specular (of luminant hue) mainly in direction (R) with little absorption. Light reflected from sub-surface irregular pigments is scattered in all directions. From these, strong coloration is seen.
2. Under diffuse lighting, the multi-directional incident white light produces countless minute surface speculars, diluting the body color reflections from all angles, so desaturating the total effective hue.
3. Matte surface reflection. Saturation changes are less with irregular surfaces, as their diffuse reflection creates desaturation under most lighting conditions. Surface smoothing (e.g. with water, varnish, oil) increases effective saturation.

is much less of that particular part of the spectrum. A curve rising at the blue end of the spectrum indicates a cold color; while a warm color would include a greater proportion of the red end of the scale.

If the curve shows just one pronounced localized peak we are probably looking at a saturated, brilliant color. Curves for pastel shades have more gradual changes, and cover a greater amount of the visible spectrum. For a light-toned surface the overall curve will be set high up on the graph, while the curve for darker colors will be low.

Real-life surfaces reflect a surprisingly wide range of hues in addition to those we think of as their 'actual color'. Green grass, for instance, reflects colors throughout the spectrum as well as this *dominant wavelength*, green. Change the proportions, and our impression of the kind of green changes. Change them further, and it may now appear blue.

Natural incident light has only limited color variations; and these principally arise from *light scatter* due to water and dust in the atmosphere – hence blue skies and vivid sunsets. Strong light

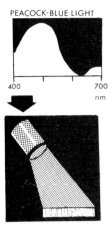

Fig. 2.15 Colored light on colored surfaces
When colored light falls upon a colored surface the resultant effect arises from their combined spectral curves.

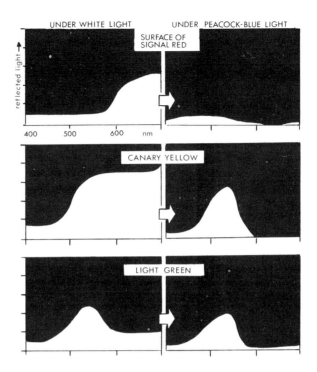

reflected from nearby surfaces can become colored through their selective absorption, so that faces are liable to become lit with sweater-yellow, foliage-green or automobile-red. In close shots, where we can see the face without knowing the origin of the spurious colorations, the strange impact can be quite disturbing.

Under white light, surface tones and colors are reproduced more or less accurately. But when colored light falls onto a colored surface the final effect depends on their relative spectral coverages. Under red illumination, for instance, a green-painted surface will appear very dark, for it absorbs all hues except green. Under green lighting it would appear to be light-toned, as it reflects most green illumination.

Assessing color

Earlier we met terms that help us to describe the tone of a surface. Further terms help us to specify and measure its *color*:

● *Hue*. This is the predominant sensation of color; i.e. whether it is red, blue, green, etc. This normally corresponds to a narrow spectral band or *dominant wavelength*. (Where a surface is entirely lacking color, it is *neutral* or *achromatic*.)
● *Saturation*. Also called *chroma*, *intensity* or *purity*, this is the extent to which a color has been 'diluted' (paled or grayed-off) by the

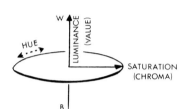

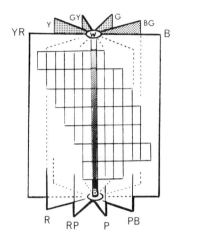

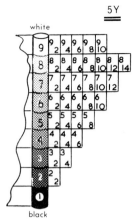

Fig. 2.16 Munsell system of colour notation

This classifies a wide range of hues for varying degrees of saturation. On a vertical gray-scale of eleven equidistant brightness steps Step 10 = white (100% reflectance), Step 0 = black (0% reflectance), Steps 1–9 being those used in practice.

A series of evenly spaced 'pages' pivot around this gray-scale, each analyzing a slightly different hue, in a series of sample 'chips'. From each gray-scale value on the inner margin pivot a horizontal branch develops with evenly numbered steps of progressively increasing color purity, to a maximum saturation (chroma) for that value.

Fullest saturation for different hues is achievable at different values; e.g. yellow reaches chroma 12 at value 8. Blue-purple reaches chroma 12 at value 3. A blue-green may never become more saturated than chroma 6.

Between each of the ten principal hues (each scaled as '5') lie sets of four subdivisions (2.5, 5, 7.5, 10) giving a total of 40 constant hue charts.

To specify a color, quote: hue deviation; value (luminance); chroma (saturation) steps. E.g. 7.5 G 7/10. (A slightly bluish green of light, strongly saturated character.)

addition of white light. As you can see in Figure 2.16, 100% saturation represents the pure undiluted color.

There are several general terms:

Tint   A hue diluted with white.
Shade   A hue mixed with black.
Tone   A grayed white.

Although we speak of colors being brilliant, pastel, deep, pale, vivid, in practice these terms are very subjective, for they are combined effects that arise from both brightness and saturation.

We can distinguish about 150 separate color differences in the spectrum, but our color judgment is often very arbitrary. Apart from the fact that color memory varies considerably between individuals, and one person's 'scarlet' is another's 'vermilion', our impressions of color can be very subjective.

It is often hard to judge whether a particular effect is caused by brightness differences alone, or is due to saturation changes as well. A deep blue sky, for example, seems to be the result of high

Table 2.4   Color terms

Achromatic values (gray scale)	Progressive scale of brightness (luminance), from black through grays to white. Usually expressed as steps in a reflectance scale.
Brightness	Our *impression* of the amount of light received from a surface. The term is often used to indicate *luminosity* (US).
Chroma	See *Saturation*. (Term used in the Munsell system.)
Complementary colors	In *light*: two colors which, when added together, produce white light; e.g. blue and yellow; cyan and red; magenta and green. In *pigment or dye*: complementary colors produce black.
Hue	The predominant sensation of color; i.e. red, yellow, orange, etc. (Term used in the Munsell system.)
Lightness	Our impression of the brightness of surface colors.
Luminance	The *true* measured brightness of a surface. Snow has high luminance (reflecting 93–97% of the light falling on it). Black velvet has a low luminance (reflecting 1.0–0.3% of the light falling on it).
Luminosity	Brightness. Our impression of the amount of light received from a light source, or reflected from a surface.
Minus colors	If the visible spectrum is filtered, to hold back one primary color (e.g. red), the result is a 'minus red' mixture. 'Minus red' = blue + green – i.e. cyan. Similarly, minus green is magenta (red + blue); minus blue is yellow (red + green).
Monochrome	Generally refers to 'black-and-white' (achromatic) reproduction. Strictly means varying brightness of *any* hue.
Primary colors	Three colors (usually red, green, and blue) which, when mixed in correct proportions, can produce any other color of the spectrum.
Saturation (chroma, intensity, purity)	The extent to which a color has been diluted with white (paled-out). If a hue is pure and undiluted, it has 100% saturation. As it becomes diluted, its saturation falls; e.g. 100% red desaturates to pink (e.g. 15%).
Secondary colors	In light, the hues resulting from the additive mixture of a pair of primary colors. (In printing, the orange-red, green, and purple-blue colors produced when pairs of *subtractive primary* colored inks are superimposed.)
Shade	A hue mixed with black.
Spectral colors	The series of color bands seen when white light is diffracted by a lens, prism, or diffraction grating. Merging continuously from red (longest wavelength; lowest frequency) through orange, yellow, green, blue, indigo, violet (shortest wavelength; highest frequency).
Tint	A hue diluted with white; i.e. a desaturated color.
Value	Subjective brightness. (Term used in the Munsell system.)

saturation (i.e. pure color, but little luminance). However, analysis reveals that this effect is actually caused by low luminance (i.e. reduced brightness). People often describe colors as 'bright' (i.e. high luminance) when in reality these are highly saturated (pure hues). In other words, it is a 'color effect' and not a 'brightness' phenomenon at all.

Color mixtures

Trichromatic color mixture

Most colors can be matched by suitable mixtures of a set of three *primary* colors. The actual primaries chosen affect the range and accuracy of possible matching, but all hues, with the exception of a few of high purity (i.e. fully saturated), can be reconstituted. This principle underlies all modern color film and television processes.

There are two fundamental systems of color mixing (see Figure 2.17):

■ *Additive mixing* For colored *light*, where the sensation of one color adds to another's to produce a new color-light mixture. The light primaries invariably used are red, green and blue.

■ *Subtractive mixing* This takes place whenever the resultant color effect arises from selective absorption processes. When light impinges on a surface the eye sees only the remaining light reflected, after this subtraction process. This is so, whether the surface color is painted, dyed, inked or pigmented (Figure 2.18). If the absorption was total, we would see nothing, or little, it would look white. Equal absorption over the whole spectrum results in 'graying' to different degrees.

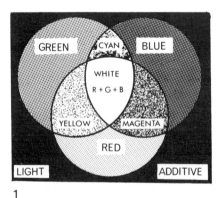

1

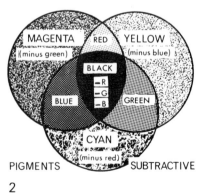

2

Fig. 2.17 Color mixing
1. *Additive mixing.* When lights of differing hues are mixed they add their respective parts of the spectrum to produce a new combination color.
2. *Subtractive mixing.* When pigments of differing hues are mixed they each absorb their respective parts of the spectrum from the incident light.

Any process is subtractive, whenever it involves a filtering action. Hence, when we shoot through color filters subtractive mixing takes place. When colored medium is placed over lamps its function is again subtractive relative to the source light, as it absorbs selectively; although whenever colored light beams are merged, the effect is additive at that point.

For pigments, the mixing together of separate colors juxtaposes a series of filtering particles, each subtracting its own corresponding region of the spectrum. If between them they absorb all colors, the result is a black surface. The corresponding primaries are yellow, magenta (a bluish-red) and cyan (a greenish-blue). Less familiar, perhaps, than our accustomed pre-blended watercolour paintbox colors, they provide the basic components offering a wider color mixture and brightness range. They form the basis (supported by black to provide greater density) of much commercial color printing.

Magenta is a mixture of the extreme red and blue limits of the spectrum, and so does not itself appear within the true spectrum. It is, therefore, termed *non-spectral*. Certain other apparently non-spectral colors are, in fact, spectral hues of low brightness set in much brighter surroundings. Thus, gray is low-intensity white. Brown is a dim orange or yellow. Others, like pink, are really desaturated versions of a spectral range (Figure 2.13).

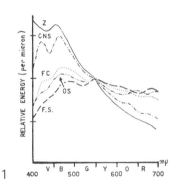

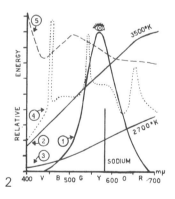

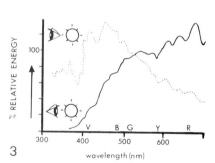

Fig. 2.18 White light
Many luminants are accepted as 'white' light although their actual spectral ranges differ considerably.
1. Variations in color quality of daylight (Northern Hemisphere): *Z* Clear zenith sky; *CNS* clear northern sky; *FC* Full sun with clear sky; *OS* Overcast sky; *FS* Full, direct sun.
2. Variations in typical luminants compared with the response of the eye. (1) The eye's response. (2) High color temperature incandescent lamp (e.g. overrun lamp, studio-type lamps). (3) Low color temperature incandescent lamp (e.g. domestic 40–75 W type). (4) Fluorescent lamp, showing peak in spectrum lines (from mercury vapor filling) within smooth spectrum from fluorescent coating. (5) Carbon-arc projector.
3. The color quality of light can vary too, with our position relative to the sun.

Color effect

It is not always obvious whether a particular effect is additive or subtractive. If you are looking at paper covered with yellow paint or ink, the color comes from a subtractive effect. But supposing, instead, it shows a close pattern of tiny green and red dots printed on it. These patterns merge and their colors blend to be seen as an additive color mixture.

Color films use subtractive principles, as they combine layers of color dye material; an inherently efficient process as color mixing takes place over the entire surface area.

Color television uses the additive process, the screen being subdivided into a permanent pattern of red, green and blue phosphor dots or stripes.

Color specification

For many everyday purposes we can identify colours by names with more or less widespread acceptance: 'scarlet, yellow, ochre, ultramarine'. But fancies like 'tartan green' or 'pixie brown' can only serve to remind the initiated. Sample cards are useful for general applications (as for paint-matching), particularly when accompanied by *spectral analysis curves* as in Figure 2.15. Of these matching-chip methods, the most sophisticated and widely used notation is the *Munsell system* (Figure 2.16).

In *trichromatic matching* color samples are matched to a light mixture derived from standard red, green and blue sources. So it becomes possible to tabulate the proportions of these known standards and from these derive chromaticity coordinates, from which the original color can be reconstituted.

Tristimulus primaries

If the three color-light primaries are set graphically at the corners of a *color triangle*, within its bounds all color-light mixtures can be plotted. Its centre of gravity represents *equal-energy white light* ($y = 0.33$ and $x = 0.33$).

By quoting the x and y coordinates we specify any color as $R + G + B = 1$ (i.e. unity). Exact color matches are achievable for hues lying within the triangle. But what if our required color falls outside these limits?

Then it will still be possible to achieve a match in color and luminosity, but only by using a higher proportion of all *three* of the primaries. Because $R + G + B =$ white, any color mixture using all the primaries must be proportionally paler (desaturated) than a mixture using only two of them.

Figure 2.17 reveals that not only can white light be derived from the *tristimulus* primaries but equally well by appropriately blending

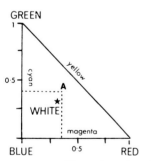

Fig. 2.19 The color triangle
The entire hue range obtainable with three given primaries can be plotted if we know the strength of any two. Using trichromatic primaries (T), i.e. primaries equally proportioned to add to unity, equal energy white is derived. (A color that is 0.4 green, 0.3 red, (0.25 blue) is shown at A.)

any complementary colors: magenta and green, or cyan and red, or blue and yellow light.

When any two colors are blended their resultant mixture will be found to lie on the line joining these two color points. Thus mixtures of red and blue light would be represented as points along the line $R - B$. No green light being present in these mixtures, the line $R - B$ can therefore be defined as a parameter in G = zero.

CIE chromaticity diagram

The international CIE system (Commission Internationale de l'Eclairage) derives from the color triangle a more precise reference system enabling any hue to be specified on its *chromaticity diagram* (Figure 2.20). Here three fictitiously supersaturated primaries are employed, the spectral hues being ranged along the 'horseshoe' boundary within. As any primary we could possibly use must necessarily be less pure, it and its mixtures must lie within these confines.

The CIE primaries are Red 700 nm, Green 546 nm, Blue 436 nm.

White light and color balance

Let us turn now to one of the most subtle aspects of color – the *color quality* of light itself. Although we might expect all 'white light' to contain similar proportions of the entire visible spectrum, in practice, such *equal energy sources* are rare. Because eye and brain adapt and interpret, we arbitrarily accept a remarkable range of

Fig. 2.20 Chromaticity diagram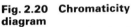

All visible hues can be classified as points falling within the curved spectral locus. Color TV primaries (NTSC) are plotted here as RGB, and reproduced TV colors lie within the triangle (gamut) formed. Subject hues falling outside the gamut are inaccurately reproduced, usually as desaturated. (Pure greens and cyans are rare in nature.)

Further gamuts could be drawn for ranges of printing inks, pigments, dyes, etc., although their shapes would be less regular. An ideal gamut is described by coordinates joining 400, 520, 700 nm points on the spectral locus.

A line from a white point to a spectral hue depicts all the possible saturation values of that hue. If two hue points are joined, any proportional blend will fall upon this line. Standard white luminants A, B, C are shown. (The purple boundary from 400 to 700 nm represents non-spectral hues. Luminance corresponds to a plane erected vertically from the page.)

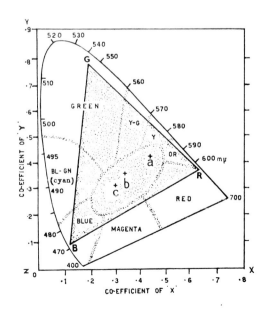

illumination as 'white light' that actually has widely differing proportions of various spectral colors. Many 'white' light sources are very lacking at the blue or red end of the spectrum, as you can see in Figures 2.18 and 2.19.

The camera does not have this property of automatic adaptation. So unless the color quality of the light and the camera system's color balance are reasonably matched there will be *discrepancies in the reproduced color values*.

How critical color matching proves to be in practice depends on whether it is important to reproduce color accurately. For many purposes, as long as the result looks reasonably realistic it is acceptable, and true color fidelity is not essential. But most practitioners aim to at least keep color quality *consistent*, even if it is not strictly correct. To enable them to do so, the color quality of the light can be measured, and expressed as a *color temperature* in 'Kelvin units' (K), as we shall discuss later in Chapter 11.

It is important to maintain good color continuity between shots taken at different times. If color values in a scene change on cutting to a different viewpoint the discrepancy can be very disturbing.

Very occasionally, we may *deliberately* introduce errors in color quality. If, for example, we want to create a warm cosy atmosphere, setting the system to a slightly higher color temperature than is strictly correct will produce that result. Conversely, a lower color temperature than normal will provide a cold bleak atmosphere.

Color filters

Color filters have several applications in lighting techniques. They are used as:

● *Color-balancing filters/conversion filters* fixed over a light source to correct or improve the match between the color temperature of the illumination and the color balance of the system, and
● *Color-enhancement filters* used to produce colored light for 'effects lighting' (e.g. simulated moonlight, firelight), and decorative lighting (e.g. colored light on cycloramas).

We shall be discussing the important subject of *color filters* in detail in Chapter 12.

Polarization

As we see in Figure 2.21, ordinary light travels in a wave motion, vibrating in all directions in a plane at right angles to its direction of travel. You may find this easier to imagine as a center-bored disc of rhythmically fluctuating size, moving alone a line which represents its travel-route. As in Figure 2.21, part 1, any polarizing filter placed in the light path will obstruct most of these fluctuations and leave it

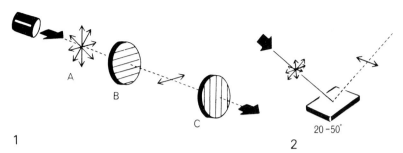

1   2

Fig. 2.21 Polarization

1. (A) Light waves vibrate in all directions in a plane at right angles to their travel path. (B) Certain transparent materials contain minute crystalline structures that admit light vibrations of only one main direction. Other directional vibrations are suppressed, so light intensity is reduced. Now the emergent light is 'polarized', according to the filter's position. (C) This plane-polarized light would pass through another similarly positioned filter. But when either filter is rotated, the polarized light is progressively suppressed, all the original light's vibrations becoming blocked when the filter planes are at right angles (crossed axes).
2. Any light falling onto a shiny surface at a critical angle (20–50°), can become polarized. (Less completely at other angles.) Being plane-polarized it can be largely suppressed by a polarizing filter. From matte surfaces the dispersed light is normally unpolarized.

polarized in a particular plane. In our analogy the fluctuating disc has been reduced to a single directional motion – e.g. is now only rhythmically fluctuating horizontally, or 'horizontally polarized'.

The polarizing filter has several practical applications:

● First, it can help to clear away or reduce unwanted reflections in glass or water, that prevent our seeing objects beyond their surfaces.
● Second, it can control flares, specular reflections and hotspots on glossy surfaces.
● Third, it provides a means of darkening sky tones.

All this the polarizing filter can do without substantially modifying color quality although there will be an overall light loss of $\frac{1}{2}$–$\frac{2}{3}$.

Polarizing screens have the property of being able to differentiate between light directions, thanks to the minute crystalline structures within the acetate sheet material from which they are formed. (To the eye this appears like gray neutral density filter.) If you place such a filter over your camera lens, only light reflected (polarized) over a restricted range of angles will pass through and be seen in the picture. Light reflected at other angles will be suppressed, so that those surfaces reproduce darker.

The polarization of light takes place naturally in everyday life, whenever light strikes glossy and polished non-metallic surfaces, glossy paint, glass, shiny leaves, calm water, etc., at around 20–50°. The actual angle varies with the refractive index of the material. Which reflections a polarizing filter eliminates depends on the angle they happen to make relative to the filter's position. Rotate the filter and the particular light reflections it suppresses alter. But this polarization is only partial, still leaving some radiation in other directions. Consequently we cannot expect under natural lighting conditions to suppress the entire reflection as with light polarized in one direction only. It may be necessary to experiment, rotating the filter for maximum effect because even polarized light can become modified on striking uneven surfaces. Though effective from one position, filtering may prove less successful from another. Nevertheless, for many purposes we can achieve an improvement that is not feasible by other means.

For more controlled situations, as when shooting subjects which are behind glass, or subjects with strong reflections, the most suitable approach may actually be to use a polarizing filter in front of lights, with a polarized lens-filter, to exclude unwanted reflections ('cross polarization').

A polarizing lens-filter also enables us to reduce haze, and to alter the apparent brightness of northern sky light, which is, to a large degree, polarized.

Polarization increases the color saturation of a shiny surface, as one would expect from Figure 2.14.

We can polarize light deliberately by putting such a filter over a light source. Stereo photography makes use of this principle, and employs separate vertically and horizontally polarized pictures for the left and right eye-viewpoints. The result is two independent differently polarized pictures which are viewed through correspondingly polarized filters and fuse in the brain, giving the illusion of solidity.

3 The eye and perception

The human eye is often likened to a camera. It does have broad similarities, but the parallel can be very misleading. As in the camera, the amount of light admitted to the eye is controlled by an iris. The image of the scene is focused by a lens onto a light-sensitive surface. But there the similarity ends, for eye and brain are, in fact, remarkably deceptive.

The eye and the camera

The eye itself has various optical shortcomings that we could not tolerate in a camera, and yet, thanks to instinctive readjustments and the brain's ability to interpret, we are left with an impression of perfect vision and overall clarity. Apart from a few puzzling optical illusions, we accept what we see as a natural and accurate representation of our surroundings.

In reality, as we look about us, eye and brain continually compensate. We make instant judgments; concentrating on some features while ignoring others. We subjectively assess tones and colors – often quite wrongly! We overlook what is plain to see – and assume what is not there.

There are often differences, too, between the real world and the way we remember it. Standing looking at a sunlit snowscape, we usually disregard how shadows are being illuminated by reflected sky light. But look at a color picture of the scene, where the camera has captured these blue shadows, and they look unreal, even artificial.

Quite often, we photograph an attractive subject, only to be disappointed with the version we see later on the screen. Or we visit the location of an impressive photograph, and feel let down when it looks less colorful and imposing than we had assumed from the photograph. A prominent feature of the scene now appears quite incidental; while something else now catches our eye in the photograph, that we did not even notice when taking the shot.

As you will see, there are many important differences between the ways *we* see and interpret the actual scene and the ways we respond to a *photograph* of it.

The camera always lies!

Let's think, for a moment, about what we are really looking at on the screen.

45

Its *flat* rectangular image shows us just a small part of the scene; typically, a segment covering from ½° to 60°. The Director's selected viewpoint provides us with a very limited amount of information. In life, we can look around freely, to assess our surroundings. But the screen gives us a few tantalizing glimpses. We can only conjecture from what the camera reveals.

Program makers continually take advantage of the screen's limitations. It allows them to make considerable economies when designing lighting, scenery, and camera treatment. The audience sees a single doorway, and infers that there is a complete room. A window shadow on the background comes from a projected light effect. No window exists.

Because we are all so familiar with movies and TV, we accept the various screen conventions and overlook how unnatural they really are; even such artifices as the close-up, cutting, the wipe, zooming.

The effect of the frame

One of the most important characteristics of the screen image that we easily overlook is the effect of the *frame*. When we look at any picture, everything we see within its frame becomes visually *interrelated* in a way that never happens in real life.

Through the lens, line and tone become transformed. A distant surface can become prominent behind a foreground subject when using a narrow-angle lens (telephoto; long focus). The unexciting lines of a nearby fence can look dynamic when shot with a wide-angle lens (short focus). Our responses are conditioned by this screen image, not by what is actually there.

Among the factors that influence our responses are:

- *Tonal balance:* the variety and arrangement of tones within the frame.
- *Line:* the shape and direction of apparent line.
- *Mass:* the effective area of tones.
- *Color relationships:* the inter-reaction of color masses.
- *Camera viewpoint:* camera height; perspective.
- *Framing:* the position in the frame of the subject, tonal masses, line, etc.
- *Shot size:* how much of the subject is visible; scale; proportions.

Altering any of these aspects of the picture can entirely change its significance, or modify our interpretation of a shot.

The critical eye

On the screen things we would normally pay little attention to in daily life can become a major distraction. Technically, the pictures may be excellent, but we can become irritated by the way the sun

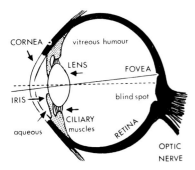

Fig. 3.1 Construction of the eye
The light from the scene enters the eye via the protective cornea, is controlled by a variable iris, and is focused by the lens (in conjunction with the cornea and aqueous). Focusing is achieved through lens-thickness adjustment (by ciliary muscles) changing the focal length of the lens.
   This focused, inverted, image falls upon the light-sensitive retina, which comprises numerous receptors. These rods and cones are joined by the optic nerve fibres to the brain. The junction region or 'blind spot' is itself insensitive to light. The fovea or 'yellow spot' comprises a highly sensitive cone-region of greatest visual acuity.

reflects in someone's spectacles. On the screen we are conscious that their eyes are in deep shadow, or the top of a person's head is very bright. A wall mirror behind them looks like a halo. We can't decipher a defocused poster in the background. Strong light reflections prevent us from seeing an oil painting properly.

A lot depends, of course, on how critically we are watching; but we will certainly notice all manner of effects in the picture that would not worry us if we were actually there. One might argue that these effects are quite *natural*; but that does not make them any more acceptable on the screen.

Everyday environmental lighting can produce some extremely unattractive effects. They appear exaggerated, crude, ugly, bizarre, when our attention is concentrated within the narrow confines of the screen. One of the aims of *controlled* lighting techniques is to minimize such features, and to enhance the appearance of the subject and its surroundings.

The eye

Our eyes, apart from a little short- or long-sightedness, seem to show us the natural world. But it can be disturbing to learn that some 8% men and 0.4% women have color-vision abnormalities. Black-and-white reproduction is an abstraction of reality, and yet we have become accustomed to accepting this as a convincing substitute – as indeed the flat color image is – for the real thing.

A brief look at the way in which the eye and the brain interpret the visual world will help forestall many of these surprises, and enable us to take advantage of their quirks.

The scene's image is focused on to a light-sensitive 'screen' coating the inside of the eyeball (Figure 3.1). This *retina* comprises layers of complex nerve cells. Most of the surface receptors (over 120 million) are *rod*-shaped. They are extremely light sensitive, but unable to

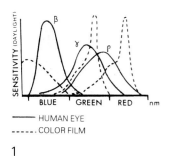

1

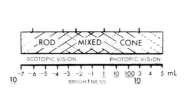

2

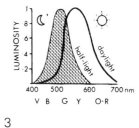

3

Fig. 3.2 The eye's color response
1. The eye is thought to posses color receptors (cones) covering three distinct regions of the spectrum.
2. Its overall response alters for lower ambient light levels, as differing proportions of cones and rods (blue sensitive) are activated.
3. Moonlight appears bluish, although actually a dim version of white sunlight, due to this 'Purkinje shift' as the eye changes from photopic to scotopic (twilight) vision, losing red-orange and gaining blue-green sensitivity.

discriminate fine detail, and are color blind save to the blue end of the spectrum. The remaining less sensitive receptors (6½ million) are the *cones*, which are concentrated in a small *fovea*. Here is the centre of our acute vision, enabling us to detect color, shape and position of objects. These cones, predominantly responsible for our color vision (Figure 3.2), are of three sorts, respectively responsive towards violet to blue light, green to yellow and orange to red regions of the visible spectrum. In passing, it is an interesting fact that even this color vision is not constant over our visual field, so that it is possible for a green or red subject to become modified to yellow and thence gray as it moves slowly from central to side vision.

What we term '*vision*' actually involves a complicated, largely unexplained, involuntary computing process between eye and brain. Our eyes really produce inverted, laterally reversed images. We can only detect sharp detail over about 1½° (2 cm wide at 2 m), within a detailless wide-angle visual field of around 240° (it is due to this peripheral vision that we see in dim light, and readily detect nearby movement). And yet, with rapid eye movements our vision hops from one point of interest to another, refocusing, readjusting our interpretation so rapidly that we are left with the impression of overall clarity. These saccadic jumps are part of an involuntary scanning process, including a tiny local 200-cycle scanning movement.

Paradoxically, a completely stationary eye-image fades rapidly. The eye continually adapts – relative to each momentary reference point – and our assessment of the image changes accordingly.

Binocular vision provides two slightly displaced viewpoints, which, with mental fusion and experience, enable us to interpret distance and dimension in the external world.

The wonders and the puzzles of vision really derive from one main factor, *adaptability*. Mostly, this adaptability extends our range of perceptivity. Sometimes it is too clever by half, and leaves us with ambiguous or incorrect judgments, as with optical illusions.

Brightness adaptation

General brightness adaptation

The eye is capable of operating under an extremely wide range of brightness levels. And with variations from the 10 000 f.c. (foot-candles) of strong sunlight to the 0.01 f.c. or less of artificial lighting, this is just as well! Two of its features enable us to do this. First, the *iris* contracts and dilates from around $f/2$ to $f/10$, giving us about 20:1 control of light reaching the retina. Second, the *retinal sensitivity* changes. Photo-sensitive pigments become reduced in bright light, increased in dim surroundings.

These combined phenomena explain why it is impossible for us to judge brightness accurately by eye, and why only a correctly used exposure meter can provide precise measurement.

Light levels of unobscured natural light remain comparatively constant for distance, whereas artificial light falls relatively rapidly with source-distance. And this, together with variations arising from source brightness and shadowing, is automatically compensated for by the eye – but not the camera.

Local brightness adaptation

As the eye scans a scene, stopping at fixation points, each fixation time, although very brief, is long enough to adapt to that point's brightness (*local brightness adaptation*). This, especially when combined with *brightness constancy* phenomena (in which we interpret object brightness relative to its background illumination), helps us to perceive shadow detail. We generally underestimate shadow density when assessing by eye.

As we ourselves move around within the scene, adaptation disguises the varying light levels and contrast that the camera itself will record, and the light-meter measure. The eye does not notice these variations so much, and 'reads into the shadows'. Shadows appear to us more luminous, i.e. less dense and impenetrable. The more we concentrate, the more we perceive through adaptation. Furthermore, where we actually know that a subject is situated within shadow (even in a photograph), we are more ready and able to adjust our assessment than if we do not.

Lateral brightness adaptation

The retinal sensitivity of the eye is not entirely local, so that when in scrutinizing a scene it adjusts to one area, its sensitivity to neighboring regions can alter, too. This gives rise to simultaneous contrast illusions.

Brightness assessment is influenced, moreover, by lighting contrast. So, although the ambient light level of dull daylight may, in fact, well exceed that of studio lighting, the latter, being more contrasty, will subjectively appear much brighter.

The *contrast range* within which the eye can perceive tonal gradation alters with light intensity. This is partly due to its internal flare (about 10%) reducing the effective contrast at higher extremes. The range of luminances that we can interpret at one time varies from 1000:1 brightness ratio at fairly high ambient light levels, down to 10:1 under dim conditions, and even around only 2:1 for very low levels.

Color adaptation

General color adaption

Provided the color quality of the incident light covers most of the spectrum, the eye will still accept very wide variations as 'white light'. Even when lighting conditions vary considerably, we still see colors looking surprisingly similar to the way they would look in daylight. Thanks to our eyes' rapid adaptation to what we are looking at, differences are far less pronounced than we would expect.

There is less visual compensation, however, when either the light or the reflected color of the object has sharp peaks or dips, or is of limited spectral range (e.g. 'pure' color).

Where two light qualities, i.e. cool and warm, are present, the eye tends to adapt to some intermediate, non-present color quality. This arises whether we are considering two lighting colors on the same subject, or two adjacent color transparency or television pictures.

Multi-color light mixtures can produce spurious color effects, and whenever the light from one source is obscured, distractingly colored shadows result. Light a white object with primary green light, and mysteriously magenta shadows can result!

Local color adaptation

After staring at a strongly colored subject the eye retains for a while a positive or negative after-image, after-sensation, of it, in more or less complementary hues. Examples of this are:

After a red image a blue/green after-image is seen.
After an orange image a peacock blue after-image is seen.
After a yellow image a blue after-image is seen.
After a green image a purple after-image is seen.
After a blue image a yellow/orange after-image is seen.

Table 3.1   Typical illumination levels

Sunlight	10 000–2 500 foot-candles	108 000–27 000 Lux
Daylight	2 500–200	27 000–2 100
Sunset	10–0.1	108–1
Moonlight	0.01–0.001	0.12–0.01
Starlight	0.0001–0.00001	0.001–0.0001
Store lighting	500–100	5 400–1 080
Office lighting	50–20	540–215
Domestic lighting	20–5	215–54
Street lighting	2–0.01	21.5–0.1

Similarly, prolonged exposure to a large strongly colored area can result in our entire color interpretation of following color quality becoming modified, until the effect has diminished. Typical examples for warm colors (red) and cool colors (green) are included in Table 3.2. We could use this particular phenomenon advantageously to emphasize a warm interior scene after a cold, bleak exterior. But again, it is an effect that can arise accidentally, after watching predominant colors.

Table 3.2   Modification of hue after color exposure

	After exposure to red	After exposure to green
Red appears	magenta, lighter, grayer	bluer, darker, intense, less orange
Scarlet appears	orange	redder
Orange appears	yellowish or greenish, desaturated	reddish
Yellow appears	very light green	slightly orange
Purple appears	violet	redder
Green appears	greener, bluer, more vivid, lighter	darker
Cobalt blue appears	bluer, green/blue	more vivid
Violet appears	Ultramarine	more purple
Magenta appears	pinkish	redder
Tanned flesh appears	yellow/brown	redder, darker
Pink flesh appears	bluish, paler	redder
White appears	blue/green	warmer

Lateral color adaptation

Akin to lateral brightness adaptation, are *lateral color adaptation* (or *simultaneous color contrast*) effects between a colored area and its surroundings. A colored subject will appear to induce a complementary hue in its neutral background, or a colored background modify a foreground subject's color. Colors appear lighter against black, darker against white.

Let us look at some examples that illustrate this common phenomenon.

An orange area against black, yellow or red will appear paler, less intense; while against a white, blue or green background it appears more strongly colored. Next to green, not only will the orange appear more reddish but the green will look more greenish than when seen separately.

A red subject against blue makes the red look orange by comparison, and the blue greener. Even *white*, when placed in proximity to a pure hue, can become modified, appearing reddish, bluish or yellowish in sympathy with its background.

Against a blue background, a blue area will seem somewhat darker than normal. Dark red or blue against large white areas may look nearly black, while the white itself becomes more vivid.

Both in color and monochromatic pictures, the effect of simultaneous contrast is most marked at the edges of the areas concerned. But it will depend considerably upon the relative sizes of areas, their brightness and hues.

Constancy phenomena

As you saw when we discussed the snowscape with its blue shadows, we regularly ignore the evidence in front of us, and interpret situations as we 'know' them to be, rather than as they *are*! This phenomenon, which is called *approximate visual constancy*, can take several forms:

■ *Approximate size constancy* We associate reduced size with distance. A series of progressively smaller objects appear further and further away; not miniature versions.

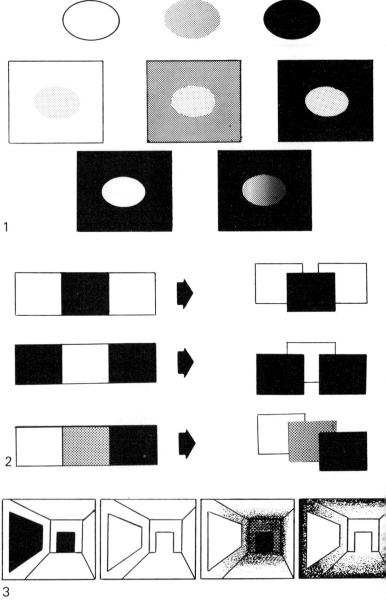

Fig. 3.3 Tonal planes
1. A light-toned object or plane generally appears larger and more distant than an identical area of dark tone. This tonal interaction is called simultaneous contrast (spatial induction). The sharper the contrast, the greater the effect.
2. A series of tonal planes can produce the illusion of advancing or receding areas.
3. Tonal gradation can produce an illusion of solidity and depth.

■ *Approximate brightness constancy*   Because we 'know', for example, that pages in a book are invariably white, we assume that the book we are seeing in a picture has white pages; even when they are dimly lit or in shadow. In reality, they may be reflecting less light than a strongly lit gray surface in the scene.

■ *Approximate color constancy*   This is an allied effect. When we assess color fidelity in a picture we subconsciously draw on memories of familiar subjects – e.g. mail box, uniforms, national flags. 'Knowing' what the subject's color should be, we tend to assume that it is 'correct', and judge surrounding colors accordingly. Similarly, we ignore what is clearly visible; for example –

● Wrongly assuming that shadows are *black*. (They are usually low-luminance areas of the colors of the surface.)
● Overlooking how light reflected from a nearby colored surface can tint someone's face.

We are often surprised when the camera reveals things as they really are.

■ *Approximate shape constancy*   Despite distortions introduced by perspective, we are still able to interpret recognized shapes correctly.

● We see an ellipse, but realize that we are looking at a circular object.
● A series of lines converge at acute angles, but we appreciate that these are planes at right angles, seen in perspective.

So disregarding the evidence of our eyes, we correctly interpret such familiar objects as normally shaped plates, table tops, coins, cubes, etc., even when seen from various viewpoints.

Color assessment

Where we cannot actually see a colored light source our eyes tend to assume the visibly modified color as actual; i.e. a gray box lit with green light becomes a green box. The exception is when we are certain that this is untrue, as with a blue face. In such cases we interpret the incident lighting as blue – paradoxically even when the effect is caused by colored make-up and white light – unless further visual clues are interpreted otherwise. Wherever possible, the eye adapts. Hence the dubious value of slightly colored key lights. The eye may adapt to the key, regard it as white, and then misinterpret other color values as abnormal.

Whenever the red and green receptors of the eye are stimulated equally we experience a sensation of yellow, whatever the actual wavelength of the light responsible (Figure 3.4). The effect may be

Fig. 3.4 Yellow light
The eye sees as identical yellow,
1. The pure yellow light from a
    sodium source.
2. The yellow light derived from two
    superimposed colored light
    beams.

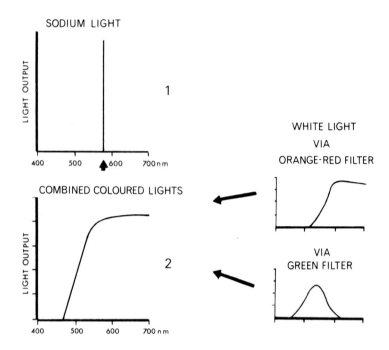

indistinguishable from the pure yellow line spectrum of sodium light. In fact, we meet two sorts of color match:

1. Where two colors contain wavelengths in similar proportions (i.e. similar spectral response).
2. Where the component energies are different, but their effect on the eye causes the two colors to appear similar.

■ *Surface texture* The color of a smooth surface will generally appear more saturated than a rough surface of identical hue and brightness. Any specular reflections will have the hue of the incident light.

■ *Light quality* When a colored surface is lit with very directional lighting (*hard light*) it usually looks brighter and more vivid (saturated) than under shadowless illumination (*soft light*) of equal intensity.

■ *Light direction* The direction of the light falling on a colored surface can also influence its apparent *saturation*. It appears greatest when the light is coming from our viewpoint, and falls off as the surface or the lighting is angled.

■ *Viewing conditions* The conditions under which we view a picture can considerably affect our impressions of color:

● A *smaller* picture (whether it is of reduced size of simply some distance away) tends to appear brighter, more strongly contrasted, and its colors more intense.

- The brightness and tone of a picture's *surroundings* can influence our color and tonal assessment. Take a small color photograph, and compare the result when mounted against white, mid-gray, and black backgrounds. The differences are often quite surprising. A carefully adjusted white illuminated border is sometimes fitted around a film viewing theater's small projection screen, to provide a more accurate appraisal of the way color and tone will appear on a television screen.
- Local lighting conditions considerably affect our judgment of picture quality. Not only can light spill onto the screen, reducing tonal contrast and desaturating color, but the viewers' eyes adapt to the general light levels. Brightness, contrast and saturation appear greatest when a screen is viewed in a dark room, but visual fatigue increases.

Color and depth

For centuries, artists have been aware of the strange way in which 'warm' colors (red, brown, orange) seem to *advance*, while 'cool' hues (blues, violet) recede from the observer. All colors appear to have their relative distances, according to the background against which we see them.

We can take advantage of this illusion when designing scenic and lighting treatment. A setting in warmer colors tends to look smaller, while surfaces in cool colors appear further away, and provide a more spacious effect.

Color and detail

The eye's ability to detect fine detail in color is limited, and varies with the hue. Our *visual acuity* is much greater for detail in orange and cyan than in green or magenta. Furthermore, because of chromatic aberrations in the eye, we cannot focus on both blue and red detail simultaneously.

As the eye cannot detect fine detail in color, modern television systems take advantage of the fact, and deliberately limit the amount of detail in the color component of the video signal. This achieves some ingenious technical economics. All detail and tonal values are transmitted in the form of changes in *luminance* ('brightness') as in any monochrome TV/video system. When luminance and chrominance are combined to form the complete color picture, all fine color detail is actually the result of a sharp monochrome image superimposed on an unsharp colored image.

Color and distance

A subject's color, and its background tones, can strongly influence our impressions of depth and distance. If you place a series of colors against a *white* background, you will find that blue-green appears nearest, then blue – purple – red – yellow, while yellow-green seems furthest away.

On the other hand, if you repeat this demonstration against a completely *black* background the situation changes. Now red appears nearest, while other colors recede in the order: orange – yellow – green – blue-green – blue – violet.

The area of a color in a picture can have a pronounced effect upon our interpretation. In *pointillism*, artists juxtapose tiny dots of paint which blend 'in the eye' by additive mixing into a new and more vibrant color than conventional methods provide.

As the real or effective size of detail diminishes, we can less readily detect colors. They seem to pale, some losing their identity before others. Blues can become indistinguishable from dark gray, and yellow from light gray. Green, blue-green and blue of similar luminance can merge. And, eventually, even the wide differences in hue between reds and blue-green are lost. Only brightness variations can be detected. And all this will affected by size, distance and light intensities. The atmospheric effects of haze, mist or smoke may cause distant colors to become desaturated, add blue to them or give them a warmer appearance.

Our assessment of fine color detail may be influenced, too, by '*spreading effects*'. A fine blue pattern tends to have a darker, more saturated appearance against black than against white; contrary to what we would expect of normal simultaneous color contrast effects. By *assimilation* the color of both the background and the pattern upon it become modified when small areas roughly equate.

When we examine *color harmonies* things become even more complicated. Color arrangements that look harmonious and coordinated when we see them close up may seem less unified at a distance. As things get smaller or further away, saturation and contrast usually appear to increase; but under certain conditions, the reverse happens – and here we reach the borderlines of the predictable!

Table 3.3   Influence of color on clarity of detail

In decreasing order of legibility:

Black	on	yellow
Green	on	white
Blue	on	white
White	on	blue
Black	on	white
Yellow	on	black
White	on	red, orange, black
Red	on	yellow
Green	on	red
Red	on	green
Blue	on	red

Color attraction

A particular feature of color is the way in which certain hues attract our attention. If, for example, there is a scarlet or orange area in shot, our eyes go to it instinctively. When we cannot see it clearly we feel frustrated; and spend time trying to discern what this defocused object really is. In a monochrome picture this does not happen, for the eye tends to pass over a defocused background.

Some colors have greater powers of attraction than others. Bright yellows and greens, for instance, are more able to hold our attention than subdued browns and purples. Saturated hues draw the eye more readily than pastel or desaturated versions. Where there are several prominent hues in a picture, we often find that the eye moves distractedly from one to another, and as a result, the unity of the shot is destroyed.

Color harmony

The effect of color is elusive, and people have long sought 'rules' to assist them in arranging color for the most harmonious attractive effect. There are no rules, but a number of guiding principles have emerged.

The most pleasing pictures tend to contain only a restricted range of colors. But we cannot be too specific about which color relationships work and which are less effective. You can achieve harmonious color schemes by interrelating *similar* hues, but equally attractive effects can be obtained by using a *contrasting* color treatment. Where most of the colors in a picture are desaturated the result can be more natural-looking than when it is full of strongly saturated areas. But too little color contrast can produce a drab uninteresting effect.

As you can see in Figure 3.5, the basic hues can be arranged in a circular diagram. Colors on opposite sides of the circle are termed *complementaries* – e.g. blue and yellow. Hues that are equally spaced around the circle are generally harmonious. These *triads* are near-complementaries. In practice, rules of thumb of this kind only serve as a general guide, because color harmony is influenced by so many other factors, such as the shape and texture of the subjects, their relative proportions, even social 'taste'.

Saturated colors, particularly when pure, are more liable to provide problems, for they attract direct attention more readily than 'impure' colors. In abundance, they can appear vibrant, forceful, gaudy, strident, obtrusive, depending upon their application. Conspicuous pattern, particularly when in pure colors, will distract attention.

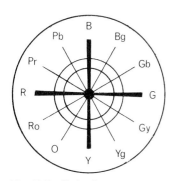

Fig. 3.5 Color harmony
Several attempts have been made to systematize color harmonies, including Rood's Hue Circle, in which complementary colors are opposite, and triads are separated by 120° (e.g. G, O, Pb). However, relative saturation, area, form and finish have a strong influence on the effectiveness of such aids.

Color memory

Although the ability to remember particular colors can be improved with practice, human color memory is unsure. And where we do not know the original color of the subject, color reproduction can be wildly wrong, yet be acceptable. It might even be preferred to an accurate version.

We think we know when the color of skin, foliage and flowers, for instance, are correct. But close inspection shows that these vary considerably between examples, and with season and light. Probably our most critical areas of assessment are flesh, food and familiar packaging.

We tend to recall colors as they would appear in daylight, although more saturated and rather warmer than they actually were. Usually we remember a light scene as rather lighter than it was; a dark scene as darker.

We tend to associate highly saturated colors with sunny lighting and so assess brightness to some degree according to the color content of a picture. Pastels and grayed colors less readily suggest bright surroundings.

Color associations

Color and emotion are inextricably interlinked, and color associations are endlessly varied. Some have universal impact, some have significances restricted to groups or nations:

Red   Warmth, anger, crudity, excitement, power, strength.
Green   Spring, the macabre, freshness, mystery, envy.
Yellow   Sunlight, the Orient, treachery, brilliance, joy.
Blue   Coolness, ethereality, the infinite, significance.
Black   Death, gloom, sorrow, hidden action.
White   Snow, delicacy, purity, cold, peace, cleanliness, elegance, frailty, mourning.
Black relieved with white   Sophistication, vigor, newness.

Even a very broad list of this kind reminds us how associative color can be, and how subtly we can evoke moods (sometimes accidentally!) by the way we use color in staging, costume and lighting treatment.

4 The principles of lighting

As you begin to work with light you will probably be surprised how many of its features are already familiar from everyday experience. You will recognize the effects, even if you have not previously considered their cause. Here you will see how they form the foundations of lighting techniques.

Perception and selection

Preoccupied with daily life, most of us give only half an eye to the world around us. We respond to its influences but seldom scrutinize its effects all that closely. We accept, for example, the well-worn concept of romantically *soft* candlelight. But if you look around a candle-lit room, it is soon obvious that a candle's small flame casts firm clear-cut shadows, and throws texture and contours into sharp relief. There's nothing softly diffused about candlelight. If light from a candle on the table does flatter, this is due solely to its low angle, as we shall discuss later.

Visual artists of all kinds are *manipulators*. They deliberately arrange line and form, light and shade, texture and color, to create carefully designed effects. They learn to recognize how features can be adjusted and controlled to create an illusion and influence their audience. Through persuasive arrangements, they express ideas.

The artist, whether working with tangible materials of paint and stone, or the intangibles of light and sound, selects, modifies, even distorts, to suit his particular purpose. He emphasizes or omits as he chooses.

When Canaletto painted canvases showing panoramic views of great cities he did not hesitate to reposition entire buildings to improve the picture's composition! Similarly, the stills photographer processes, retouches, dodges his prints to develop an idea. The Lighting Cameraman and the TV Lighting Director are more restricted in their opportunities, but they, too, have their deceits.

The way ahead

If your aim is simply to get pictures from a camera, then flooding the scene with light will do the trick. But to use light as a *persuasive tool*, you need to appreciate its various properties, and become adept at controlling the way in which it illuminates the scene. So we shall be exploring these aspects of the craft in some detail:

- *Light quality:* from hard shadows to diffused illumination.
- *Light direction:* the effects of light direction on the appearance of the subject.
- *Light coverage:* the ways in which adjusting the spread of light guides the attention and influences the mood.
- *Light intensity:* the subtle changes that develop as we control the balance of lighting (relative intensities).
- *Color:* the influence and opportunities that color offers.

Light quality

Between the extremes of sunlight's clear-cut shadows and the flat shadowless illumination of an overcast day lie infinite blends of light quality.

Hard light

Hard light is the highly directional illumination produced by any small-area light source. The light radiates from such a *point source* in straight lines, so casts sharp, clearly defined shadows, revealing surface contours and texture. The more concentrated a light source is, the 'harder' its illumination will be; the sharper its shadows.

Fig. 4.1   Light quality
1. Hard light comes from many natural and artifical sources. The smaller the effective light-emitting area, the harder the source. (Hence a distant localized soft light can produce distinct shadows.)
2. Soft light, too, comes from natural and artifical sources, and from scattered hard light. For well-diffused, shadowless light, large-area sources are necessary. Light diffusion reduces source efficiency.

A hard light source does not have to be *powerful*. It can equally well be a single match, or a candle, or an arc. The 'hard' quality of light comes from its effective *concentration*, not its intensity (brightness). Although our most familiar hard light source, the sun, is in fact enormous, it behaves as a concentrated source, due to its great distance.

If a hard light source is close, or large relative to the subject, a certain amount of its light will be diffracted round the edges, causing the outline of cast shadows to be softened and less clear-cut.

Because it is very directional, hard light can be restricted and localized. You can easily confine it to selected areas. So you will use hard light (usually from spotlights) –

● Whenever you want to reveal (or even emphasize) the modeling and texture in a subject;
● When you aim to light only a restricted part of the subject or the scene;
● When you wish to prevent light from spilling over adjacent areas.

If you use hard light *appropriately*, the result will be a well-modeled effect, in which contours are clearly revealed. Carefully controlled shadows help to define the subject, and relate it to its background.

If the lighting angle is unsuitable, hard light can produce some very unattractive effects:

● Harsh crudely defined modeling;
● Coarse textures;
● Important areas of the subject may be left unlit;
● Distracting shadows may form;
● The subject's shape may even appear distorted.

As we shall discover, although hard light has its drawbacks it is the light quality necessary to provide controlled, well-modeled lighting treatment.

Texture

If a surface has many extremely small irregularities (e.g. as in rough concrete) each bump or projection on it will cast its own tiny shadow, and create an overall random pattern. It is this effect we call the surface's *texture*.

Texture is a matter of scale. If you look at a stone wall close to, you can see the individual pieces from which it is built, each well modeled and casting its own shadow. But look at that same wall from a distance, and details merge to form an overall texture. Moving even further away, this texture is no longer clear, and the wall may become a plain overall tone.

Soft light

Soft light is the diffuse, shadowless illumination that results from scattered light. We meet 'soft light' under natural conditions, when the sun is obscured by thick layers of cloud, by fog and mist, or reflected from the ground and nearby surfaces.

Lighting equipment produces soft light in various ways; by scattering through diffusion, reflection, or multiple sources. The particular properties of soft light are that:

● It can be used for shadowless illumination.
● It will illuminate detail in shadowy areas, without casting additional shadows.

- It can be used to create subtly graded shading.
- It can reduce or suppress unwanted texture or modeling.
- Where a large area is to be illuminated, a single key light can be supplemented with soft light to produce a single-shadow effect.
- It can be used to simulate natural lighting from sky light or reflected light, or cloudy skies.

At first sight, the even overall illumination that soft light provides might seem ideal. Surely it shows us everything clearly, and avoids distracting shadows.

Well, if the soft light is positioned at an angle to the subject, that side will be illuminated, and we shall see subtle modeling as light progressively falls off in graded half-tones on its far side. At best, carefully applied soft light can achieve attractive delicate modeling; although the system may not reproduce this delicate shading accurately.

Table 4.1　Comparison of hard and soft light

	Advantages	Disadvantages	Advantages	Disadvantages
	As a key light		*As a fill-light*	
Hard light: shadow-creating, directional				
Readily restricted and controlled	Gives well-defined, clear-cut modeling	Modeling can be harsh	Coverage can be controlled	Casts additional shadows
Beam focusing and localization readily achieved	Shadow-formation reveals texture and form	Shadow-formations may be unattractive or distracting	Effective over long distances	Destroys subtleties of modeling
Compact light sources possible		Local hotspots likely	Does not fall off rapidly	
Soft light: diffused, non-directional				
Not readily restricted	Can provide many half-tones, and delicate tonal gradation	May suppress modeling with excess use	Provides subtle half-tones	Not readily localized
Relatively large bulky light sources			Does not cast spurious shadows	Spreads over adjacent areas
	Does not cause distracting shadows	May cause lack of clarity and crispness, making subject flat overall	Reduces harshness of shadows, making shadow detail visible	Brightness falls off rapidly with distance
	Does not emphasize modeling and texture	An inefficient light source for high lighting levels		

It is extremely difficult to localize soft light, however, and prevent it from spilling around. Because diffused light fills shadows, reducing texture and modeling, it can produce a very uninteresting overall 'flat lighting' – particularly if it is positioned near the camera.

Light direction

The lighting angle is important

The angle at which light strikes a subject can have a major effect on its appearance. The optimum direction will depend on the subject itself, and what you want it to look like. The lighting angle you would use to emphasize the surface detail in a close shot of a coin, or the texture of cloth, would produce a very unattractive portrait! The lighting that enhances a woman's hair style may make the top of a bald head embarrassingly bright. We may deliberately cast the shadow of a tree branch across a wall to provide an attractive pattern. But the shadow of a microphone boom becomes an ugly distraction.

Classifying direction

Few of us are particularly good at estimating odd angles accurately. Ask someone to point over 90° and their guess will be reasonable enough. But what of 30°, 60° or 135°? We are lighting in a three-dimensional space, and have to make rapid off-the-cuff judgments about the angle and direction of lighting. So we need some kind of system for classifying lamp positions, simply and unambiguously. The one we shall be using here is straightforward and universal, based on the *clock face*; a method that has proved to be remarkably accurate. Figure 4.2 shows you the idea.

Fig. 4.2   Classifying light direction
Using two imaginary clock faces, lamp positions are easily designated. The camera is always shown as at 3 o'clock V (3V), 6 o'clock H (6H). (Hours are 30° steps, minutes are 6° each.) Intermediate positions between hours are shown by + (clockwise) or − (anticlockwise) signs.

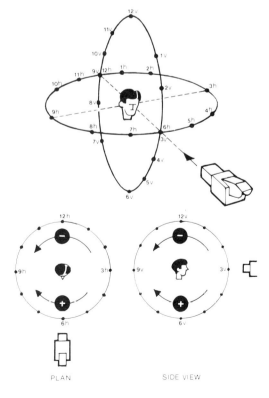

PLAN   SIDE VIEW

Imagine that your subject is at the center of a *horizontal* clock face.

● The camera is located in the 6 o'clock position – which we shall call **6H** (i.e. **6 o'clock horizontal**).
It is looking straight at the subject.
● As we stand beside the camera, 12 o'clock (**12H**) is straight ahead of us; immediately *behind* the subject.
● To the camera-left of the subject is 9 o'clock.
● And to the right is 3 o'clock.

We can refer to any horizontal direction simply by quoting the nearest 'hour' on the clock, rather than having to estimate angles in degrees. (If 12 o'clock is taken as 0° then the *hours* represent steps of 30°, and each *minute* represents 6°.)

Having established the *horizontal* direction, we can extend this idea and imagine a second clock face. This is *vertical*. Again, the subject is located in the center. Any positions on this vertical clock face are called 'V'.

● 12 o'clock is immediately above the subject, at **12V**.
● 3 o'clock is in front of the subject, where the camera is located (**3V**).
● 9 o'clock is immediately behind the subject (**9V**).
● 6 o'clock is directly below it (**6V**).

So we have only to specify that a lamp is at e.g. 4H/1V (i.e. round horizontally at 4 o'clock, then up at an angle to 1 o'clock) and this has pinpointed its position in space.

From the camera's viewpoint

As you saw, in our system the camera is always at 6H/3V, and we shall always consider what things look like *from the camera's viewpoint*, irrespective of the direction the subject itself is facing.

When you stop and think about it, this is the only rational approach. Assuming that the lighting remains constant, the effect of any light changes as we alter the camera's viewpoint. Suppose, for

Fig. 4.3 Light direction
We classify the direction of light relative to *our viewpoint*; irrespective of the subject's position.

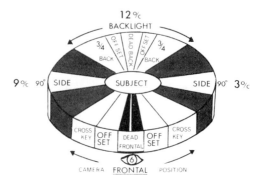

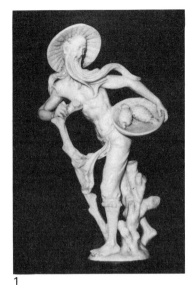

1

2

3

4

5

Camera viewpoint
Here we see the same objects from different directions under constant lighting.
1. This statuette is flatly lit by a level *frontal key* (notan).
2. But when seen from the side, the key has now become a *side light*, which effectively reveals the modeling and shape of the subject. We see a chiaroscuro effect.
3. If we move the camera round to the rear of the subject, the lamp now becomes a *back light*. The subject is silhouetted, and we can see that it is translucent. But there is the danger of lens flares from this low back light.
4. Against a back light, this appears to be a glass vase with a speckled pattern. Details in the glass are silhouetted.
5. But when seen from the opposite direction, this light becomes frontal, and it reveals that these specks are tiny flakes of gold embedded in the glass, and the object's appearance is quite transformed.

example, someone is looking towards a single lamp at 6H/1V. If the camera is facing them (at 6H), this light is shining straight at the person – frontal lighting as far as the camera is concerned.

Leaving the light alone, move the *camera* round them until it is in line with their shoulders. They will now appear in profile, with the light coming from the side (*side light*).

Move the camera round further until it is behind the person, and the original lamp now appears as *back light*. Because we have changed the camera's viewpoint, the *effective* lighting angle has altered, and the subject's appearance varied.

Although in practice we seldom move lights around the subject, we regularly switch camera viewpoints, and as a result the effect of the lighting changes. One of the skills of the craft of lighting is to

Fig. 4.4   Light direction and viewpoint
1. A subject's appearance will vary with the direction of the lighting.
2. The effect of a particular lamp will depend on your viewpoint.
So if either the direction of the light or your own viewpoint change (e.g. switching to another camera position), the subject's appearance will alter.

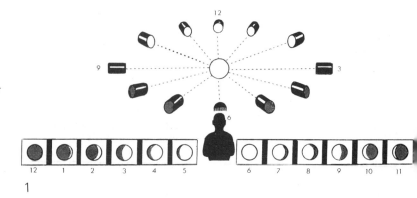

1

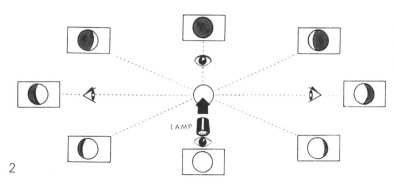

2

ensure that the subject's appearance is compatible from varying positions.

Frontal lighting

If you place a lamp very close to the camera lens shining forward onto the subject (from 6H/3V) everything that the camera can see will be lit. The subject's shadow will fall immediately behind it, and so will not be seen. (Look from another viewpoint and it will be clearly visible.)

Although the camera will show any surface decoration (i.e. color and tone), no surface contours, irregularities or texture will visible, for there are no shadows. In other words, there is no *modeling*. The subject looks flat. The closer the lamp is to the *lens axis* (the forward line from the lens to the center of the shot), the flatter the illumination will be.

Theoretically, 'dead frontal' lighting of this kind would avoid shadows, suppress modeling, and prevent bumps, wrinkles and other surface irregularities from being seen. In practice this lighting angle has severe drawbacks. If, for instance, you attempt a flattering portrait of a person this way, aiming to 'youthen' their appearance by suppressing creases, bags and skin folds, you will simply dazzle them.

1

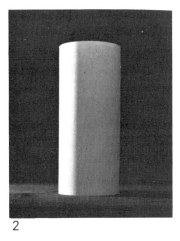

2

3

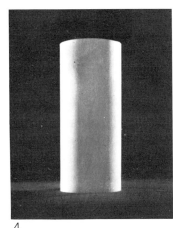

4

5

Tubular forms

These pictures show clearly how our interpretation of shape can be influenced by the lighting.

1. Here the only clue we have to the shape of this flatly-lit object is its curved ends.
2. *Side lighting* seems to reduce the width of the subject, as the shadow increases.
3. Shading on both edges reduces the subject's apparent width further. We tend to associate this kind of effect with the specular reflection from a convex metallic surface.
4. Double-rim lighting has a broadening effect.
5. But at lower intensities it can produce a bisected effect instead.

In general, if the surfaces face-on to the camera are highly reflective, the light will bounce straight back into the lens. Even when the surfaces are fairly rough, central hotspots are often clearly visible; although this may not be obvious on broken-up surfaces. The size and distance of the lamp will affect how *evenly* the subject is illuminated.

Where you want flat lighting it is usually sufficient to place the lamp (preferably a soft light source) just beside or above the camera. However, where you need absolutely shadowless illumination (e.g. in multiplane graphics) it may be necessary to use the somewhat cumbersome Pepper's ghost device shown in Figure 15.16.

Angling the lamp

Moving the lamp away from the lens axis in any direction, the subject's appearance progressively alters – depending on its shape, contours and surface finish. If your subject is symmetrical (e.g. a

The impression of space

The way subjects are lit will influence
our impression of their solidity,
depth, space.

1. With diffused light alone (ceiling
   bounce light), there is no sense of
   depth or form.
2. When we add a side key (9H 3V),
   depth and form are noticeably
   improved.
3. An additional key (at 2H 2V)
   increases the illusion further.
4. Kill the soft light, leaving the two
   hard keys, and the effect is
   harshened and quite dramatic.

1

2

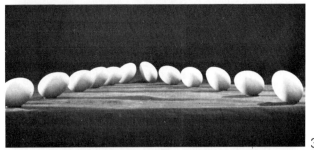

3

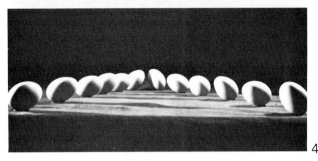

4

ball), it will probably look pretty similar, whatever the direction of
the light. But if the subject is strongly contoured, has a very irregular
shape, with a variety of textures, its appearance can vary remarkably
as the lighting angle is altered.

As you know, when you angle any light, illumination on the side of
the subject furthest from the lamp falls off. Shadows form beside any
bumps or projections. Increasing the angle, these shadows grow;
slowly at first, then more quickly as the lighting angle becomes
acute. These shadows spread across the surface, obscuring surface

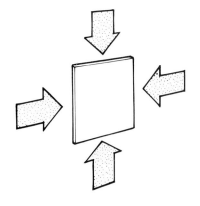

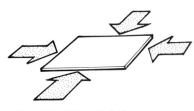

Fig. 4.5 Edge-lighting
Whatever the position of a surface, if light skims along it at an acute angle any texture or unevenness will be emphasized.

detail. Texture becomes more prominent. The shadow of the subject falls on nearby planes.

We saw earlier (page 26) that when a plane is angled to the light its apparent brightness falls. (Although it now appears brighter if viewed from the direction of the lamp.) Where a surface is curved, each point on it is at a slightly different angle to the light, and this creates the *tonal gradation* or *shading* through which we interpret the subject's contours.

- Under *soft light* the fall-off in brightness will be gradual – blended 'half-tones'.
- Under *hard light* the transition is more abrupt. The shadows have harder edges, and shading is coarser.

Side light and edge lighting

If you place a lamp to the side of a subject at 3H or 9H this *side light* will be at right angles to any surfaces facing the 6H camera viewpoint. Theoretically, if that surface was perfectly flat it would remain unlit (for it is in line with the light rays); although any parts protruding from the surface would be brightly illuminated.

However, in practice most surfaces are not completely flat, and, especially where the lamp is fractionally forward of the side position, its light will skim along any frontal planes at a very shallow angle. The slightest unevenness will cast abnormally long shadows and emphasize contours. We see this effect in natural lighting, of course, when the sun is setting.

When we deliberately arrange lighting in this way to reveal texture it is generally referred to as *edge lighting*. This 'textural lighting' is an excellent way of demonstrating the characteristic surface of wood, grain, stone, leather, paper, fabric, carving, embossing, and similar low-relief treatment.

Whether you choose to have the light skim along the surface from the side or from the upper or lower edges depends on the subject. One direction may be more convenient than another. You may want to emphasize certain features, or need to avoid shadows falling onto the surface.

Back light

When you move a lamp round to a position directly behind the subject (12H/9V) the source itself will be hidden and generally ineffectual.

- If the subject is *solid*, all that will catch the light will be any irregularities around its edges (e.g. hair, fur, feathers).

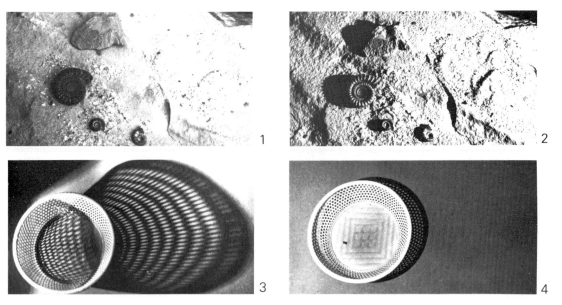

1

2

3

4

Edge lighting
1. Under *soft lighting*, texture and form of this subject are lost. So we have difficulty in seeing details clearly.
2. Edge lighting throws detail into sharp relief, and emphasizes the texture and form of the sand, rock surface and the ammonites.
3. Here oblique lighting produces strong shadow patterns. They may be attractive, distracting, or quite inappropriate. Those falling across the basket actually prevent us from seeing details of the weave.
4. With the key light slightly offset, we can see the basket's texture and form, and the shadow gives us an idea of its depth.

1   2   3

Texture and form
See how our impressions of form, texture, contours and materials change under different lighting.
1. Here the silver vase and the lay figure are displayed most effectively, but the ball is less successful.
2. Edge lit from the *left* (9H 3V) shadows obscure some objects, but texture is emphasized.
3. Edge lit from the *right* (3H 3V) different aspects of the articles become visible. The large ball is seen more clearly, but the vase is almost invisible.

1   2   3

Appearance changes with light direction
Here the main light moves round the statuette, and its appearance alters:
1. Using dead frontal lighting, all planes are illuminated equally.
2. Where the dominant lighting is from behind the object, its outline is emphasized, and we now see that the material is translucent.
3. Lit from the side (3H 3V) some features of the object are sharply defined, while others are in shadow.

1   2   3

Subject position
1. Lit from another angle (9H 3V), the emphasis changes. In this case, unimportant aspects of the subject are highlighted, while important features (e.g. the face) are lost in shadow.
2. If we turn the object round, these details become visible, but the subject now appears distorted from this viewpoint. (Key still at 9H 3V.)
3. With the object positioned for the best effect, two *side lights* (at 3H 3V and 9H 3V) provide a well-modeled result.

- If the subject is *translucent* (e.g. frosted glass) we shall see its structure and surface details silhouetted – probably with a central hotspot.
- And if the subject is *transparent*, the camera will be looking straight into the lamp, and see little else.

If you move the light from this dead-back position, towards 1H or 11H it will begin to illuminate the side edges of the subject, progressively emphasizing its outline there. Within an angular range of around 10–11H and 1–2H the edges of the subject are more effectively lit, and here we have the regular positions for *three-quarter back light*.

Lamp height

So far, we have been talking about positioning a light source at around lens height – i.e. *level* lighting. There will be times when you have little option but to light a subject in this way; e.g. with a lamp attached to your camera or held in the hand. But apart from closer work, such as a model on a table, level lighting can have real disadvantages:

- Frontal lighting at camera height will cast the subject's shadow onto the background behind it. This shadow will increase in size with the background distance.
- A person will be dazzled by level frontal lighting.
- A lamp located at around 10H to 2H will shine straight into the camera lens unless you take care to mask off the light.
- A lamp supported at camera height is vulnerable. It is easily knocked out of position, and people (or cameras) can inadvertently move in front of it.

Apart from these limitations, you will usually find that lighting from a level position does not provide the optimum effect. Instead as you will see, most lamps, whatever their direction, need to be some distance above the floor, shining down at the subject.

Let us look, now, at what happens when a lamp is *raised* above lens height. As you would expect, the visual effects are similar to those we have just discussed when moving the lamp *round* the subject; except that now shading and shadows progressively grow *downwards* as the lamp is raised. Its shadow is cast down towards the floor.

If the light is falling on the subject from above it will be edge lit; contours and texture being emphasized. With odd exceptions, overhead or *top lighting* is seldom attractive, and only suitable when you want to isolate the subject in a localized pool of light.

When you *lower* a light from a level position to shine *upwards* onto the subject, shading and shadows progressively grow upwards. Shadows are thrown up onto the background. Although this effect may be appropriate enough when simulating a low light source (e.g.

firelight), results from low-level lighting are usually strange, even bizarre, and only really suitable for highly dramatic situations.

Basic lighting principles

Lighting a flat surface

At times you will want to light flat surfaces unevenly, for shading or patchiness can make them look more attractive. But there are situations where any unevenness would be unacceptable – e.g. when lighting a title card, graphic, photo blow-up, skycloth.

The *edge lighting* that so effectively showed surface relief on a coin, or the texture of a brick wall, would reveal all wrinkles, bumps, and creases in a cyclorama!

The *frontal lighting* that can produce a flatly lit cyclorama would reflect in a shiny background or an oil painting.

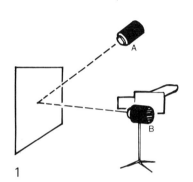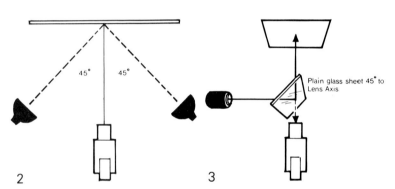

Fig. 4.6   Lighting a flat surface
1. Light from a frontal position near the camera (B), may reflect to produce hotspots, flares. Offset at (A) will overcome reflections, but may cause uneven illumination.
2. For critical work, dual lighting from either side (30–45° offset) is preferable. Soft-light sources reduce spurious surface texture or irregularities (blisters, wrinkles).
3. Shadowless lighting (along the lens axis) is necessary for multi-layer graphics (Pepper's ghost).

In Figure 4.6 you will see typical methods of lighting fairly small flat surfaces. Larger areas present different problems, which we shall discuss when lighting *backgrounds*.

Three-point lighting

Let us just recap. As you move a lamp round a subject, shading and shadows form to the side. If you raise the lamp, they move downwards. So if you do both, modeling moves diagonally. For some reason, the *combined* effect is usually more pleasing than either 'raised frontal lighting' or 'angled level lighting'.

As a rule, the most attractive lighting effects derive from an approach called *three-point lighting*. The technique is not sacrosanct; it can be varied as needed. But it is certainly a good starting point.

■ *Key light*   First we position the main light – usually a hard light source – at an angle of around e.g. 4–5H or 7–8H, and about 1–2V elevation. This *key light*:

● Establishes the light direction;
● Creates the principal shadows;
● Reveals form, surface formation, and texture;
● And largely determines the *exposure*.

■ *Fill light (filler)*   Then carefully positioned diffused light is added:
● To illuminate shadow areas.
● And reduce the overall tonal contrast.

■ *Back light*   Finally a hard source is placed behind the subject, facing towards the camera – but *not* shining into its lens and causing *lens flares*! This *back light* is usually around 11H to 1H at a vertical angle of e.g. 10V to 11V.

Fig. 4.7   Lighting an object
Effective lighting treatment usually involves four basic lighting functions:
Key light. Usually one lamp (spotlight) in a cross-frontal position.
Fill light. Usually a soft-light source illuminating shadows and reducing lighting contrast. But may not be required, or alternatively a reflected hard light is used (rarely a direct spotlight).
Back light. Usually a spotlight (or two) behind the subject, pointing towards the camera. (Occasionally soft light is used.)
Background (setting) light. Backgrounds are preferably lit by specific lighting, but they may be illuminated by the key or fill-light spill.

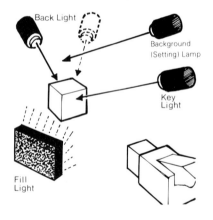

Back light creates a rim of light along the top and an edge of the subject. Without it, the subject may merge into its background and appear flat. In addition, back light helps to reveal its edge contours and give it solidity. If the subject is translucent or has tracery this light will show up the details.

Again, there is nothing mandatory about these arrangements. You may want to use a single back light at 12H/9V, or very rarely, even on the ground at 12H, shooting upwards towards the rear of the subject. Sometimes *two* backlights either side at e.g. 11H and 1H will attractively *double-rim light* the subject.

The arrangement you decide on will depend on a number of factors, including:

● The nature of the subject;
● Its form and tones relative to its surroundings;
● The way it is facing;
● The camera viewpoint;
● The particular atmosphere or ambiance you are aiming at.

The key light

Let us look in more detail at our choice of key light. First, there should only be *one* key, and this will normally be a focused hard light source (a spotlight), casting a single shadow and well-defined modeling. If you use two or more frontal keys on the same area (e.g. crossed frontal key lights) they will create two sets of conflicting shadows and modeling. One lamp dilutes the shadows cast by the other, and this situation is best avoided.

Where shadows (or modeling) from the key light are too prominent, a diffuser (scrim) placed over it will soften the light to some extent, but will also reduce its intensity. Occasionally, one may use a soft light source, to reduce modeling and minimize shadows. But as this diffused light is likely to spread uncontrolled over the rest of the scene its value is limited. Arguably, as a rule of thumb, one might say that the closer the key light is to the subject, the softer it should be.

The *angle* of the key light will be influenced by the subject and what you want it to look like. You will get a better idea of how crucial this angle can be when we look at portraiture.

Many types of TV production, including talk shows, interviews, newscasts and games shows, use stylized settings and action is limited. Here you can usually position the key lights for the most effective results. In more complex situations the key light's position may have to be something of a compromise; in order, perhaps, to avoid a camera shadow, or to get round an obstructing piece of scenery.

There are often visible or *implied* light sources in realistic surroundings. Then, ideally, the direction and angle of the key light should relate to these sources. If someone is standing near a window in daylight, and you have keyed them from another direction, the effect will look false and unconvincing.

In practice, compatible light directions are most important in longer shots. For waist shots or closer you can often 'cheat' the key light position, placing it at an angle that achieves the most attractive effect; even if it is environmentally inaccurate. This is done regularly in film making, where close-ups are shot separately and are lit 'down the nose-line' for maximum effect.

When there are no obvious sources of illumination in a natural scene (i.e. the camera does not show a ceiling light, or window, or decorative lighting fittings) the problem of compatible light directions does not normally arise.

Fill light (filler)

The modeling and shadows cast by the key light usually appear too harsh and strongly contrasted. So we need to add *diffused* light that will illuminate these areas and reduce the overall contrast, without casting new visible shadows. This 'soft' light is termed *fill light; filler; fill-in*. The illumination from some 'soft light' sources is insufficiently diffused, and casts faint shadows. But these are not usually obvious on-camera.

The character of fill light

As you know, soft light has two main limitations:

- It is liable to spread over the nearby scene.
- Its intensity falls off quite quickly with distance.

These characteristics can be an embarrassment in

- *Low-key situations.* The soft light introduced to improve portraiture may over-illuminate dimly lit surroundings.
- *Distant action.* Sometimes it is not possible to place a softlight fixture reasonably close to a subject; e.g. as local fill for someone in a pool of light on a darkened theater stage. Then you may have to use a diffused spotlight as a fill light; keeping its intensity low, and hoping that the key-light is sufficiently strong to wash-out the additional shadow.
- *Strong sunshine.* When shooting in bright sunshine, you may have little option but to use 'hard fill' from a high intensity arc to fill major shadow areas. Nothing else is sufficiently powerful.

It is best, however, to consider this 'hard fill' technique as a first-aid measure rather than a regular practice.

The intensity of fill light

How bright should the fill light be? There is an oft-quoted rule-of-thumb of '*one-half to one-third as bright as the key light*'. But we often need no fill light at all! There is no hard and fast rule. The intensity of the fill light has a considerable effect on the picture impact. It should *never* exceed the intensity of the key light, and most would agree that it should rarely equal it. The purpose of the fill light is to augment.

How much is needed will vary with the amount of tonal contrast you want in the picture. A highly dramatic situation may require no fill light at all. A high-key comedy scene can need a considerable amount.

If you use an excessive amount of fill light, shadows and shading will be weak, modeling slight, and the overall effect flat, even lifeless. Dynamic lighting requires a careful balance between the relative intensities of the key and fill.

If you can switch off the *key light* and can still see fully illuminated shots, then there is too much fill light!

- The fill light should not be strong enough to modify the *exposure*.
- It should not suppress the modeling created by the key light.
- Fill light should not create its own spurious shadows, or modeling.
- It should not be so powerful that it establishes a different light direction from the key light.

The level of fill light is normally adjusted to suit portraiture. If it doesn't happen to reveal sufficient detail in the background, light those areas separately or arrange extra lighting there, rather than increase the fill light's intensity.

It is worth stressing these points, for strong fill light is too often used to iron out mistakes; to disguise the ugly modeling that comes from badly placed key lights – e.g. to illuminate the deep eye shadows produced by oversteep keys. Admittedly, there are occasions when this is the only way to cope with unplanned action or impromptu shots, but it should only be used as a first-aid measure.

Fill light ratio

Figures quoting 'ideal' proportions of fill light intensity to key-light intensity can be misleading, for choice is affected by so many factors. Two conventions are used when describing the relative intensities of key and fill light:

- The 'key-to-fill' ratio (their separate intensities), or
- The 'key-*plus*-fill ratio'

Provided we realize which convention is being used, either approach is useful.

Factors affecting fill light intensity

As you will see, it depends on a number of factors, including:

- *Significance*
- *Subject*
- *Time of day*
- *Interiors*
- *Mood*
- *Dramatic effect*

■ *Significance*   Why are you introducing fill light?

● Is it to avoid the lop-sided effect of an unfilled offset key?
● Is it to control the density of shading, and build up a three-dimensional effect?
● Is it to make all information in the shaded area clearly visible (i.e. drawn detail, lettering, texture)?
  The answer here influences the strength of the fill light.

Suppose the subject is a decorative vase. An offset key light reveals its form, and a back light rims its outline. Without fill light, the deep shadow cast on one side of the vase is likely to make it look unbalanced. All surface detail and texture there will be hidden.

By adding a low-intensity fill light you can show the overall form, and the vase begins to look more three-dimensional. Surface detail is hardly discernible in the shaded area.

Increase the fill light a little more, and we can now see decorations in the '*luminous shadows*'. Make the fill light brighter still, and the vase loses some of its roundness, but now all details of the decoration are clearly visible.

In this example the intensity of the fill light is determined by whether you want a dramatic image, a solid-looking effect, or a display in which all of the vase's surface is sharply defined.

■ *Subject*   Where subjects are well textured and strongly contoured they are shown at their best if we light to create strong vital modeling. On the other hand, where the subject has a smoothly shaped surface and/or a delicate texture, this treatment would look harsh and crude. Instead, medium- to high-intensity fill light would be more appropriate to prevent texture becoming exaggerated.

■ *Time of day*   The amount of fill light you need can vary with the *time of day* action is taking place. When shooting at sunrise or dusk the prevailing atmosphere will be ruined if you use too much fill. Shooting in strong sunlight, though, you will often need intense fill light to prevent shadows becoming dense.

■ *Interiors*   Strong fill light conveys an impression of 'openness' and space. If you use an excess of fill light when lighting an interior scene (location or studio setting) it will destroy any feeling of enclosure. Where action is taking place near a window, more intense fill light is a natural effect. But when the action is well within a room (particularly if the windows are small), contrast would usually be greater, so less fill light is needed. Shadowy interiors require very little fill light.

■ *Mood*   The intensity of fill light can vary with the *mood* of a scene. One might light happy light-hearted action to a low contrast, while using much less fill light for a sad or violent scene.

■ *Dramatic effect*   There are times when you will deliberately leave shadow areas *un*filled, for dramatic effect.

It is night. The camera moves closer and closer towards a wayside notice. Moonlight casts strong shadows of an overhanging branch onto it, hiding its message. A gust of wind blows the leaves aside, and we discover the warning that it is an unfenced cliff-edge!

A silhouetted intruder sits in a darkened room, a large-brimmed hat casting a shadow over his face. He strikes a match, and by its light we recognize who he is.

Familiar enough visual clichés perhaps, but in these situations strong fill light would have ruined the dramatic moment.

Fill light positions

There are various opinions on the 'ideal' position for the fill light. Let us look at these in some detail (Figure 4.8). We can summarize the main locations as *frontal fill*, *offset fill* and *wide-angle fill*.

- *Frontal fill.* Filling from around the camera position is probably the most obvious method; a soft light on or above the camera. One could argue that from this position we can fill any shadow areas seen by the camera, and there is little point in doing anything else. However, a frontal fill light does have disadvantages for certain key light positions, when it further illuminates the area already lit by the key light and reduces the modeling.
- *Offset fill.* An offset fill light is less likely to nullify the effect of the key light or add to exposure. It will not reduce subtle half-tones or flatten modeling produced by the key.
- *Wide-angle fill.* If you place the fill light at a greater angle to the key it lights only the shadowed area. However, there is a chance that it may produce secondary modeling there.

Diffuse lighting

The sharp modeling that a hard key light provides is far too harsh for some kinds of subjects; particularly when you want to convey softness, delicacy, subtle contours; e.g. when lighting babies, children, elderly faces. You can approach this situation in three ways:

- Simply increase the intensity of a frontal soft fill light, so that shadows are extremely 'transparent', leaving modeling visible but slight.
- Heavily diffuse the hard key, so that it is less strident, and produces 'softened off' modeling.
- Use fairly strong soft light to illuminate the subject, while back light or side light strongly defines its contours.

The last technique is capable of most beautiful results, with subtle shading and delicate half-tones. We see the effect in nature when

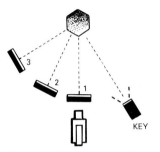

Fig. 4.8   Fill-light positions
The effect of the fill light will vary with its position relative to the key light:
1. Frontal fill.
2. Offset fill.
3. Wide-angle fill.

shooting towards the sun, and sky light alone illuminates the face (*contre-jour* lighting). Badly used, however, this technique can produce quite uninteresting, flatly lit pictures.

It usually relies for its success on

- Angled soft light – e.g. from 4H or 8H;
- Little or no fill light from other directions;
- Carefully located lighting from behind the subject;
- A sensitive balance of intensities; avoiding either excessive back light or over-strong frontal illumination.

Back light

There seems to be quite a lot of confusion about the need for back light.

- The amateur tends not to use it at all, except accidentally.
- Certain 'pictorialists' seem to use it abundantly, as an 'essential' atmospheric effect.
- Some 'realists' regard back light as false and unnatural (presumably overlooking *contre-jour* shots into the sun).
- Many add back light by rote, as a routine for all lighting treatment.

In reality, back light is a persuasive tool that, when appropriate, makes a valuable contribution to pictorial lighting.

Hair and clothing are frequently of similar tone to the background, and will appear to merge with it if there is no back light. Color differences alone are not sufficient to isolate the subject and make it stand out. Tonal separation is a particular problem with dark hair and clothing.

Even when tonal values are quite distinct, and there is a marked contrast between the subject and background (e.g. a dark suit against a light background), back light will usually enhance the picture. Particularly with dark clothing, back light catches its folds and edge contours, and gives it shape and solidity; preventing its being reproduced as a silhouette.

Without back light, translucent subjects would lose their entire visual impact. They would appear opaque. Where a subject has openwork (tracery, lattice, mesh) back light helps to model details, and prevent it being lost against the background.

In most cases back light is only *distracting* when it is too bright, or when it is totally inappropriate. The strong *double-rim* back lighting that creates a glamorous aura around a beautiful girl can look incongruous when used to light a weatherbeaten tough.

You will not always need back light; especially where the subject is keyed from the side (e.g. for a profile shot). When back light is steep, or widely offset (e.g. at 10H or 2H), results can be worse than having none at all. If a person is lit with badly angled back light the effect can appear artificial, phoney, or just downright ugly.

If the lamp used for back lighting is too low, or some distance away, it is liable to cause *lens flares*. These appear as spurious patches of light, or overall veiling or fogging that grays-out the picture. (This is due to inter-reflections within the lens system.) It is good working practice to have a well-designed *lens hood* (*lens shade*, *sky shade*) fitted to the camera lens to shield off stray light, and keep lens flares to a minimum. The effectiveness of this lens hood varies with the lens angle.

The regular solution to lens flares is to lower the top flap of a barndoor to keep light off the camera. Alternatively, you can raise the lamp so that it is lighting from a steeper angle. Another solution is to strategically hang a vertical cloth or board (a *gobo*, a *flag*, or a *teaser*) to shield off the light.

There are situations where you may want to avoid the shadow from a person's back light falling forward onto something they are demonstrating. Then it may be preferable to use a soft light source as a back light instead. There are drawbacks, though. It is not easy to localize this diffused back light, so it may spread around, and cause lens flares. Also, its intensity can vary as the person moves around.

Lighting balance

When *balancing* lamps we adjust their relative intensities to achieve a particular effect.

- In a 'coarse balance', high tonal contrasts dramatize the play of light and shade.
- A 'fine balance' provides subtle tonal variations, with delicate modeling.

Lighting balance can have a considerable effect on the attractiveness of the picture, and on the prevailing mood; and is clearly a matter of personal judgment and experience. Intensities are always relative, for as the viewpoint changes, the effective balance will change too. This is an intrinsic problem in multi-camera production.

Summarizing

Let us pause at this stage and look back over our journey so far.

We have in fact covered the underlying basics of lighting. We have looked at the essential differences between the character of hard light and soft light; and how we can select and blend different light qualities to influence the appearance of the subject. We have seen how texture and modeling can be controlled through lighting treatment.

We have seen, too, how the angle at which light falls onto a subject directly affects its appearance. As a single lamp is moved round the subject the effect of the lighting gradually alters. We have seen how frontal lighting displays the subject . . . how side lighting empha-sizes, even exaggerates its modeling . . . and back light gives your subject a third dimension, causing it to stand out from its background.

Lastly, we have seen how in *three-point lighting* one combines light of differing quality from different directions, to develop an overall effect.

Shadows

Paradoxically, imaginative lighting is as much a matter of creating and distributing shade and shadow as it is of selective illumination. All shadows come from an absence of light, but they originate in several ways.

There is the shadow formation we see on a half-lit object; where part of it has been left unilluminated – a side-lit globe, for example. This type of shadow is often termed *shade*.

Shadows are also formed wherever the incident light has been interrupted by a solid object:

- *Primary shadows* are the shadows thrown onto a subject by contours and projections on its own surface – e.g. a nose shadow falling onto the face.
- *Secondary shadows* are those cast by the subject onto its immediate surroundings – e.g. a person's falling onto the floor or a nearby wall.
- *Tertiary shadows* are cast onto the subject by other nearby objects – e.g. a leaf pattern falling onto a person beneath a tree branch.
- *False shadow*. This is not really shadow at all, but the *impression* of shadow that results when you leave part of a surface unlit. It happens mostly when illuminating a large area such as a cyclorama with a series of lamps. If adjacent lighting overlaps, it double-lights a portion and creates a hotspot. If part of the surface is left unlit by adjacent lamps. this can appear from a distance as a false shadow.

■ *Primary shadows*   These are a natural effect and help to reveal the subject's form. If they are too prominent, however (e.g. a long nose shadow), they may be distracting.

■ *Secondary shadows*   Secondary shadows tend to unify a subject with its surroundings. We recognize from the size and position of a person's shadow how far they are from a wall. Their floor shadow reveals the proximity and the nature of the floor surface. Without secondary shadows, people may appear to 'float' (see *chroma key*).

When a subject is positioned very close to a background it will normally cast a strong silhouette beside it. This secondary shadow is liable to intrude or confuse the picture. To reduce the shadow's prominence, you can:

- Use soft light to illuminate the subject.
  (This will reduce modeling and may result in flat lighting.)
- Move the subject away from the background.
  (This is often the simplest solution.)
- Move the background away from the subject.
  (Sometimes impracticable in small set-ups.)
- Raise the key light to reduce the shadow height, or hide it behind the subject.
  (This increases modeling; perhaps over-emphasizing it.)
- Move the key round the subject to reduce the shadow width.
  (A straight-on key will provide minimum shadow, but may flatten modeling.)

If you try to 'light out' a background shadow by adding an extra localized lamp to dilute it there is a danger that its light will fall onto the subject and overlight parts of the background.

■ *Tertiary shadows* When shadows fall onto a subject they help to relate it to its surroundings. As the pattern of light through prison bars or the grille of a confessional falls onto a person they emphasize the environment.

Shadow density

Where shadows are used decoratively they generally need to be strong, well defined and unambiguous. But in practice this is easier said than done. All too often, shadows are distorted, interrupted, broken up, obscured, or diluted by other lighting.

Fig. 4.9 Shadow formation
1. Shadow density is greatest on matte, light-toned, flat, undecorated areas placed at right angles to the light, and undiluted by spill light from other sources. The density depends, too, upon the hardness, size, power and distance of the light source.
2. Shadows generally look denser when they have sharply defined edges.

Fig. 4.10 Shadow sharpness
Shadow sharpness is influenced by
the opacity of the subject and
firmness of outlines. Shadow
sharpness decreases:
1. With a larger-area source.
2. As the subject moves further from
    the background.
3. As the subject/lamp distance
    decreases; and as light diffusion is
    increased.
4. Where the subject is small relative
    to the light source, and where the
    lamp is near the subject, the
    shadow cast by one part of the
    source becomes lit by another
    part. This illuminated shadow
    (half-shadow or penumbra) forms
    a graded border to the main
    shadow (umbra). Sharpest
    shadows arise from point sources,
    and parallel-ray light beams.

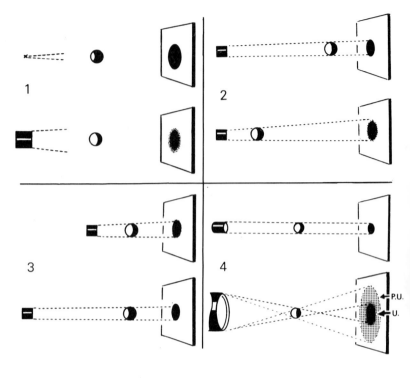

Fig. 4.11 Shadow size
1. The size of the shadow is always
    larger than the subject. Shadow
    size increases as the lamp/subject
    distance is reduced. Close lamp
    positions give greater size
    changes, and exaggerate shadow
    perspective.
2. Shadow size increases with the
    subject/background distance:

$$\frac{\text{Shadow size}}{\text{Subject size}}$$

$$= \frac{\text{Lamp/shadow distance}}{\text{Lamp/subject distance}}$$

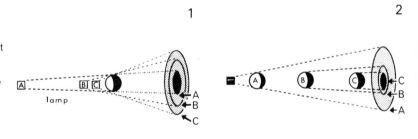

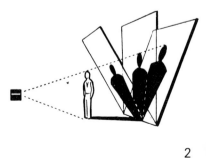

Fig. 4.12 Shadow length
1. Shadow length increases at
    shallow light angles, and
2. With obliqueness of the lamp to
    the background.

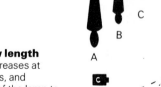

Fig. 4.13 Multiple shadows

1. Two or more lamps casting shadows on the same background give deep shadow (umbra) where their respective shadows coincide, half-tone shadows (penumbra) where the shadow from one lamp is diluted by the light of another, and a background lit by their combined light.

   Colored lamp filters can supply multi-color shadows and subtle color mixing. By fading between lamps with individual cut-outs, background shadows can be changed at will. Multi-shadows can suggest multiplicity of objects (e.g. numerous tree shadows on to a sky-cloth).

2. The *penumbrascope* consists of several open lamps on a rotating fitment. This casts multiple shadows which weave from side to side, continuously changing in tone.

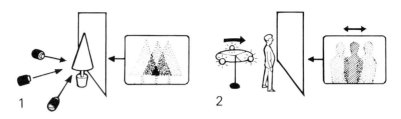

Dilute vague shadows are invaluable for breaking up surface tones that would otherwise be plain and uninteresting, or developing certain atmospheric effects.

It is best to avoid the conflicting multiple shadows that can arise when two or more separate lamps light the same area, or where the beams of lamps from the same direction overlap ('dirty light').

Where you cannot place a lamp in the optimum position to achieve a particular shadow effect – e.g. a patch of light falling onto a wall from a nearby window – it is sometimes preferable to *simulate* the effect by painting/airbrushing the wall instead! Rather than use a misshapen shadow pattern, it may be better to leave it out altogether.

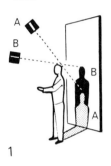

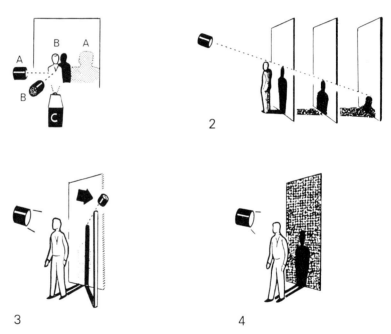

Fig. 4.14 Shadow control

Where background shadows are obtrusive, their prominence may be reduced in several ways. Perhaps a lower camera position will hide the shadow, or the pictures may be recomposed to omit it. Shadow-free lighting may be the solution.
1. The lamp may be raised or moved round the subject to throw the shadow off the background. But this can degrade subject lighting.
2. Subject-to-background distance can be increased.
3. The offending shadow may be diluted, but the extra light may cast spurious shadows or overlight the surface.
4. Disguise the shadow with background break-up, or hide it in a foreground object (e.g. foliage).

Fig. 4.15 Shadow-free background
To display a small object free from shadows, arrange it on a transparent sheet spaced from a lit background, or on an illuminated panel.

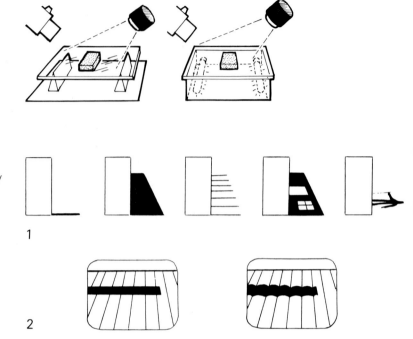

Fig. 4.16 Revealing shadows
Shadows can reveal various aspects of the scene that are not immediately obvious:
1. The rectangular plane is revealed here as a flat, a solid object, a series of parallel planes, a building, a flat concealing a person.
2. A shadow falling across the floor reveals that it is flat, or is contoured.

1

2

Lighting opportunities

By now, it's obvious that lighting is about a lot more than enabling the camera to get good pictures! Lighting is a persuasive tool. If you use light casually, you are liable to get some very unexpected results. Selective lighting enables you to influence the audience in many subtle ways. Light can not only

● Guide *what they look at*, but
● Affect how they *interpret* what they see, and
● Influence how they *feel* about what they are seeing.

Unbelievable? Then imagine the following scene.

> It is a darkened room. The door opens, and just a narrow shaft of light from the doorway pierces the blackness. The light falls onto one corner of a large object . . . It appears to be an old chest . . . We are intrigued. What is in the chest? Where are we?

Compare that treatment with the audience response to . . .

> The screen is dark. Someone enters the room and switches on a ceiling light, and the entire room is flooded with light. There happens to be a chest in one corner, among the bric-à-brac.

We shall follow up this idea of persuasive lighting in greater detail in Chapter 8.

What is the aim?

Ideally, before lighting any subject you should pause and carefully consider exactly what you want your lighting to achieve. A fundamental aim of all lighting is to try to compensate through light and shade, for the loss of the third dimension; to overcome the limitations of the flat picture. To clarify, and help the audience to interpret what they are seeing.

But beyond that, you need to assess the picture's *artistic purpose*; for your final techniques will spring from that decision.

■ *Structural lighting* Very often, your main aim is to present the subject clearly and unambiguously, conveying an impression of what it really looks like. The lighting is *structural*, concentrating on displaying a subject's construction, form, shape, contours, texture.

When lighting specific objects, antiques perhaps, you want the audience to appreciate the grain of the wood; the beauty of a curved surface; the skill of intricate carving; the subject's harmonious proportions. When lighting people, you will usually be seeking to make them look as attractive as possible; to show their costume to advantage.

■ *Atmospheric lighting* On other occasions, particularly in drama productions, *atmospheric* treatment would be more appropriate, in which the light creates a mood, an emotional attitude to what is being seen. You are using light to dramatize, beautify, transform. Emphasis is on effect.

You may use light to soften, to harshen, to exaggerate. You may use shadows to give the subject an unusual mysterious quality. Or gentle shadows may cast a delicate tracery onto the subject, to imbue it with an added charm.

'Structural' and 'atmospheric' are not mere labels here, but a reminder that we frequently use light for quite dissimilar *purposes*, that require different techniques.

Assessing the subject

Whether a picture's purpose is to sell something, or to demonstrate how a piece of equipment works, or to reveal the beauties of an *objet d'art*, the form of lighting used directly affects its success. If the subject is a dish of luscious fruit you may want the audience to feel they can almost taste the juice oozing from it, as it forms a glistening pool on a silver dish . . . Carefully angled back-light can reveal its translucent beauty, and persuade the audience of its mouth-watering qualities.

On the other hand, if the shot is part of a botanical program, demonstrating how the skin texture of the fruit changes when attacked by a parasitic fungus, then selective edge lighting could be much more appropriate. Your aims determine your approach.

Supposing, for instance, that the subject is a larger-than-life bronze statue.

- Which are its most important features?
- Do you want to emphasize its form or texture?
- Are there aspects that should be concealed or underplayed?
- Does the tone or finish of the statue pose any technical problems (exposure, specular reflections)?
- Should the light isolate it from its surroundings?
- Is the statue delicate and ethereal perhaps . . . elegantly poised as if in flight? It may be lit best by soft light revealing subtle half-tones and gentle curves.
- Or is it a dramatic dynamic work . . . a forceful imposing structure standing four-square and solid on the ground. It may need harsh angular hard light; coarse modeling with areas of deep shadow.

Inappropriate lighting can diminish the subject itself.

If you were to use the dynamic lighting for the first statue it could appear hard and rigid, and lose its delicate qualities. Under soft light, the dynamic statue could lose much of its strength and vigor.

Directing attention

The most obvious way of concentrating the audience's attention on a particular subject or a selected area is to use a *spotlight (spot)* to isolate it in a pool of light. If the light follows the moving subject, it becomes a *follow (following) spot*.

You can use the spotlight alone, while everything else remains unlit, or you can have the spotlight emphasize the subject, while the surroundings are illuminated to a lower intensity. Whether you use a spot with a sharply defined edge or a soft edge that merges into the surrounding lighting will depend on the circumstances.

There are several variations on this theme. You could, for instance, have the main subject in a group frontally lit, while others nearby are left silhouetted against a light-toned background; a singer, perhaps, backed by a choir.

Occasionally, you may deliberately leave the main subject *in shadow*, to arouse audience interest; e.g. the shot opens with someone silhouetted against a light background. As they begin to sing, a spotlight is slowly faded up to reveal the artiste. At best, this is an intriguing, fascinating approach. But it can become someting of an anti-climax if the audience is disappointed by what they see!

Lighting and composition

Through the way in which you distribute light you can create compositional forms and interrelate subjects. You can see this quite easily with a very simple experiment.

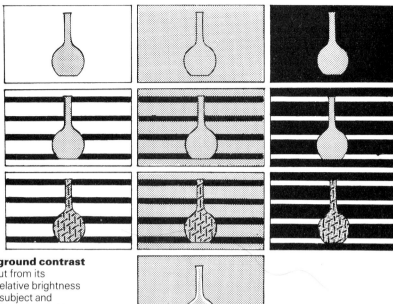

Fig. 4.17 Subject-to-background contrast
How strongly a subject stands out from its
surroundings depends upon its relative brightness
(tone), colors and clarity. Where subject and
background might merge, back light can provide
isolation by rimming the subject and emphasizing its
outline.

1

2

Tone and compositional balance
The way in which light is distributed over the
background can influence the final compositional
impact.
1. The central area of light provides a balanced,
   unified, but rather conventional and dull effect.
2. In this particular subject, the main areas of interest
   lie to the right, and this would be emphasized by
   asymmetrical background lighting.
3. Where the background shading is against the
   subject form, it creates unbalance. You might
   introduce such unbalance deliberately, to create a
   feeling of uneasiness, tension, unrest.

3

The impact of tonal shading

1. Faces will often stand out more clearly against a shaded background. But shading may inadvertently direct attention to the lower parts of the shot.

2. Cycloramas and backgrounds are often lit by lamps hidden behind a cove or a groundrow. Brightest at the bottom, the surface shades progressively towards the top. Although generally effective, unattractive effects can develop. In this low-key shot, the light appears to be cut off behind the lowest parts of the subject, and the legs merge with the dark floor. (This could be avoided by placing the person on a raised area.)

- Take three objects and place them in front of a plain background, with a similar patch of light behind each. The result is a display of three *isolated* items.
- Increase the size of the central light area, and that item appears *more important*, supported by the others.
- Join the light patches to form one large area, and the group is *unified*.

Using quite straightforward devices you can modify the apparent structure and form of a setting:

- By casting (or projecting) shadows or light patterns onto a background (page 249);
- By shading backgrounds, you can confine the eye;
- Light streaks can serve as a pointer to direct the attention to selected subjects;
- Deep shadows can add to the mass of scenic pieces (columns, arches, screens), and give them greater prominence;
- By adjusting the relative tones of surfaces, you can modify their apparent size and distance.

Visual continuity

When the camera cuts to a new viewpoint, and the entire appearance of the subject or the setting changes, you have problems! Yet it can happen quite easily. Unless you want to create a visual shock there needs to be a feeling of visual continuity as pictures are intercut. It may be the result of lighting treatment and/or scenic design.

A performer may look strongly lit when shot from one direction and silhouetted from another.

They may have a light-toned background in one shot, and a moment later, a cut shows them against dark surroundings; the visual jolt can be quite disrupting. At worst, it can even look like day in one shot and night from another angle!

So this is a reminder to look out for that particular problem, equalizing tones perhaps to improve the visual flow.

Technical limitations

Both film and television systems have their technical limitations, and we need to bear these in mind when staging a production. Scenic design, lighting, costume, make-up are all restricted in various ways. Today these limitations are fewer, but they do exist (Chapter 13).

When a film camera shoots into a lamp, or lights reflected from a shiny surface, it simply reproduces the strong reflection as a bright white blob. But many television cameras, when faced with specular reflections or light sources, are liable to produce spurious 'comet tails', 'streaking', or a 'stuck-on' image of the source. So one avoids large speculars wherever possible, by judicious lighting or using dulling spray on shiny surfaces.

As we discussed earlier, the film medium can generally handle a more extended tonal range than the TV camera. The contrast range the TV system can usually accept and reproduce accurately is limited to around 20:1 to 30:1.

The final tonal contrast range presented to the camera partly derives from the inherent contrast within a scene, and partly from the way the scene is lit. Lighting can all too easily increase the contrast beyond the camera's range. If you allow darker areas to fall into shadow and strongly illuminate lighter areas, this would greatly increase the overall contrast; perhaps by as much as tenfold. Sometimes we want to do just that, to build up a dramatic, low-key effect. But normally it is advisable to control the lighting contrast range carefully, to prevent it from exceeding the +20:1 tonal range overall.

If the scene itself has a wide contrast range all one can do is to keep deep shadow to a minimum, and ensure that darker areas are strongly lit – unless, of course, you actually want them to merge into solid black.

It may be necessary to shade off light-toned areas, or even leave them unlit if there is already a high level of ambient light around.

These steps will lower the overall contrast. But, in practice, it is not always possible to illuminate darker areas without some of the light also spilling onto light-toned surfaces. Similarly, shading a light-toned area may leave a darker surface nearby underlit.

Lamp functions

While lamps have a single function (e.g. a decorative pattern on a wall), the purpose of others may vary with the camera viewpoint, or the action.

● From one angle a lamp may serve as a frontal key light, while from another viewpoint it becomes a backlight.
● The 'sunlight' coming through a window may provide an atmospheric effect in one shot and also serve as a key light, when someone stands looking out in a reverse shot.

Table 4.2   Lighting terms

Accent light	Loose term for light emphasizing a particular feature of a subject. Also as *modeling light*
Back light	Light from behind the subject. Usually illuminating its edges
Base light, foundation light	Diffuse light flooding the scene uniformly, to reduce overall lighting contrast and prevent local underexposure
Bounce light	Diffuse light obtained by random reflections from a strongly lit surface, such as a ceiling or reflector board
Camera light, basher, headlamp, camera fill light, spot bar, Opie light	Small light source mounted on a camera to reduce contrast for closer shots, improve/correct modeling, and for localized illumination
Clothes light	Spotlight specifically revealing form and texture in clothing
Contrast-control	Soft fill light from camera position, illuminating shadows seen by camera, so reducing contrast
Cross-light, counter key	Additional angled frontal key at any height (at about 4–5H, or 7–8H)
Edge light	Light skimming along a surface, emphasizing its texture and contours. (From around 8–9H, or 3–4H, 12–1V, or 5–6V)
Effects light	Light producing specific highlight areas on the background – e.g. around a practical lighting fitting
Eye light, catch light	Eye reflections of a light source (preferably only one). Often from a camera-light specifically for that purpose
Fill light, filler, fill-in	Spot light illuminating shadows cast by the key light
Hair light	Localized lighting to reveal hair detail
Key light, key	The main lamp illuminating the subject; casting principal shadows, and establishing the prevailing light direction
Kicker, cross-back light, 3/4-back	Back light from directions 1–2H, or 10–11H
Modeling light	Loose term for any light revealing texture, contour, and form
Rim light	Illumination of the subject's edges, usually by back light
Set light, background light	Light illuminating the background alone
Side light	Lighting at right angles to the lens axis (at 3H or 9H). Strongly emphasizes subject's contours. Creates ugly portraiture for full-face shots, but effective for profiles
Three-point lighting	Term for rudimentary three-light set-up with key, filler and back light
Top light	Overhead lighting. Undesirable for portrait lighting
Underlighting	Lamp angled upwards, from below the lens axis (e.g. 4–6V). Used as an effect, or to relieve downward shadows and modeling

Table 4.3   Typical lighting treatment for common materials

Fabric	Cotton	S, mF, sa/sl, (z)	9
	Wool	H, wF, sa/sl/la, (z)	
	Tweed	sH, wF, laK	
	Silk	S, msl	3, 5, 8, 9
	Damask	sS	2, 5, 9
	Velvet	sH, sF, saK	
	Filmy	wK, sB	7, (9), (11)
Fur		sH. sF, saK/laK	
Feathers		H, sF, wK, sB	(7), 10, (11), 12
Canvas		sH, mF/nF, vlaK	
Leather	Smooth	mS, sl(m)	2, 4, 5, 6, 8
	Grained	S, sl(s)	
	Rough	H, wF, vla	5, 6
Wood	Natural	H, wF, sa	(14)
	Polished	S	1, 2, 3, 4, 5, 6, 8, (14)
Paper	Matte	S, sa	(z, 7), 13
	Shiny	S, sa	3, 4, 5, 13
Metal	Cast	H, mF, sa	(14)
	Machined	S	2, 3, 4, (14)
	Polished	S	1, 2, 3, 4, 5, 6, 8, (14)
Stone	Natural	sH, wF/mF	9, (14)
	Tooled	sH, wF/mF	(14)
Plaster		sH, mF	(14)
Plastics		S	1, 2, 3, 4, 5, 6, 7, 8, (z, 11), 12, (14)
Glass	Sheet	S	1, 2, 3, 4, 5, 6
	Formed	S	1, 2, 3, 4, 5, 6, 7, 8, 10, (z, 11), 12
	Solid	S	1, 2, 3, 4, 5, 6, 7, 8, 10, (z, 11), 12
Pottery	Unglazed	H, wF/mF	(14)
	Glazed	S	2, 3, 4, 5, 6, (7), 8, (14)
Liquid		S/H	1, 2, 3, 4, 7, 8, 12
Flowers		sS, sl, sB	(7), 10, 12, (14)
Foliage		sH, mF/sF	3, 6, 7, 10, 12, (14)

KEY

Light character relative to camera position:

K	Key (hard light)		H	Mainly hard light
F	Filler (soft light)		S	Mainly soft light
B	Back light		/	Alternative choice
s	Strong		sa	Slightly angled
m	Medium		sl	Side light
w	Weak		la	Low angle
			vla	Very low angle
			nF	No filler

1. Surface is revealed by reflections (lights, nearby objects)
2. Avoid excessively large areas of reflection
3. Avoid too many reflected highlights
4. Avoid spurious flares reflected from surface
5. Avoid reflections concealing surface pattern
6. Dulling surface by spraying, waxing, etc., masks character
7. Back light reveals structure and translucency
8. Overall sheen or gloss best revealed by large broad sources
9. Avoid hard sculptural effect
10. Avoid confusing shadows
11. Silhouette subject relative to background
12. Defocus background to isolate subject
13. Preferably use symmetrical lighting
14. Pronounced background lighting on shadow side of subject
z. Depends on material's purpose

(Entries in parentheses are not always applicable.)

A range of both general and specific terms have come into use over the years and these are summarized in Table 4.2.

Distorting reality

Lighting is frequently a compromise. The kind of lighting that shows one feature to advantage may hide another. The light direction that helps to reveal the texture of a surface can produce shading that makes it hard to read what is printed on it. Light shining through a translucent screen may reveal not only its delicate beauty but also the construction of the framework that supports it. Shadows that define a subject's contours may disguise its overall shape.

As we saw earlier, if you light a textured surface with a hard light near the camera it can appear smooth. Edge light it, and you can so exaggerate its surface texture that it becomes unrecognizable.

Surface interpretation
A surface's appearance can change according to the way we light it.
1. The piece of flock wallpaper has been arranged both horizontally and vertically, and lit from directly above. The horizontal plane shows only slight detail, for the lighting is at right angles to its surface. But the light shines onto the vertical plane obliquely (edge-lighting) and its surface pattern stands out in sharp relief. Our interpretation of the surface can vary with the lighting angle.
2. Under a spotlight this jug appears dark and glossy, but surprisingly smaller than in (3).
3. Here the jug is lit by soft bounce light. See how different it looks. It appears larger, shinier, and of lighter tone, as it reflects its light surroundings. We are more aware of its surface markings.
4. When the surface is dulled, or it reflects dark surroundings, its finish appears quite matte. The overall effect is far less interesting.

1

2

3

4

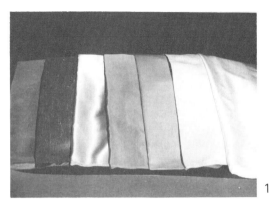

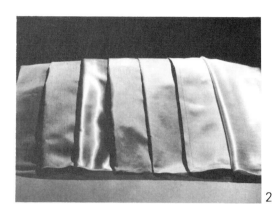

Textiles

Here several different materials have been lit (1) frontally from the camera position, and (2) by back light at 12H 11V. See how the appearance of some materials changes noticeably, while others remain reasonable similar under different light directions. Back light can create hotspots on horizontal planes, which reproduce as white detailless areas ('bloom, crush-out, block-off').
The textiles are (*left to right*): velours, slub rayon, satin, pure silk, cotton, linen, crêpe de Chine.

Light a vase from above, and its shape and proportions become distorted. Lower parts are thrown into detailless shadow. Place the vase on an illuminated panel, and its surface detail is underlit and clearly visible; but its overall appearance is strangely different from usual.

Lighting common materials

Because the way you light a surface can influence its appearance it follows that various types of material ideally require different lighting treatment. You would not expect the set-up that displays the full beauty of a filmy fabric to reveal the multi-faceted texture of a rough-hewn stone statue to greatest effect.

There are, indeed, optimum approaches that show each material most effectively. These are not rigid, there are many subtle variations. But those outlined in Table 4.3 are a general reminder of the most successful.

When you are lighting an *isolated display* there is no reason why you should not selectively light to suit that particular subject. You can bring out those special aspects that suit the occasion: revealing the translucency of a piece of jade, the sharply etched designs in engraved glass, the brush-strokes of an oil painting, the barely visible structure of an embedded fossil.

In most situations, you will find yourself faced with a wide range of materials. You cannot hope to light each item in a room individually. But that does not mean everything has to share a common lighting treatment. You will often be able to pick out major features of the location to bring out their particular qualities; as you will see when we talk about lighting 'places'.

5 Lighting people

Show the hero in a highly dramatic situation, with a half-lit face and deeply shadowed eye sockets, and your audience may accept the lighting treatment as 'powerful', 'moody', 'exciting'. Light a newscaster that way, and you may be out of a job!

'Bad' lighting

When most people talk about *'bad lighting'* they are usually criticizing *inappropriateness*. Ugly shadows, distracting hotspots, flat or excessively contrasty picture quality, burned-out faces . . . all are acceptable enough *when they are appropriate*. They may even be part of the 'atmosphere' of the occasion. But when they are quite unjustified, even the mildest of viewers can become incensed.

Table 5.1 Lighting people

	Front light	*Back light*
Lamp's vertical angle too steep	Harsh modeling which gives a haggard and ageing appearance. Gives black eyes, black neck, and long vertical nose shadow. *Emphasizes*: forehead size, baldness, and deep eyes. Figure appears busty. Casts hair shadows onto forehead. Hat brims and spectacles produce large shadows.	Nose becomes lit while face in shadow ('white nose'). Effect most marked when head tilted back. 'Hot top' to forehead and shoulders.
Lamp's vertical angle too shallow	Picture flat and subject lacks modeling. Produces shadows on background.	Lens flares or lamp actually in shot.
Underlighting	Inverted facial modeling. Shadows of movement beneath head level (e.g. of hands) appear on faces. Shadows may be cast up over background. 'Mysterious' atmosphere when underlighting used alone. Useful to soften harsh modeling from steep lighting. Reduces age lines in face and neck.	Largely ineffectual but can be used for back lighting women's hair. Shadows from ears and shoulders cast on face.
Light too far off camera axis	In full face, nose-shadow across opposite cheek. An asymetric face can be further unbalanced if lit on wider side. One ear lighter than the other.	Ear and hair shadows are cast on cheek. One side of nose is 'hot'. Eye on same side as back light appears black – being left in shadow while temple is lit.

Note: All lamp positions are relative to camera viewpoint.

Fig. 5.1 What is 'bad' lighting?
Various unattractive effects can arise
from unsuitable portrait lighting.
1. *Steep frontal key light.* Black eyes.
   harsh modeling, long nose
   shadow.
2. *Two frontal keys.* Twin nose
   shadows and shoulder shadows
   (latter can also arise from two
   back lights)
3. *Steep back light.* Hot head top;
   bright nose tip, and eartips; long
   bib shadow on chest.
4. *Oblique frontal key.* Talking profile
   seen on shoulder. Bisected face.
5. *Side key.* In full face bisects the
   face. Half of it crudely edge-lit, half
   unlit.
6. *Dual side-keys.* Create central
   'badger' shadow effect, and
   coarse modeling.

'Bad lighting' makes people look unattractive. They can appear tired, older, haggard, even ill. 'Bad lighting' draws attention to thinning hair, outstanding ears, a large nose. It can emphasize wrinkles and bags, scrawny necks. It can make a face look clownlike, bizarre, strange, ugly. It can cause it to look lopsided or formless.

Although a good make-up can enhance a person's appearance – even glamorize them – poor lighting can destroy its effect entirely. Unsuitable lighting can nullify all the skills of the make-up artist, creating a pathetic, disillusioning, even grotesque result.

Bad portrait lighting is an anathema! You can modify the appearance of most objects and locations very considerably without audience comment, but pictures of people are another matter.

Poor lighting is a visual insult to a guest appearing in front of the camera. But the irony is that in daily life, people are badly lit much of the time and we don't notice it! When the sun is at a steep vertical angle, or side-lighting a face, or when lights in a ceiling shine straight down onto a person ('top light'), we regard the results as 'natural'. Light someone that way in a studio interview, and the effect looks unattractive and quite unacceptable.

There is, it seems, a sense of 'appropriateness' that we apply critically when watching the screened image. In life we compensate, and make allowances. We don't even notice the hot tip to someone's nose, or the way their ears stick out, or a double neck shadow. But point a camera at them, and we find it *distracting*.

Styles in portraiture

The human head has a basic shape, and although facial characteristics vary, and individuals have their own preferences, the fundamental problems of lighting people are pretty constant.

Admittedly, some people are easier to light than others. It is a challenge to achieve an unobtrusive nose shadow when lighting a Cyrano de Bergerac character, and deep-set eyes pose difficulties.

There are preferences in the ways we like to see people lit. Generally speaking, the strongly contrasted lighting that 'suits' a rugged outdoor character would be regarded as unsuitable for a beautiful woman. Conversely, the defocused delicate half-tones that enhance the latter would seem out of place in a portrait of a coal-miner or a lumberjack.

On the whole, it would be fair to suggest that the images people prefer are usually stylized, and the result of their exposure over the years, to traditional motion-picture lighting. In film making it has long been the practice, where attractive close-ups are required, to light these separately from the long shots in order to achieve the optimum effect. Although the lighting treatment may not match, this is not considered too critical.

Apart from photography that is intended to provide a factual record (as in e.g. medical and archaeological work), most picture making is more concerned with *effect* rather than accuracy. Tests have shown, for instance, that people generally prefer color reproductions in which faces are rather warmer and somewhat yellower than in life, and consider this color quality more 'natural' than true color values.

The 'ideal' portrait

Conventional portraiture should not be a predictable 'rubber-stamp' treatment that follows a performer around wherever he goes. Like the screen fight in which the hero somehow never gets disheveled or dirty, the result would be unconvincing. That is not to argue, though, that it should fluctuate from burned-out faces to detailess murk, on the grounds that this sort of thing happens under true-life conditions!

Unorthodox treatments may occasionally come off brilliantly, and create arresting pictures. But don't rely on it.

In an ideal lighting arrangement each lamp has its particular function. The fill light does not destroy the modeling created by the key light. Back light makes its own separate contribution to the overall effect.

In practice, it is impossible to avoid unattractive modeling from time to time. As people move about during action, back light turns into disturbing side light, they lower their heads and develop black eyes! Sometimes you can improve a shot by rebalancing the relative intensities of the lamps, introducing some sort of supplementary key light, or relying on a camera-light to make the defects less obvious . . . but that is a dilemma that must wait for the moment.

Lighting faces

In principle, lighting a person is no different from lighting any other subject. In reality, though, there are strong emotional overtones. We can accept a shot of a jug, where the shadow of the spout spreads downwards over its surface; but a similar shot of a person with a nose shadow over the lips and chin is 'ugly'.

- Certain parts of the face must be seen clearly. The eyes are all-important. If there is no light reflected in them ('eye-lights', 'catch-lights') they seem to have a dull, 'dead' look.
- Other parts of the head can be allowed to fall into shadow. They may even look better if they are of a lower tone than the main features (e.g. the neck, and ears).
- If hair is too bright, it will look unnatural, frizzy, over-brilliant.

Begin by lighting yourself

As you will soon discover, when lighting people, even a slight change in light direction can alter a person's appearance considerably. Key them from the left instead of the right, or put the key directly over the camera, and they will look different. Not necessarily better or worse, but certainly different.

You can learn a lot about the effects of lighting angles by looking at yourself in a mirror! All you need is a small flashlight, a mirror, and a darkened room.

Looking in the mirror, point the flashlight straight at yourself, adjusting it until your nose shadow disappears. The light will be directly in front of you – the equivalent of a lamp beside the camera lens.

Notice how flatly lit your features are. When the light is dead frontal, shadows and surface modeling are absent. The resulting picture can be very uninteresting. A certain amount of shading and shadow are desirable, to give the head and features form and dimension. As you will have noticed, light from that direction is *dazzling*; a point to bear in mind when using lamps attached to a camera.

Raise the flashlight, and shadows will appear under your eyebrows, nose, chin, lips, ears. The steeper its vertical angle, the longer these shadows will grow. Modeling on the cheeks will become more pronounced. So will any frown-lines, wrinkles, and other facial irregularities.

If the lighting angle is too steep it will emphasize and exaggerate facial modeling as the lower parts of the head shade off. For a thin, gaunt face the overall effect starts to become less attractive at around 40°. A rounder face with less defined features may still look fine – even improved – when lit from quite a steep vertical angle; e.g. 45–50°.

With your lamp at about 30°, slightly tilt you head up and down, and notice how the shadow-lengths change. Under steeper lighting these variations are much more noticeable. Light that is well angled when you are looking straight ahead effectively becomes much steeper as you look down to read from a book, or speak to a child. Conversely, if you look upwards, the same lighting becomes effectively shallower, and more flattering. (Watch out for the way experienced broadcasters have a habit of keeping their heads up to avoid steepening the key light angle!)

Look at the effect as you move your head slowly from side to side. Your nose shadow spreads across the opposite cheek, and the side of the face further from the lamp becomes shaded. Turn your head a little further, and your nose shadow will meet the shadow on your cheek at around 45° off-axis. At first, there is a triangle of light on that side of your face, but as you continue turning, the whole area falls into shadow, leaving it half-lit.

Next, watch what happens when you move the light upwards *and* sideways at the same time. The shadows move diagonally downwards in the opposite direction.

Of course, you can go on to see the effects of 'side light', 'top light', 'underlighting' even 'back light'; depending on your agility, and how long your flashlight lasts out! The great merit of this particular exercise is that you can repeat it over and over, critically examining how light behaves.

Fig. 5.2 The effect of the lighting angle
As the angle of the light alters, it changes the person's appearance in various ways.

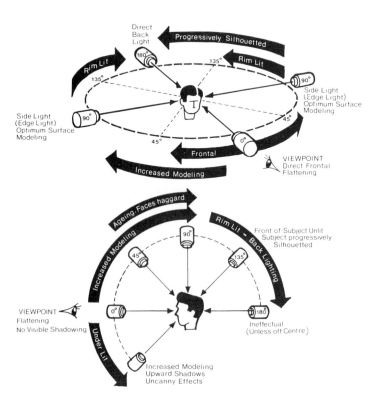

With a little patience you will soon get to know the 'feel' of different lighting angles; and that needs to become instinctive if you want to use light as a precision tool.

Now that you have seen the broad effects of different lighting angles you will find it helpful to go back over the exercise, carefully concentrating on *one* aspect of the face at a time, noting how it changes:

● The nose shadow – check its length and shape.
● The eyes – watch how the appearance of the eyes changes (reflections). See how shadows fall across the eyes.
● Wrinkles, frown-lines, and laughter-lines around eyes – varying in prominence with lighting angle.
● Lip shadows – well-molded lips vary in prominence, and their apparent shape alters.
● Neck modeling – note how the neck shadow can create a 'dirty neck' effect.
● Ears – check the prominence of ears. See how shadows grow within ears (making them dark inside), and streak across the head.
● Eyelash shadows – falling on the upper cheeks and the side of the nose (particularly when false eyelashes are being worn).
● Cheek-bone and cheek modeling – can become over-prominent.

Lighting zones

There is nothing mechanical about lighting people. In an ideal world you 'custom-build' all lighting, styling it to suit each person's facial characteristics. You can, for example, design the lighting for a regular announcer, host, or guest to compensate for any imbalance or irregularities in their features, so complementing their make-up treatment.

More often, though, you will find that the same lighting arrangements have to suit a number of *different people* who sit in the same chair, or stand on the same spot. But that does not mean that you have to resort to 'general illumination'.

A good way to begin lighting people is to think about the light as coming from a series of *zones*. Within each zone you will achieve effective lighting. The lighting angle you choose within a zone is selected to suit an individual person's requirements.

These zones were derived originally by studying many heads and analyzing the effect of angular changes in the light. Where the angle of a lamp is too slight it may be ineffective. Where the angle is too steep the result may look crude, and unattractive. There are usually optimum positions that suit most people, and you will see these in diagrammatic form in Figure 5.3.

Although we move our heads around in all directions, the general effects of directional changes can be summarized as a series of plan (horizontal) views:

Fig. 5.3 Lighting zones
For each basic head position there
are optimum light directions that
produce attractive portraiture.

Angles have been rated here as:
White = good
Gray = fair
Black = poor

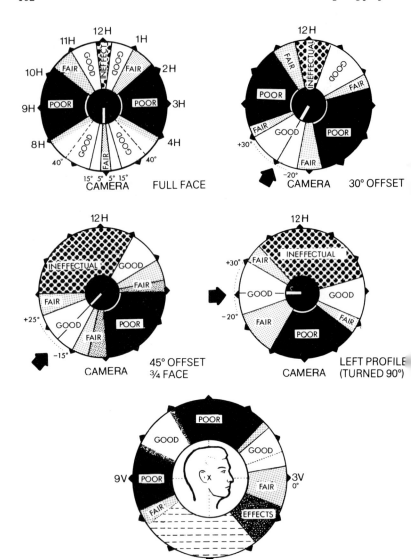

- Facing the camera ('full face')
- Slightly turned to the side ('3/4 frontal')
- Half-turned to one side ('1/2 frontal')
- In profile ('side view')

The effects of the *vertical* angle of the light, which you can see in
the last figure, will apply to each of these head directions.

Classifying light directions

Now you've seen the general results of changing the lighting angle
let's take a closer look at these characteristic effects. To do this, we

need a method of specifying lamp positions, so we'll use the '*clock method*' you met earlier. Just to recap, the person is located at the center of a couple of clock dials.

- On the *horizontal* dial, which shows where the lamp is located *round* the subject, the camera is positioned at 6 o'clock – i.e. 6H.
- On the *vertical* dial, which shows the *steepness* of the lamp, the camera is positioned at 3 o'clock – i.e. 3V.

For general directions we shall refer to 'hour' positions (30° intervals), although 'minutes' will give you more precise (6°) positions if you need them.

The basic effects of lamp positions

When working under pressure there is too little time for experiment and correction, and inspiration has to be coupled with down-to-earth practicality.

When lighting people you need to know, for instance, that if you light the human head from certain angles it will result in one eye socket falling into deep shadow, or the nose will have a bright tip, or a distracting nose shadow will stretch across a cheek. When there is a problem you need an immediate solution. When you see an unattractive effect you need to immediately recognize what is causing it, and how it can be remedied. *Analysis* helps you to anticipate, to avoid problems, and to diagnose the causes. As we look in detail at the effects of light direction in portraiture you will become increasingly familiar with its potential pitfalls.

Full face: frontal lighting – central

■ *6H/3V: dead frontal: level* A key light within about 10° of the lens position is going to dazzle anyone looking towards the camera.

Table 5.2   Lighting balance

	Too bright	Too dim
Frontal light	Back light less effective. Skin tones are high and facial modeling lost. Lightest tones tend to be overexposed. Gives a harsh pictorial effect.	Back light predominates and often becomes excessive. Darker tones are underexposed. Can lead to muddy, lifeless pictorial effect.
Back light	Excessive rim light. Hot shoulder and top of heads. Exposing for areas lit by excess back light causes frontal light to appear inadequate.	Two-dimensional picture which lacks solidity. Subject and background tend to merge. Picture appears undynamic.
Filler	Modeling from key light reduced and flattened.	Produces excessive contrast and subject too harshly modeled.

The head – vertical lighting angles

1. With a lamp near the camera, facial modeling is not very pronounced.
2. As you raise the lamp to a slightly steeper angle, downward shadows grow longer and various features (eyes, nose, chin, cheekbones, etc.) become more strongly modeled.
3. This double-exposure shows that, although the nose shadow grows slowly, the chin shadow quickly spreads down the neck as the angle increases.
4. As downward shadows lengthen, they seem progressively to 'age' a person. If someone has plump or less-defined features, their appearance may not change much, even under quite steep lighting. But where a person has angular or strong facial contours (e.g. deep-set eyes), the modeling soon appears crude and harsh as the lighting becomes steeper. Finally, at very steep angles, the head becomes skull-like.

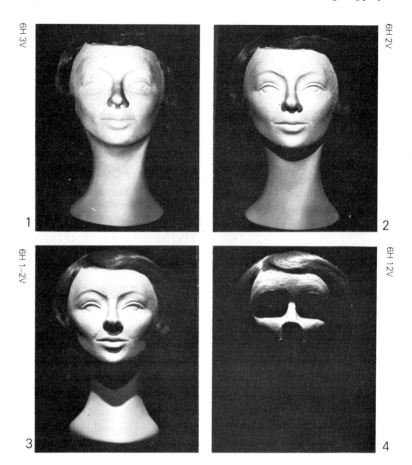

They are likely to half-close their eyes, and will probably find it impossible to read cue-sheets or prompters. Even a low-intensity camera light has to be used discriminatingly if you are not going to discomfort the performer.

Frontal keys are very 'flattening'. In a shot lit by a dead frontal key even strong facial contours and deep wrinkles will be suppressed. If you use a *soft* light-source for the key the bags and wrinkles in an ageing face can be 'lit out' to a remarkable degree. Whether the effect is flattering, or whether it produces a portrait bearing little resemblance to the real person, is a matter of judgment. If you use a dead-frontal key on a smooth unwrinkled face it is liable to appear somewhat flat and blank.

A dead-frontal key light also has the disadvantages that

● It casts a person's shadow onto the background immediately behind them.
● It produces strong light reflections in spectacles.

■ *6H/5V up to 6H/4V: underlighting* Now let's move the frontal lighting down *below* the lens. All shadows are cast upwards, forming strangely unfamiliar modeling on the face. Areas under the brows,

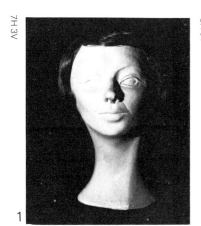

The head – horizontal lighting angles

As a key light moves *round* the head (1 . . . 2 . . . 3), a person's appearance is considerably altered. One side of the face becomes increasingly lit, while the other is progressively shadowed.

If you light from both sides (4), the result is a centrally shadowed 'badger' effect, with dark eyes, and a prominently-lit nose.

When the key is raised *and* offset (5), as it usually is, the shadows and the facial shading will be diagonal.

Fig. 5.4 Changes in effective light angle

The fundamental effect of light direction remains, but the strength of modeling alters with changes in the subject's position.
1. A key light is correctly angled for good facial modeling.
2. When the head tilts down the facial modeling steepens.
3. Tilting the head upwards, it looks more directly into the light and modeling is reduced.

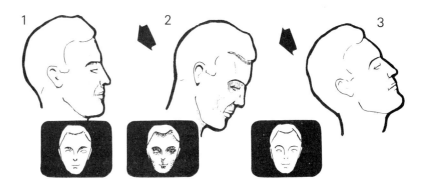

the nose and the chin, where we normally see shadow, are fully lit. Nostrils can become quite prominent. Strange shadows appear on the bridge of the nose, above the top lip, and on the brow. The person's eyes are very strongly lit.

We usually associated underlighting with bizarre, horrific, uncanny situations. Dramatically, this is a very useful visual cliché. But underlighting does arise naturally enough in daily life, wherever any light source is lower than head height (e.g. a table lamp, a bonfire).

Underlighting has other applications, too. As an emergency measure you can add low-level soft light to reduce the effect of the harsh modeling from a steep key light. You can use soft underlighting to glamorize and youthen. Atmospherically, under-lighting can suggest the warm, comforting fireside.

The practical disadvantages of underlighting include the way in which it is liable to spread upwards over the background, cast shadows onto faces, and overlight legs.

■ *6H/3V to −2V: from level 0° up to e.g. 45°*   As the vertical angle of the key light increases, modeling improves. Shadows become more noticeable under the brows, nose, lips, cheeks and jaw. We have the impression that the face is getting narrower, and the nose appears to lengthen.

Central lighting tends to draw attention to the proportions of the face, and a high forehead (or receding hair) can look more pronounced.

■ *6H/2V to 1V: e.g. 45–60°*   The steeper the key, the harsher the modeling. At these angles you get a number of unattractive effects. The eyes are thrown into shadow, and the eye sockets (orbits) become black and skull-like. If there are catch-lights reflected in the eyes, the effect can be strange and mystical!

A long nose shadow spreads over the lips and chin. The neck is thrown into deep shadow, and a bib-like shadow area covers the upper chest. Thinner faces take on an emaciated look. The forehead becomes over-pronounced. Eyelash shadows sweep down over the cheeks, and bushy eyebrows bristle.

Steep keys are only suitable for occasional dramatic effects (60–75°).

■ *6H/1V: top light*   Lighting from this angle flattens the top of the head, and gives the shoulders prominence. The front of the face is barely lit; except as the head is tilted upwards, when a long nose-shadow spreads over the lower part of the face. This is an ugly light direction with little pictorial value.

■ *6H/12V to 11V: top light*   Overhead lighting produces quite crude, ugly effects. The top of the head and shoulders are strongly lit. The top of the nose and the ears are illuminated. The brows may be lit, but the rest of the features are in shadow.

Avoid the temptation to use this kind of top light as a *hair light*. It flattens the hair, and invariably produces a disturbingly bright nose!

This is essentially a lighting effect to be used sparingly, if you are simulating a particular type of environment. Where illumination from above is unavoidable (e.g. ceiling lights, sky-lights) the only solution is to keep its intensity to a minimum, and augment it with more attractive lighting treatment.

7H 2V

6H 2V

5H –2V

1

2

3

The head – frontal key light
Here you can see the effect when the raised central key light at 6H 2V (2), is moved to the *left* as in (1) at 7H 2V, or to the *right* as in (3) at 5H 2V.

Because most faces are asymmetrical, a person's appearance can change considerably with the key direction. Most have an optimum key position: left, center, or right.

Full face: frontal lighting – horizontal offsetting

■ *+4H to −6H; +6H to −8H*   As you offset the key light to one side the face seems to become increasingly lop-sided. The greater the amount of offset, the greater the apparent unbalance. This happens because shading causes the far side of the face to look narrower, while the more strongly lit half nearer the key remains prominent.

Several other things happen as you increase the lamp's offset angle. The nose shadow spreads across the face onto the opposite cheek. At first it appears as a shadow outline on the cheek and the corner of the far eye. Then as the lamp is offset further, this shadow joins with the increasing cheek shading to form a light triangle. The triangle gets smaller as the horizontal offset angle increases, until finally the side of the face opposite the key is entirely in shadow.

After a while, most tyro lighting directors develop a phobia about *nose shadows* – not without cause. They either try to avoid a nose shadow altogether (by lighting along the nose line, or using soft light) or spread it across the face to form a 'triangle-on-cheek' effect. There are those who favor a *cross light* key position (45°H/45°V) for dynamically lit full-face shots, but the effect is somewhat mannered, and not altogether satisfactory.

Full face: frontal lighting – diagonal offsetting

In most practical lighting your key will be both raised and offset, so all shadows will be cast *diagonally*, in the opposite direction from the key. For most purposes you will find that:

● The optimum *vertical* angle for the key lighting a full face is well within the range of −3V to −2V, i.e. 10–45° from the horizontal – preferably avoiding these extremes.
● It is best not to *offset* the key light beyond the zone −6H to 4H, or +6H to 8H (5–60° from center). Preferably keep well within these limits.

Nose shadows

The closer the key is to the lens position, the shorter will the nose shadow be – but at the expense of reduced facial modeling, and the prospect that you will dazzle the person.

1. The nose shadow can be quite obtrusive.
2. Moving the lamp round the subject will lengthen the nose shadow, until at around 45° horizontal, 45° vertical, it will join with the cheek shading to produce a cheek triangle, which many consider less obtrusive – even attractive.
3. However, if the head turns *away* from the key (or the key is moved towards the shaded side of the face), the triangle breaks, and the long nose shadow appears.
4. *Cross frontal* key lights result in double nose and chin shadows. These are not only distracting, but can apparently broaden the tip of the nose, and sharpen the chin.

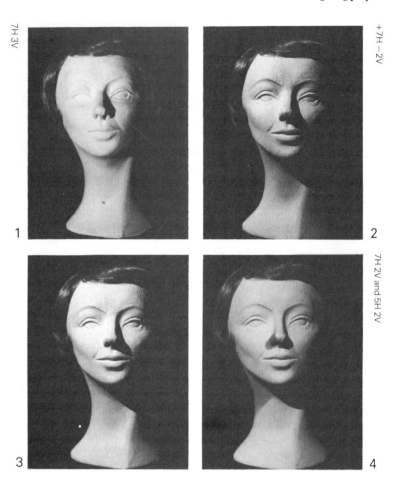

Try to avoid extreme angles. If the key light's vertical or horizontal angle is too small, the result can be disappointingly flat. When the angle is too great, modeling can look too harsh.

For most subjects a useful working region when lighting the full face is within

- A *vertical* angle of around 20–40° and
- An *offset* angle of 15–35°.

Full face: side light

Moving the key round to either side of a person (3H or 9H) results in *side light*. Side light produces extremely poor portraiture for a person facing the camera. It bisects the face. The inner corner of the near eye socket is shaded, the nose contours are emphasized, and a shadow of the profile falls on the far shoulder. Where the key is steep (e.g. above 2V or 10V) results become even more grotesque.

If the full face is lit from *both* sides (3H and 9H) the result is a curious 'badger' effect, with a broad dark band of shadow down the center of the face!

Turning the head

■ *From full face towards camera right*   Let us assume that someone facing the camera is lit with an offset key light on camera right (30° offset and vertical angles). As a head turns, the effects of the key light alter.

● Turning towards the light, the nose shadow and facial shading diminish.
● Turning away from the light, the nose shadow grows. Lighting that provided good modeling for the full face may no longer suit the new head position.

■ *The head angled towards camera right*   We can think about the partly turned head in two stages:

● '3/4 frontal' – slightly turned to the side (30° offset); facing 7H or 5H.
● '1/2 frontal' – half-turned to one side (45° offset); facing −8H or −4H.

At the same time, it is convenient to consider the face as divided vertically at the nose to provide the 'near side' and the 'far side' of the head. The direction in which the head faces is often called the *nose line*.

There are two important things to notice when the head is angled. Unless the person is looking directly at the key light there will always be a nose shadow. Its vertical length will depend on the lamp's height.

● If the key is on the side of the head *nearer* the camera, the nose shadow will fall on the far side of the face and may not be visible in the shot. The near side of the face will be quite strongly lit, so there may be no need for fill light.
● If the key is on the *far* side of the nose line the nose shadow will be clearly visible in the shot. (Obviously, if the person has a small

Head direction
The lighting that is effective for one head direction may not suit others.
1. When this head is turned three-quarters right, *facing the key* at +4H +1V, modeling is optimum.
2. But when it turns towards camera, we see a strong nose shadow (cheek triangle) and neck shadow.
3. Turned further, looking away from the key, modeling is decidedly unattractive.

1

2

3

nose, the shadow is less obtrusive.) The head has a much more 'solid' appearance from this key position, as shading reveals its shape and surface contours. If you feel that the tonal contrast is too great for the situation you can add soft fill light.

When you move the key further round the far side of the head (9–11H) its near side becomes progressively shadowed, until finally it is only rimming the head as back light. So that you can see the effect, try lighting the '3/4 frontal' to '1/2 frontal' positions from a lamp much nearer the camera (e.g. 6H). You will find the result uninterestingly flat.

Move the key further round towards 5H, so that the face is *looking away from* the key, and you will see how ugly modeling forms around the near eye and the nose, the front features becoming fully shadowed at about −4H.

At 3H the key is edge-lighting the head; strongly modeling the front of the face and neck, and casting forward shadows of the ear and hair.

From 1H to 2H the lamp becomes a 'kicker', catching the near side of the face and attractively delineating the edge of the head.

Remember, in all these examples we have been assuming that the head is looking towards the *left* of the screen ('camera left'). If the head faces *right*, the directions all become reversed, so that 7, 8, 9, 10, 11, for example, become, respectively, 5, 4, 3, 2, 1 and vice versa.

Profiles

When the head is turned to one side, in a *profile* position, emphasis is on the outline of the face. If the person is facing camera left (9H) the most attractive key positions are within about 10° of the nose line. If you light from the far side at level up to about 9–10V you will see the edge light skimming along the near side of the face, with a short nose shadow. The jawline and the cheek are shaded towards the ear.

Raise the key towards 11V and, as you would expect, the face becomes increasingly gaunt. Shadows become longer under the brows, cheekbones, nose, lips and chin.

If you light the profile from the *near side* of the nose line (e.g. at 8H) the side of the face nearer the camera is more fully illuminated, but less well modeled. The nose shadow is out of sight, on the far side of the face. Some Lighting Directors feel so strongly about avoiding the nose shadow that they are happier to accept the reduced modeling of this lamp position. Where you want a *high-key* effect (i.e. predominantly light tones), or where the person has a well-contoured head, this nearer key position is often more successful.

For a left-facing profile you will find the key positions from 7H to 6H pretty unrewarding, for the facial modeling is poor. Move the key further round from 6H to 4H, and the effect becomes increasingly unattractive.

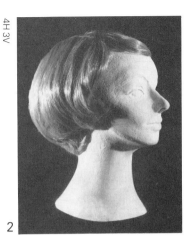

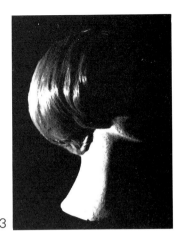

1

2

3

The angled head

When the head is at an angle, both the facial modeling and the nose shadow can become more critical.
1. Where the key light is offset relative to the head direction, a prominent nose shadow forms.
2. If the person is looking straight towards the key, the nose shadow disappears, and general facial modeling is good.
3. A side light from around 9H 3V gives form to the hair and the neck.

The head – the profile

1. If someone turns to look *away* from the key light, the effect can be very unattractive: a shadowed eye, white nose, shadowed face. A brow shadow defaces the nose. Only the hair is lit successfully.
2. Frontal light produces flat, unmodeled features.
3,4. As the key moves round the profile head, modeling is emphasized (edge lighting), and the nose shadow grows.
5. You cannot compensate for ugly modeling by adding soft frontal fill light.

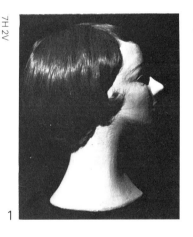

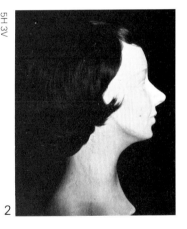

1

2

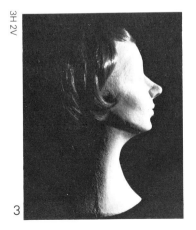

3

4

5

1

2

3

4

5

6

+4H +1V (×6)

The head – filler position for the full face

This series of shots shows the results of using different positions for the fill light.

The top row (1, 2, 3) shows the final effect of a key on camera right (4H 1V) plus fill light. The next row (4, 5, 6) shows the effect of the fill light alone.

1,4. Using a *frontal* fill light.
2,5. The fill light is offset 45° left.
3,6. Left side fill-light. Generally unsatisfactory for full-face portraiture.
 7. Frontal fill light cannot compensate for the harsh modeling caused by a steep key light.

7

    Nearer to 3H at an elevation of from 3V to 2V, a lamp will produce a useful rimming side light, which will supplement the key light at around 9H.

    Move the side light slightly forward of 4H, and you will see that the cheek and ear are skimmed with strong hair and ear shadows. A lot depends, of course, on the nature and style of the hair.

The head – filler position for the profile

1. High frontal fill light (e.g. from hung soft light) produces unattractive modeling. (The key is at 1H 2V.)
2. Filler from 4H 3V destroys modeling from a key light at 3H 3V.
3. When the subject is keyed from +4H 3V, a frontal filler controls contrast.
4. However, it does not correct the harsh modeling from an *oversteep* profile key (4H 1V).

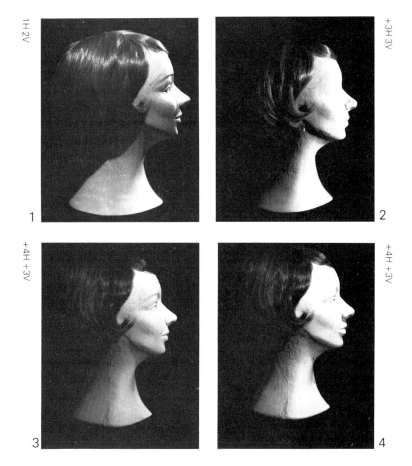

If, instead, you move the side light towards 2H you will see that the light falls off as the right-hand rim narrows. Move it still further back, from 1H to 12H, and only the top hair is tipped with light.

As before, if the head is facing right, these various light directions will be reversed.

Fill light in portraiture

Is it necessary to have fill light? Well, it will not usually be needed if the key light is quite frontal to a camera (e.g. ±10°). Most of the subject is fully lit so there is little shadow area to fill.

Where an appreciable amount of the face is in shadow, fill light is usually desirable, even if it is of relatively low intensity. So you will normally need it, for example, when the key is offset by 25° or more, or the camera has a profile shot with a key near the nose-line. Then fill light prevents the final portrait from becoming over-contrasty. In a highly dramatic situation, however, any fill light can spoil the tension of high tonal contrasts.

While a male with rugged features may need little or no fill light, shots of women and children will usually be enhanced by a medium to high level of fill light.

The tone of skin can influence the amount of fill light you need. While facial modeling is often much better defined for darker skin tones, shadows are more likely to require fill light; particularly when someone is wearing light-toned clothing. Lighter skin tones, on the other hand, all too easily reproduce as over-pale and poorly contoured, so may need proportionally less fill light.

Fill light position

The optimum position for the fill light is influenced by where you place the key light, and the angle of the head. If, for example, you fill a left-facing profile (keyed at 9H) from around 7H or 8H it will not wash-out the shading at the back of the head, on camera right. Fill from the front (6H), and it will reduce this shading to some extent. If your subject is moving around, then you might need to position the fill light to suit varying head directions.

Fig. 5.5 Portraiture fill-light position
The best position for the fill light will vary with the head position. Here the variations are shown for full-face, three-quarter face, and profile positions.

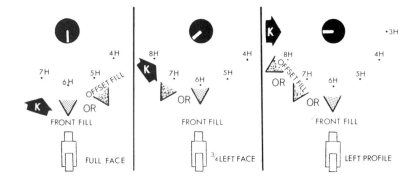

Back light

Back light, as we saw earlier, serves to rim the subject with light to reveal its edge contours and to help to separate it from the background.

Dead back light

With the lamp directly behind the subject at 12H/9V, looking towards the camera (at 6H/3V), the lamp is hidden by the subject and the light only outlines the hair and ears. Raise the lamp, and it will shine over the subject straight into the camera lens! Raise it further to the 10–11V position (30–60°) where it is out of shot, and the shoulders together with the top of the head become progressively lit. Gradually the contours and texture of shoulders and arms become clearer.

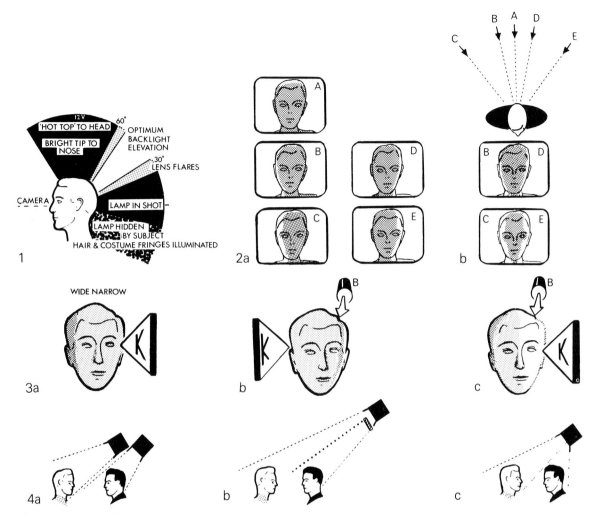

Fig. 5.6 Back light

1. *Effect of vertical angle.* The effectiveness of back light changes with its vertical angle.
2. *Offset angle.*
   (a) Back light rims the subject's sides, to an extend depending upon its horizontal angle. The width of the illuminated rim broadens as the light moves from behind the subject towards a side position. Any subject contouring or protuberances (e.g. ears, hair) will cast long shadows forward over the side of the face when slightly offset back light is used, these shadows shortening as this back light moves towards ¾-back position.
   (b) Combined back lights produce double-rim lighting, which exaggerated, becomes a 'badger' effect. Slight double-rim lighting at the subject's edges can produce a glamorous, attractive visual effect. If inappropriately applied, however, it can overemphasize head outlines (i.e. ears, coiffure) and apparently exaggerate head-width, neck thickness. The shaded centre-stripe effect that results from broadly angled double-rim or from side-lighting is seldom appealing. It often arises in horseshoe-grouped shots where cross-light for facing speakers bisects a centrally-positioned person.
3. *Facial balance.* The direction of back light may modify the apparent balance and width of the face.
   (a) When the key is directed on to the narrow side of the face, the wider side may be visually narrowed by shading.
   (b) When the key lights the wider side of the face, a back light on the narrow side rims it, effectively widening it.
   (c) A back light on the same side as the key will tend to nullify the key's modeling and overlight one side fo the face. Increased compensatory fill light would encourage flat results.
4. The amount of back light required by one subject may be less than the key light intensity for another nearby subject. To obtain a suitable balance one may:
   (a) Use separate lamps where space permits;
   (b) Use a localized diffuser;
   (c) Arrange a single lamp so that the key is fully positioned on one subject, while its beam-edge fall-off serves to back light the other subject.

If the back light is too steep (e.g. 11V–12V) a dark bib shadow falls on the chest, and the top of the nose is strongly lit, making it extremely prominent – especially if the person tilts their head up or leans back.

Low-level back light, from e.g. 7V to 8V behind the subject, will underlight filmy costumes, and the underside of coiffures and wide-brimmed hats; but it is seldom used.

Offset back light

If you offset any back light so that it is around 11H/10V or 1H/10V the effect is asymmetrical, lighting one edges of the subject more than the other. A rim of light appears on one side of the head and neck, and one shoulder is brighter than the other.

If you offset the back light even further (e.g. to 10H or 2H) this lop-sided effect is very obvious. More of one side of the face is lit, with a large forward shadow of the emphasized ear. The eye on that side is rimmed around the brow and temple, so that its socket looks shaded. From the front, the person appears to have one black eye!

Back light – vertical angle
As a back light directly behind the subject (dead back) is raised, the amount of light falling on top surfaces increases, progressively rimming the subject.

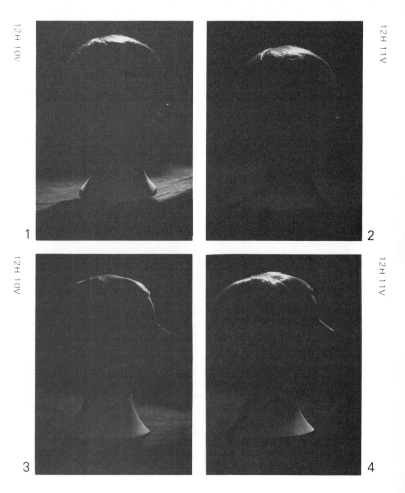

1H 10V

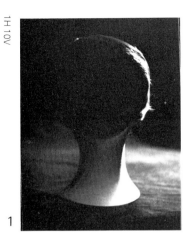

1

2H 10V

2

+2H 10V

3

Back light – horizontal angle
As a back light is moved round a subject, one side of it becomes increasingly lit. From a slight rim at (1) (chiefly on the hair), the side of the face becomes increaingly modeled (2), illuminating the nose tip, and eventually producing a 'black eye' effect (3) with crude facial modeling.

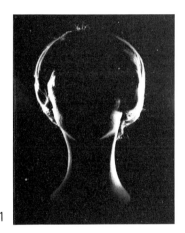

1

2

3

Back light – double rimming
1. Lighting with two angled back lights, either side of the subject, produces a *double rim* effect. The final result largely depends on the person's hair style.
2. Where hair is swept back or short, double-rimming reveals the facial structure.
3. Glamorous portraiture results when double rimming is combined with soft frontal lighting.

Offset and the key

If you offset the back light on the *opposite* side of the head from the key it will attractively rim the head on that side. But if you offset it on the *same* side as the key it will increase the amount of light falling on part of the head (perhaps causing localized over-exposure), and will tend to cancel out modeling created by the key.

When you want to enhance the impression that all light is coming from one direction (e.g. a person standing at a window in daylight) it may be more realistic to have the back light on the same side as the key. But it is a technique to use carefully, for it can make the head look unbalanced.

Fig. 5.7 Lighting a single person

The plans and elevations show typical range of lighting angles within which good portraiture is achieved.

1. Full face.
2. Offset head (¾ front, 45° turn).
3. Profile.
4. Vertical angles for each are shown.

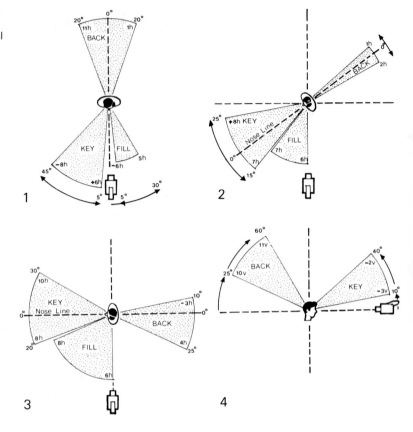

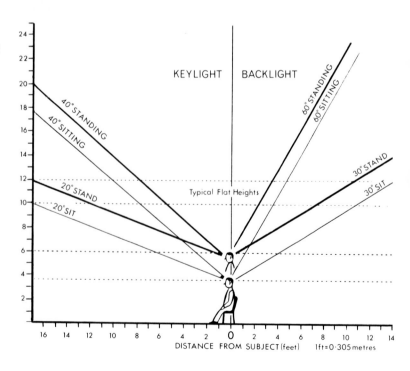

Fig. 5.8 Vertical angles

A quick way to check the vertical lighting angles is to use the heights of flats as a guide.

The head – effects lighting

Occasionally, you may want to create a strikingly dramatic effect when lighting the head:

1. *Underlighting* is associated with uncanny, horrific situations.
2. Where the underlighting is steeply angled, the eyes become shadowed and the face becomes mask-like. Hand movements can cast shadows on the face.
3. *Side lighting* bisects the face, and creates strong coarse modeling.
4. When lit from *overhead*, the head becomes crudely modeled and skull-like.

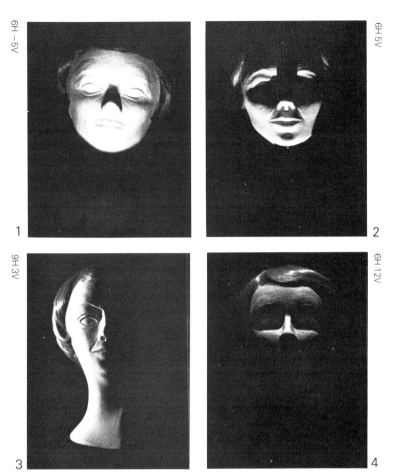

Double-rim back light

Another approach that is very effective *when appropriate* is *double-rim* lighting. Here you have a pair of back lights (e.g. at 11H and 1H) to create a rim on either side of the subject. The effect can be quite beautiful, but it is liable to broaden the face, and make the forehead look higher. Think twice before using it when lighting men.

There are Lighting Directors who use double-rim lighting simply because it achieves symmetry. Unlike single-point back light, both shoulders are equally bright. But the pair of forward-pointing head shadows are distracting, so any benefits are really outweighed.

The realities of portraiture

Portraiture dynamics

When you light a *scene* it stays put! *People move around*, looking great one moment and downright unattractive the next. What they do in rehearsal can vary during the 'take'. As we saw earlier, your audience is a lot more critical about what *people* look like than any other subject appearing in front of the camera.

There is nothing more frustrating and more time occupying than seeing an ugly effect such as a 'black eye' or a 'white nose' and not knowing how to cure it. With portraiture know-how at your finger tips you will immediately recognize the solution. Whether you have the opportunity to do anything about it is another story!

Types of portraiture

The way you light a person can vary considerably with the occasion. You will change the emphasis and character of your approach to suit the general effect you are seeking:

● *Natural portraiture.* This is the sort of artless effect you achieve when shooting with available light on location. It is often unflattering, but one's impression is of totally natural illumination, unsupplemented by extra lighting treatment. The illusion can often be surprisingly difficult to simulate in the studio.
● *Formal portraiture.* Here the shots and the lighting have a self-conscious air about them, as in a regular studio interview. Lighting is arranged to provide an attractive effect.
● *Character portraiture.* Here lighting is used to emphasize the character or personality. Lighting may draw attention to the subject's personal features: wizened, unshaven, weatherbeaten, delicate, soigné, etc.
● *Glamorous portraiture.* Here we use light to add emotional overtones, and to beautify. This is the realm of soft frontal, haloing back light, and soft-focus lenses.
● *Environmental portraiture.* Lighting treatment will only be convincing if the way people are lit is compatible with their environment. If a shot shows a group of card players lit by a low central light the treatment should create exactly that effect. Whether you actually use a single central lamp or a series of localized keys to produce the harsh steep lighting is unimportant. It is the result that counts.
● *Bizarre portraiture.* Deliberately strange facial lighting to provide uncanny, ugly, grotesque effects.

Corrective lighting

When you meet for the first time someone you have previously only seen on the screen, they often look puzzlingly different. They may, of course, be taller or shorter, older or younger than you expected, but there are other indefinable differences, too, from the way they look on camera.

These dissimilarities are not just a matter of make-up treatment. A girl who seems very homely off-screen appears attractively vivacious on the screen; while a beautiful woman may look puzzlingly ordinary. The lens may portray a handsome man as 'foppish', and a quite run-of-the-mill face as 'characterful'.

It is often said that 'there are those whom the camera loves', but the answer probably lies in their *facial proportions*.

Lighting can emphasize or reduce variations in a person's features. Take a series of shots of someone you know well, with various key light positions; left, center, and then on the right. You will certainly see changes in their appearance. In some cases, the results can be startling!

Looking closely at a number of faces, you will rarely find them symmetrical:

- The forehead looks disproportionately large or small.
- The two sides of the face have different widths.
- The eyes are not in line; one is higher in the face than the other.
- The nose appears long, short, bent, angled . . .

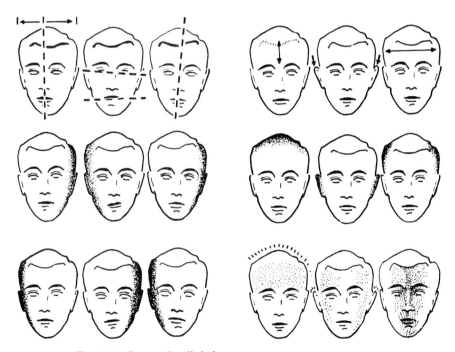

Fig. 5.9 Corrective lighting
The human face is rarely symmetrical. Most faces have characteristic unbalance (*top left*) or disproportion (*top right*).

On camera, such irregularities may be emphasized, but can be disguised with varying success by make-up and, where the subject is stationary, by lighting (*center*).

Where the subject is mobile, corrective lighting is impracticable, and irregularities may even be exaggerated if incorrectly lit. For example, by the key light on the wrong side (*bottom left*), by top light, combined top light and back light, and double-rim lighting (*bottom right*).

- The mouth is set at a slant.
- Mouth and eye-line converge on one side.
- Left and right profiles can look remarkably different.

We overlook many of these characteristics in daily life, although if we see someone in a mirror the discrepancies may be more apparent in the reversed image.

As you can see in Figure 5.9, light can exaggerate or reduce these effects. We may need to play them down in order to make a person appear as we normally recognize them. If, for instance, you key a person on the broader side of the face it will emphasize the difference, and unbalance the face further. Key them on the narrower side, and the shading on the broader side will narrow it, apparently evening-up the features.

Unfortunately, there are times when in 'correcting' (disguising) one fault you may draw attention to another. Steepening a key to improve a bulbous nose may make the eyes become small and 'piggy'. Adding filler to the eyes can then reduce modeling, and cause the neck to thicken!

You will soon learn to avoid emphasizing a broken or generous nose with coarse or strong highlight shading. You will recognize how a long nose becomes longer with a shadow beneath it, and how the tip of a pointed nose appears to twist with a diagonal side-wing shadow.

You must always anticipate changing head positions. Even when a person is sitting still, there will always be a certain amount of animated head movement. Facial modeling varies, and momentary defects may go unnoticed. But you will find that, for example, a delicately poised triangle of light on the cheek which looks fine when the head is still can disintegrate into a long nose-shadow or a half-lit face as the head moves.

If you light a full face with an offset key at 8H, when the head turns towards the key, the lamp is now pointing straight down the nose line, and the modeling is excellent. But if the head turns the other way, towards 4H, some very unattractive things happen. The features are virtually flattened on the key side, with ugly cheek and eye-socket shadows appearing on the nose. The important frontal area of the face is shaded (lit by filler alone), and the cheek, ear, and neck areas are over-emphasized by the resulting side light.

And this, as you will see, is the sort of lighting quandary that can arise when a subject moves about, and is shot from several viewpoints.

Pitfalls in portrait lighting

You have now met a number of the unattractive visual effects that can bedevil portrait lighting. These can considerably affect a person's appearance. Worse still, they can also moderate how the

Table 5.3   Static portraiture – corrective approaches

		Lighting	*Camera*	*Subject*
Hair	Light	(11), 15, (18), (20)	B	
	Dark	(10), 21		
	Thinning	11, 15, 16, (18), 20		
	Bald	2, 13, 15, 16, 18, 19, 20	B	
Forehead	Prominent	2, 13, 14, 16, 18	B, D	d
	Wide	(3), 5, 8, 10, 13, 14, 18	B, D	d
	Narrow	2, 4, 6, 7, (9), 11, 12	(A)	
Eyes	Deep-set	2, 7, 22	B	d
	Protruding	16, 18	D	e
Nose	Large	2, 4, 6, (7), 9, (11), 14, 16	D	a, d
	Small	1, 3, 5, (8), 10		e
	Broken	2, 4, 7, 9	D	a
	Long	2, 4, 7, 9	B, D	a, d
	Bent	2, 4, 7, 9	D	
Mouth	Large	2, 4, 7, 9, 16, 20	D	b, (c), d/e
	Small	1, 3, 8, 10		
Chin	Large	2, 4, 7, (11), 13, 14, 16, (17), 18, 20	D	a
	Small	1, 8, (10), (22)		
	Narrow	(12)		b, (c), d
Neck	Thick	4, 13, 14, 16, 17, 18, 20	(A)	e
	Wrinkled	2, 7, 11, 16, 17, 18, 20, (22)	(A), C, D	e
	Double chin	1, 7, (11), 16, 17, 18, 20, (22)	A	d
Ears	Large	3, 13, 14, 15, 16, 17, 18, 19, 20	D	b, (c)
	Prominent	3, 13, 14, 15, 16, 17, 18, 19, 20	D	c
Features	Wrinkled	2, 4, 7, 11, 14, 22	C	b
	Unmodeled	1, 3, 5, 8, 10, (12), 18, 20	D	b, (c)
	Broad	5, 8, 10, 13, 14, 18, 20	A, D	b, (c)
	Narrow	2, 6, (9), 11, 12		(c)
Spectacles		1, 3, (9), (11), 16	D	d, e
Figure	Heavy	2, (3), 5, 8, (10), 13, 14, 16, 19	D	b, (c)
	Slight	(11), 12		a
Disfigurement		14, 16, 17, 18, 19		b, (c)

KEY

Lighting:

1	Raise key light	12	Double rim light
2	Lower key light	13	Avoid double rim lighting
3	Offset key light	14	Avoid offset back light giving emphasis
4	Key more frontally	15	Reduce back light
5	Key far side of ¾ face	16	Draw emphasis to other characteristics
6	Key near side of ¾ face	17	Hide in shadow
7	Avoid hard key	18	Shade off area
8	Avoid soft key	19	Merge area with background
9	Watch nose shadow emphasis	20	Use localized scrim
10	Increase lighting contrast	21	Increase hair light
11	Decrease lighting contrast	22	Include some soft underlighting

Camera:

A Use higher camera angle
B Use lower camera angle
C Perhaps use lens diffuser
D Particularly avoid close wide
   lens angles

Subject:

a Use full-face position
b Use ¾ face position
c Use profile position
d Tilt head upwards
e Tilt head downwards

Note: Items in parentheses may apply in certain cases

audience reacts to that person. Here is a gruesome summary of lighting defects:

Nose
- The shadow cast by the nose can make it look longer or distorted.
- It can emphasize or exaggerate the nose shape.
- As the head moves, the nose shadow will vary in length or shape. A triangle of light on the cheek may make and break distractingly.
- If a person has two key lights(!) there will be twin nose-shadows, forming a pseudo-mustache on the upper lip or 'wings' either side of the nose tip.
- Highlights on the tip of the nose can cause it to look bulbous or up-tilted.
- Highlights on the side of the nose can make it appear broken, bent, twisted, or misshapen.

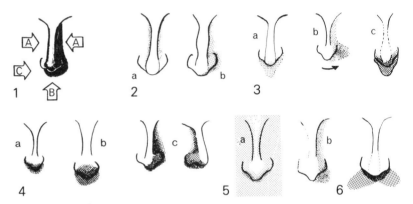

Fig. 5.10 Shadow and nose shape
1. Nose structure is revealed by shadow formation: (*a*) by the shadow-edge along the bridge of the nose; (*b*) by the tip-shadow on the upper lip and cheek; (*c*) by shading on the nose tip.
2. Shadow reveals the nose's bridge structure, and can (*a*) emphasize the thinness and length of the nose; or (*b*) the unevenness of a broken or irregular bridge structure.
3. The tip shadow can (*a*) lengthen the nose, emphasizing long or large noses; (*b*) twist the nose; (*c*) bend it down.
4. Tip shading (*a*) shortens the nose length; (*b*) emphasizes retroussé noses; or (*c*) turns down the nose end.
5. Where light falls upon part of a nose it may (*a*) appear bulbous (from steep back light); or (*b*) broader or bent sideways (from side light).
6. Double nose shadows are not only distracting but can broaden the nose at its tip.

Eyes
A person's eyes are usually the center of our attention, and so their appearance is all important.
- Eyes without reflected highlights look dead, expressionless, wan.
- Very large eye-lights, or a multiplicity of reflections, look strange and unnatural.
- Contact lenses can result in large, abnormal eyelights.

- Eyes can be partially or totally shaded by the nose or the brow.
- Widely angled back light, or side light, can produce a single 'black eye' as the light catches the brow.
- Eyelashes can create strange-looking patterns on the upper cheek or the side of the nose.
- A pair of 'black eyes' is a typical result of steep lighting, or a down-turned head.
- When eyes fall into shadow they appear to recede, and the head has a skull-like appearance.
- Spectacles can create various problems. Lights reflected in spectacles are extremely distracting.
- Where the shadow of spectacle frames cuts across the eyes it can even give the wearer a shifty look; as if they are unwilling to look straight into the lens.

Hair
- Dark hair without reflected highlights can appear as an untextured sculpted mass.
- It can merge into a dark background, leaving only the isolated face visible.
- Excessive back light can cause some quite unattractive effects. Apart from the 'hot shoulders', hair can appear fuzzy, greasy, overbright.
- It can inappropriately over-glamorize the subject.
- A hot or steep back light can emphasize thinning hair, and seemingly flatten the top of the head.
- It can even cause a fair-haired person to look bald.

Ears
- When insufficiently lit, the insides of ears appear dark and formless.
- Excess or steep back light can emphasize ears, making them seem to stick out or look larger.
- They can appear translucent.
- 'Hot tops' to ears draw attention to them, as do forward ear-shadows falling on the cheeks. (Particularly noticeable in profile shots.)

Chin
- Chin shadows are normally quite unobtrusive, but a steep key will throw a long shadow down the neck. So, too, will very steep back light. The effect is particularly disturbing on bare shoulders.
- Diagonal chin shadows produce a lop-sided effect, but shadows from side-light are even worse, as they result in a 'talking profile shadow' on the shoulder.

Forehead
- A hot top to the head can emphasize the size of the forehead, creating a bulbous effect.

- Where the temples are too bright, the face can appear over-widened.

Facial modeling
- If there is insufficient modeling the features will lack definition, appearing rather blank, and larger than normal (*key too frontal, or too soft*).
- If a thin face is lit with a steep key it can take on a gaunt appearance.
- A face can appear lop-sided or unevenly balanced if keyed from the wrong side (wider side), or lit by a widely angled key.
- Where the key angle is too shallow the face may appear over-broad.
- A face may appear much wider than normal if keyed from the wrong side, or if back-lit on the broad side of the features.
- If a key is too steep, or too hard for the subject, it will emphasize eye bags, wrinkles, flabby skin, skin defects (e.g. scars, blemishes) and facial texture ('orange peel effect').

Sometimes one may deliberately create unattractive effects for dramatic reasons; e.g. as the suave villain leans forward to threaten the victim, the lighting angle coarsens his features, giving him a grotesque appearance.

The important thing is to recognize these lighting pitfalls, and know how to remedy them.

Fig. 5.11 Clothes light
The amount of light required by clothing may differ from that needed for portraiture. To adjust relative light intensities use: (*a*) a scrim or half diffuser to shade off the overlit area; (*b*) separate lighting; (*c*) augmented lighting.

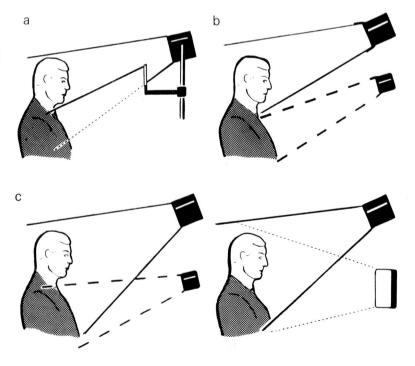

General maxims

Let's summarize some of the accepted principles we have now met:

1. Place the key light within about 10–30° of a person's nose direction.
2. Have them look *toward* rather than away from the key light, if re-angling their position.
3. Avoid *steep lighting* (above 40–45°) or oblique angles.
4. Avoid a very wide horizontal or vertical angle between the fill light and the key light.
5. Do not have more than one key light for each viewpoint.
6. Use properly placed soft light to fill shadows.
7. Avoid side light on full-face shots.
8. For profiles it is generally better to place the key light on the far side of the face, if you want maximum modeling, and to avoid casting shadows of a boom microphone on faces.

Lighting groups

So far, we have been talking about lighting a single person. What happens when there are two or more people together?

Fig. 5.12 Lighting two people
1. Treatment is developed from basic portrait lighting.
2. Individuals may be given their own key and back lights (taking care that one person's lighting does not fall onto the other). When people are close together it is more practical for them to share lighting; one person's *key* serves as the other's back light. Relative angles and intensities may be a compromise.
3. This arrangement provides a three-dimensional effect and avoids microphone shadows falling on the people.
4. Here the key light is on the same side of the faces; providing better visual continuity on intercutting.
5. This method projects nose shadows away from the camera, making them less prominent, but solidity is reduced.

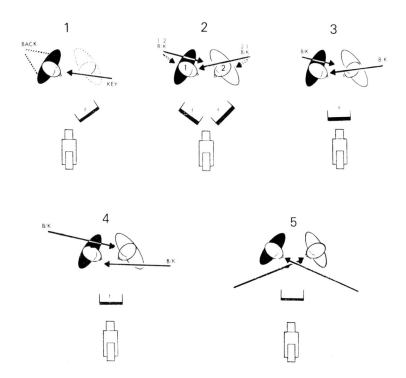

Two people

The obvious technique is to light them as two separate people, each with their own key light, back light and filler. Even when you have sufficient lamps you will often find that people are too close together to individualize their lighting. The fill light for one person is liable to illuminate their neighbor. One person's key light may produce a strong side light on the other.

The solution will often be to use *shared* lighting. Here you arrange cross lighting that provides a key light for one person and effective back lighting for the other. You can use half-height diffusers/scrims ('bottom-half jellies') to reduce the intensity of the back light where necessary. Whether you key on the far side of the faces or the near side will depend on the features, and how you feel about nose-shadows.

Three people

In television the three-group is a very common arrangement for discussions and interviews. At first sight, it would seem to be a straightforward situation, in which you light the people as three separate individuals. You can do just that. Difficulties often arise, though, when the person in the center turns over a wide angle to speak to the others on either side, and the director takes cross shots.

Let's look at this in some detail, for it is a typical enough example of everyday problems that may have no ideal solutions.

In Figure 5.13(a) you have an arrangement in which each person is lit with a separate properly positioned key light and back light, and a communal central filler. Mechanically, that is simple enough, but check out how the *fill light* relates to each position in turn:

● It is quite acceptable for the central person – although liable to flatten out the modeling from their key light.
● However, you will find that it is very poorly positioned for the other two people. It fills them from the *side* in each case. When the filler is suspended at e.g. 3 m/10 ft the overall effect can be particularly unattractive.

Next, let's take a closer look at the *central person's key light*. He will look fine when facing the front; but what happens when he turns his head 45° or more either side towards the others? On a cross-shooting camera the key light is now effectively at about 50–60° off the nose axis, and the back light becomes a 'side-rear light'(!). The face is bisected, and there is no fill light on the camera left side of the face! Portraiture in cross shots from either side is poor. However, if the central person only turns a little, and there are no widely angled cross shots, the treatment may get by.

Figure 5.13(b) shows another approach. Here the filler comes from two soft sources at improved angles. (You need to ensure that

Fig. 5.13 Lighting three people

1. Individuals may be given their own key and back lights, with a communal fill light (taking care that one person's lighting does not fall onto the other).
2. Treated as a wide two-group, with a separately lit central person.
3. As an overall group, with supplementary lighting.

Each arrangement has its limitations:

(a) Where a sound boom is used, microphone shadows may fall on the end speakers.
(b) When people sit close together, separate lighting may be impracticable.
(c) Cross-frontal keys are necessary when people are sitting in-line. In horseshoe layouts keys tend to move upstage, away from the camera.

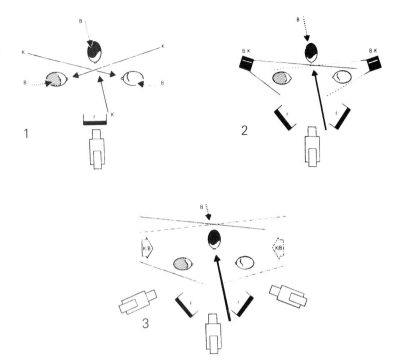

each lights only the person intended.) The result is slightly better for the people either side.

For convenience or economy, the people on either side may have shared key/back lights. The central person is lit individually as before. You might introduce additional localized fill lights either side upstage, and alter the back light positions for the center person as he re-angles.

In Figure 5.13(c) you have another treatment, in which the entire group is communally lit with two key/back lights either side of the group. Now the central person is satisfactorily lit for cross shots when angled to face someone at either side. But when he faces forward, the two side lights bisect his face. You might fade up a localized frontal key to override this effect. It may work. Any attempt to increase the frontal filler instead will flatten the overall modeling.

You might decide that the optimum answer is to cross fade unobtrusively from the lighting that suits the group shot to a separate treatment for the frontal viewpoint. You could use soft frontal lighting, and add frontal or angled keys for the center person as the shots change.

The quandary of providing good portraiture for a person who looks to either side as well as frontal has no foolproof solution; but this exercise does help you to get accustomed to rationalizing such problems.

Panel group

In Figure 5.14 we have the ubiquitous grouping that is a regular feature in many TV games shows and discussions. A panel of guests, questioning journalists, team of experts, or what-have-you are seated in a line. They may face a host, anchor, or chairperson, or a further opposing group.

You can approach this situation in several ways, including:

● Individual keys and back lights with a communal fill light;
● Paired keys with paired back light;
● Overall shared key light, and shared back light.

Fig. 5.14   The panel group
1. Individual
2. Paired
3. Overall

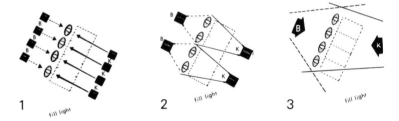

Individual. Clearly, this method gives the greatest control. But if people are sitting close together and the lamps are some distance away it is not always practicable. There may not only be difficulties in isolating individuals but they may move into each others' lighting set-up.

Paired. The 'paired' approach gives a certain amount of flexibility. You can use half scrims/wires where necessary, to reduce the light intensity for one person of the pair – e.g. on a bald head, white hair, pale face – while fully lighting their partner.

Overall. When the group has a single communal key light, and a single back light, one just has to hope that this arrangement will suit them all! It may. Although this simple approach has the advantage that it uses few lamps, it is necessarily a compromise. You may find, for example, that a single lamp barely spreads over the group, and the people at the ends are of slightly lower intensity than those in the middle. If you take the lamp further away (to increase the beam spread) the lower light level may be an embarrassment.

Where there are considerable differences in the tones of hair, skin or costume it may be possible to adjust exposure in closer shots to improve the results.

Large groups

When you have to light a large group of people, such as an audience or an orchestra, treat it as methodically as possible. Do not let hope triumph over experience, and just pour on light. It is all too easy to achieve overlapping hotspots (overlit areas) and dead areas (unilluminated areas).

Sometimes you can use the 'overall' technique, with generous fill light. A distant high-power HMI arc or 10 kW quartz light can be quite effective. It may be central, or located over to one side. If the group is spread over a wide area you will find that the lighting angle varies considerably, so that some people are frontally lit while others near the edge of the group are lit quite obliquely, and this becomes very obvious in closer shots.

Arguably the best approach is to sub-divide the group into several sections, each with suitably angled keys and back lights. Filler can be communal or sectionalized. Certainly, when lighting an orchestra where there will be close shots of instrumentalists this method offers the closest control.

Fig. 5.15 Large groups
1. Lit *en masse.*
2. Sectionalized.

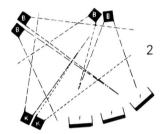

Where a director wants *close shots* within a large group it is often impracticable to move the camera in among them. Instead, it is usually necessary to shoot with a narrow-angle lens (long-focus, long focal length, telephoto) on a distant camera. This can result in a certain amount of perspective distortion, in which faces appear somewhat flattened. To improve the modeling, and compensate for unavoidably compressed features, you may find that reducing the filler improves the modeling, so that the effect is less apparent.

Lighting action

The problem of movement

In television and motion pictures, lighting has to be arranged so that there is good *visual continuity* from shot to shot, however much the subject or the camera moves. People must not change appearance from moment to moment, or the background vary with each new viewpoint. The mood, atmosphere, time of day should look similar in successive shots of the scene.

Now this is all very obvious, but it is not necessarily simple to achieve in practice. The lighting treatment that produces great results from one viewpoint may be less successful from another.

Fig. 5.16 part 1 The effect of moving the head

In Fig. 5.3 we saw the light directions that suit different head positions. If the head turns, how effectively is it then lit?

Top row. Lighting arranged for a full-face shot to camera.
1. Full face.
   Result: Good.
2. The head turns half-left (same lighting).
   Result: Good/fair.
3. The head turns to profile (same lighting).
   Result: Fair to poor.
Conclusion: The head turning towards the key continues to be acceptably lit in all positions. (Check out what happens when it turns *away* from the key!)

Center row. Lighting arranged for a three-quarter face shot (half-left).
1. Full face.
   Result: Potentially poor. (Too dramatic.) May be good for character lighting.
2. The head turns half-left (same lighting).
   Result: Good.
3. The head turns to profile (same lighting).
   Result: Good to fair.
Conclusion: Potentially good, but may be too offset for full face.

Bottom row. Lighting arranged for a profile shot.
1. Full face.
   Result: Very poor (bisects face).
2. The head turns half-left (same lighting).
   Result: Fair.
3. The head turns to profile (same lighting).
   Result: Good.
Conclusion: Unacceptable when head turns to camera.

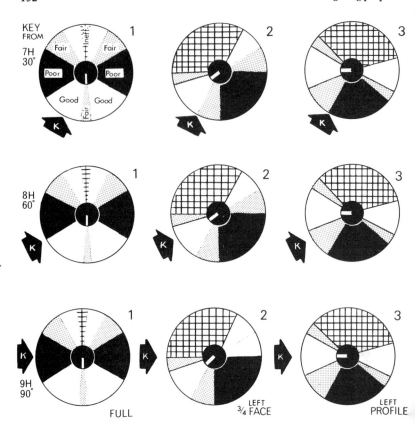

How difficult is it to light for action, and to suit several camera viewpoints? That will depend on a number of things:

- *The camera treatment.* In longer shots where people are seen in full-length, facial modeling is far less critical, and the appearance of the surroundings becomes important. Where cameras are taking waist shots or closer, we can see the person's face clearly, and appropriate portrait lighting is essential. The surroundings are less visible, and may even be out of focus (due to limited depth of field).

- *The extent of the action.* The more people move around, the less time there is to see any lighting imperfections. If the audience is watching a fight sequence they are not going to notice two nose shadows! Where the camera is taking critically close shots of action over a large area it can call for very extensive localized lighting treatment.

- *The nature of action.* Certain head positions may degrade the lighting; e.g. a back-tilted head causing back light to illuminate the nose and forehead; a forward-tilted head causing eye shadows and harsh facial modeling.

- *Where the action takes place.* If people are moving around obstructions such as arches, trees, columns, screens, suspended

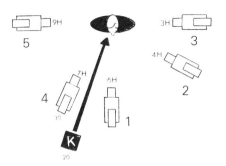

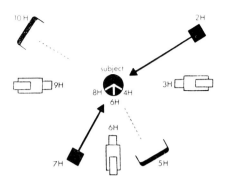

Fig. 5.16 part 2   The effect of moving the camera (see Fig. 5.3)
As the camera viewpoint changes, the effect of the key light's angle alters:
Position 1 – Good; Position 2 – fair; Position 3 – fair;
Position 4 – good; Position 5 – fair.

Fig. 5.16 part 3   The effect of changing subject and camera position
A set-up may sufficiently suit various positions.

panels, etc. it may be difficult to position lighting for optimum portraiture. There will be problems of scenic shadows falling onto the performer, or his lamps overlighting the scenic pieces.

- *Whether defects are acceptable.* Some distractions are more acceptable than others. Where, for instance, a hat brim casts a shadow across the eyes an audience accepts it readily. But if a person stands with the shadow of an overhanging branch falling across the eyes, it will not. Sometimes a defect is not obvious. A bushy mustache will hide any nose shadow and a beard can disguise facial modeling.

Action and background lighting

Lighting any scene really involves two coordinated treatments:

- *Action lighting* – the lighting for the performers.
- *Background lighting* – the lighting for the setting (scenery) or environment.

It is best, of course, if you are able to light the people and the setting independently, for this will give you the greatest flexibility. But that is not always feasible. You may not have sufficient lamps (or power) to rig two separate sets of lighting in the area. Quite often, people are positioned too close to the background anyway, for you to be able to light them separately.

In those circumstances you will have to combine the functions of lamps. The same lamp may, for instance, have to serve as a key light for a person *and* light the wall behind them. Where you need to do this, the lighting angles, coverage and intensity must be something of a compromise, for the treatment that is optimum for the person will not necessarily provide the best modeling on the background.

Types of movement

Lighting opportunities vary with the kind of movement:

- In large-area performance a *group of people* moves around; dancing, marching, parading, performing feats (acrobatics, etc.).
- A *single person* may move around either at random or to specified spots – e.g. a singer on stage, a lecturer in an art gallery.
- In a drama or situation comedy *people* can be very mobile; moving to different parts of a room during the source of action, facing in various directions, sitting, standing, etc. Camera viewpoints can vary considerably.

Methods of lighting movement

There are three fundamental ways of coping with movement:

- *Area lighting*
- *Localized lighting*
- *Specific lighting*

Which you use will depend not only on the action itself but also on your lighting facilities (the number of lamps available; the lighting board), the amount of advance information you have, the time available to set the lamps and rehearse the action, and so on.

Area lighting

'General lighting' for areas is necessary when you have full-stage action (e.g. a choir), widespread movement (e.g. ballet), or situations where you know little about a last-minute item and the time is short. It is m*ethodical* illumination. There are several techniques:

1. Soft frontal plus back light – Figure 5.17(1)
Flood the area with soft light as a high-intensity *base light* or *foundation light*; adding *back light* to rim the performers and make them stand out from the background.

This rudimentary approach can result in poorly modeled, high-key pictures. But there are occasions when it does work for widespread movement (e.g. dancers), particularly where the soft light is angled either side of the acting area, or confined to one side. Central soft light produces flat results from frontal cameras but acceptable pictures on cross-shooting cameras.

When large areas of the floor are visible most purists would contend that there should only be a *single shadow* from the back light, for a multiplicity of shadows is muddling and unattractive. A wide-angle high-intensity back light can be central, or offset to one side, balancing offset frontal soft light.

Fig. 5.17 Area lighting

1. *Soft-frontal area lighting.*
   Soft overall base/foundation light, with modeling 3/4 back lights.
2. *Side keys area lighting.*
   Carefully angled soft light as filler for spotlights over the side walls of the setting, lighting selected areas.
3. *Overall three-point lighting.*
   (a) *Total treatment for smaller areas.*
       The back-light arrangements may be (2) dead back, (3) one 3/4 back, or (1 + 3) two 3/4 back lights.
   (b) *Total treatment for large areas.*
       Broken down into a series of sections; taking care to avoid 'double-keying' due to large overlaps ('doubling'), or unlit areas ('dead areas').
4. *Dual key lighting.*
   The area is split into two sections with their own keys; (a) overlapping, or (b) separate.
5. *Localized three-point lighting.*
   Selectively arranged keys and back lights for each position. Fairly localized fill light. Furniture, windows, doors, etc. are used as *locating points.*

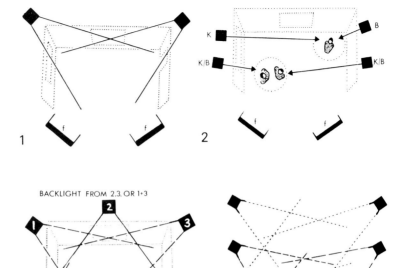

BACKLIGHT FROM 2,3, OR 1+3

USED EITHER OVERALL(a) or SECTIONALISED(b)

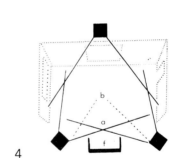

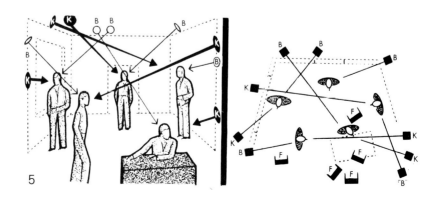

2. Soft frontal with side keys and back light – Figure 5.17 (2)
Here you use soft frontal lighting as before, but with fresnel spots along the sides of the setting, providing key lights and back lights where needed; mainly to suit cross-shooting cameras.

At best, the results can be very attractive, but the success of this arrangement depends on carefully positioning and controlling the soft light. At worst, it produces haphazard, mediocre illumination, with a very arbitrary filler angle.

Again, the overall impact is high key, as the soft light illuminates the entire setting; particularly the side walls. This method has advantages when sound booms are being used.

3. Overall three-point lighting – Figure 5.17(3)
In this approach overall 'three-point lighting' is applied to the entire acting area. It is a mechanical solution that may, with luck, get surprisingly good results, if cameras do not cross-shoot too obliquely. The back light is arbitrary. You can use one or two dead back lights, one three-quarter back, or balanced three-quarter back lights.

Where there is a large acting area it may be necessary to subdivide it with a series of 'follow-on' keys and back lights, to provide sectionalized three-point treatment.

4. Dual key – Figure 5.17(4)
Here the area is keyed from both left and right, with central soft light providing the fill light. Where separate action takes place in either the left or the right sections it receives three-point treatment. Although there is the continued hazard that a subject will be double-keyed, this treatment works for most camera positions.

Localized lighting

Ideally, of course, one would like to be able to light each individual camera shot for optimum effect. In motion-picture making this is axiomatic. It is possible, too, in television, given favorable conditions. But where action is continuous, or is shot from several angles using intercut cameras, one is usually obliged to compromise. However, if skillfully handled, this compromise need not be at all obvious.

You will encounter many situations that appear to be quite complex, and yet nothing prevents your putting every lamp exactly where you want it to get precisely the effect you're aiming at. Then there are others which are seemingly quite straightforward, where you are extremely restricted in your lighting treatment, and have to be satisfied with the mediocre.

Quite often you will find that, although there is a fair amount of action, the performers are actually going to a number of localized spots. They enter a door, look out of a window, stand beside a fireplace, sit at a table or in various chairs. Each of these *locating points* (*action points*) can be lit independently with three-point lighting tailored to the subjects' directions there. One lighting set-up will often suit various shots or action sequences in that area; perhaps with slight re-balance of light intensities.

Given enough lamps, dimmers, etc., and adequate time to readjust them, you could light each individual shot. But this degree of finesse is not normally feasible.

If the action in a particular position becomes too general, and the shortcomings are unacceptable, you may need to sneak in some additional 'first-aid' improvements for certain shots (additional filler, repositioned key light). Sometimes you might provide the optimum quality for certain 'key shot' positions, and accept that others will be rather less successful.

Specific lighting

Even with quite long continuous shooting, it is possible to provide high-grade portrait lighting where the action is fairly limited. There are a number of approaches.

■ *Supplemented area lighting* – Figure 5.18   Here the acting area is lit using any of the techniques you saw earlier, and additional lamps introduced as necessary to suit specific shots or action.

■ *Bisected key* – Figure 5.18(1)   In this treatment the lighting is arranged to suit two or more different viewpoints. If, for instance, some one seated at a desk turns through 45° to face another camera viewpoint, you may position the key so that that they appear reasonably well lit from both directions.

■ *Preferred position* – Figure 5.18(2)   Here the best results are achieved from the main camera angle (Cam 2), but still quite satisfactory when shot from another direction (Cam 1).

■ *Separate set-up* – Figure 5.18(3)   In this arrangement the person moves from one lighting set-up to another.

■ *Dual set-up* – Figure 5.18(4)   With anticipation, it may be possible to arrange a set-up that suits two different situations equally well.

■ *Dual-function lamps* – Figure 5.18(5)   You can use the same lamp for different purposes as the camera angle changes. A key light can become a back light for another shot, and vice versa.

■ *Supplementary light* – Figure 5.18(6)   As the person moves away from the hung fill light an additional source (usually a camera light) can be added to compensate for the loss; perhaps with re-balance.

■ *Auxiliary key* – Figure 5.18(7)   As a person moves between two spaced keyed areas a small supplementary key takes over at their crossover point to maintain the light level. (In a dramatic situation

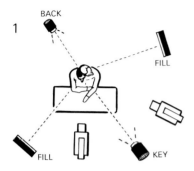

1

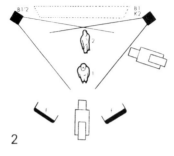

2

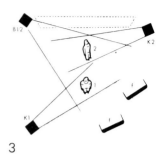

3

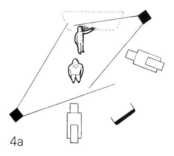

4a

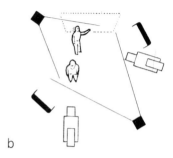

b

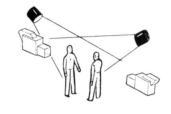

5

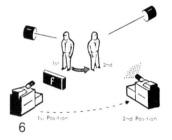

6

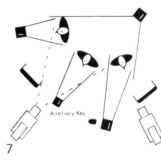

7

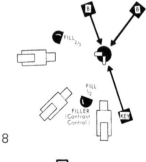

8

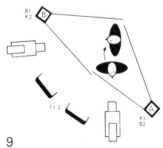

9

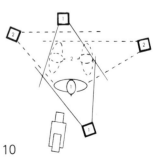

10

Fig. 5.18 Light for movement

1. *Bisected key.* A key midway between two head positions may serve each equally well. However, the side of the face lit changes as the head turns.
2. *Preferred position.* The main subject position is well lit (2), while another (1) is only adequate.
3. *Separate set-up.* The person moves from one lighting set-up to another. Some lamps may have dual functions.
4. *Dual set-ups.* Lamps are arranged to suit both head positions.
   (a) Portraits are good, but distracting shadows of the person stretch across the background map.
   (b) This set-up anticipates the problem, avoids the shadow, and produces good portraits.
5. *Dual-function lamps.* Lamps serve as a key light from one angle and a back light from another.
6. *Supplementary light.* An additional lamp is faded up to augment the second position. Here, as the person moves away from the filler, a camera lamp is faded up to provide fill light.
7. *Auxiliary key.* Here a localized extra key is added for an intermediate position between two main areas.
8. *Auxiliary fill light.* A compromise solution, in which fill light is boosted to act as a 'key' when the camera or subject moves.
9. *Rebalance.* The intensities of the lamps are rebalanced for each position; e.g. For Cam 1. Lamp A is strong, B is weaker. For Cam 2. Lamp B is strong, A is weaker.
10. *New lighting set-up.* Switching or cross-fading to a different set of lamps lighting the same area; e.g. from Group 1 to Group 2.

we might deliberately omit this extra light, so that they go into partial shade when moving between positions.)

■ *Auxiliary fill light* – Figure 5.18(8)   In certain situations, where keys are widely angled and the person turns over 90°, the modeling can become unacceptably crude. As an expedient, a strong auxiliary fill light may be added to help disguise this problem.

■ *Rebalance* – Figure 5.18(9)   This is a dual-function situation (6) in which the lamps' intensities are rebalanced as the person turns.

■ *New lighting set-up* – Figure 5.18(10)   Here we have switched or cross-faded from one set of lighting to another to suit the different positions.

■ *Following key*   You can follow the subject with its own specific lighting; a follow-spotlight or a mobile hand-held lamp. In certain types of show this is a method of concentrating the audience's attention on the performer or a localized group within a large-area display – ice-skating displays, stage acts. It is also used dramatically to simulate a person walking around with a lighted lamp or candle.

6 The production process

Originally, the techniques of motion picture and television program making were worlds apart. But as technology develops and production techniques change, they share more and more common ground.

The hybrid arts

Today, the differences between the film and television/video media are becoming so blurred that it is often difficult to discern whether a particular production was developed using film or video camera. On the cinema screen you see material shot with TV cameras and transferred to film for projection. For years films made for theater distribution have been shown on the TV screen, as well as films created specifically for the medium. Program inserts and commercials may be shot on film and transferred to videotape, or shot on tape and transferred to film. Films contain video effects. All-video productions are confusingly referred to as video 'films'. Little wonder that it becomes increasingly difficult to judge which you are actually seeing on the screen!

There are still important distinctions between shooting for film and TV. Film always has to be processed before you can judge the final results. The TV/video camera lacks the general mobility and rugged simplicity of the film camera. But both are catching up!

A standard film camera can be fitted with a video-camera accessory that shows the Film Director on a nearby TV picture monitor what the Camera Operator can see in the optical viewfinder while shooting. This video duplicate can be taped, and replayed immediately to monitor various aspects of each take. There are even systems which allow the Film Director to sit in a nearby remotes van, directing two or more film cameras: using inter-camera switching, remote stop–start, automatic shot timing and marking. The result is Film, the technique is derived from TV production methods.

Modern videotape editing facilities have now developed to a point at which they can be manipulated as easily and with as much sensitivity as film editing systems.

Shooting conditions

TV and film cameras shoot under a remarkable variety of conditions; in some circumstances relying on existing illumination, in others supplying full lighting. Even where some sort of illumination already exists it is not always suitable, and for technical and artistic reasons has to be augmented or substituted.

The extent and the complexity of lighting needed varies considerably with the type of project. In some aspects of program making (news gathering, for example) the TV cameraman on an ENG (electronic news gathering) assignment probably has more in common with a film cameraman shooting a news event than with his counterpart on a drama production.

News events are shot impromptu in the most unpredictable circumstances – from the multi-media gathering where everyone seems to be adding their own illumination to the most adverse conditions where the cameraman is relying on his single camera-light to get any picture at all. Shooting in fog, rain, snow, unrelieved darkness, lighting is often a matter of improvization, of adaptation, making the best of it. Equipment is often minimal and has to be highly portable and unobtrusive. There are times when the best viewpoint and the most effective light direction are quite incompatible. Time is short. There are often problems with crowds and weather.

A few strategically placed stand lamps or a battery-powered hand lamp usually form the basis (or the totality!) of lighting treatment. Yet even here the principles you have met will help you to choose the optimum approach.

Fortunately for the cameraman working under these conditions, there is a high audience tolerance when things go wrong. Even downright bad lighting, half-lit faces, burned-out tones are readily accepted as they are presumed to be due to the difficult conditions. Losing focus or poor framing is one thing on a news event, and quite another if it happens when shooting drama!

Although *documentary shooting* has many of the elements of the 'news event' there is the unconditional need for the best possible picture quality. Subjects range from conventional interviews shot under the most unpromising conditions to the downright 'impossible'. Yet high standards are generally maintained.

Public events of many kinds from ceremonials to operas, from sports meetings to vox-pop street interviews have their own characteristic problems. Lighting treatment spans from the impromptu to large, full-scale systematic layouts. In many locations lighting treatment is very much a compromise, due to restrictions in lamp positioning, facilities, power, access, safety problems, the lack of rehearsal, etc . . . but that is the name of the game.

Location exteriors and interiors are extensively used in drama. Shooting in real surroundings avoids the complications and expense of building a studio replica; but at the same time, it does mean that all facilities have to be installed for the occasion. As we shall see, each site poses its specific problems and opportunities. In some cases the very realism of the location can make any uncharacteristic lighting look false and artificial.

Studio productions offer their own particular kinds of challenge.

From daily soap operas to cooking demonstrations, from game shows to newscasts, comedy to classical drama, a wide gamut of techniques are called into play.

Although one finds common lighting methods used in them all there are many subtle differences in emphasis:

- The blatant color treatment suitable for a pop group.
- The delicate ethereal qualities of classical ballet.
- The harsh illumination of sordid drama.
- The vibrant razzmataz of a giveaway show.
- The quiet formality of a piano recital . . .

Different approaches, from the blatant to the sensitive understatement – yet all aspects of persuasive lighting techniques.

The Lighting Director must adapt to the particular idioms for which he is working. He must suit the methods and temperaments of the production group with which is he involved. Shortly we shall look in some detail at one approach to studio drama production; for this arguably represents the peak of lighting opportunity.

In addition to orthodox forms of TV production there is the ever-growing use of chroma key and various video effects developments within TV programs, and these require extended techniques that we shall look at in Chapter 15.

Basic film mechanics

The Lighting Cameraman

Film making takes many forms, and people have their own ways of organizing and going about the task. But in general terms the *Lighting Cameraman* or *Director of Photography* is responsible for production lighting; for creating the visual image we see on the screen. He interprets the film director's pictorial aspirations.

In some working relationships the Director explains the treatment, and draws on the Lighting Cameraman's experience to convert this into camera mechanics. The Lighting Cameraman decides on the most suitable camera set-up: its positions and movements, shot size, picture composition, etc. In other production teams the Director has well-defined ideas about what he wants, which the Lighting Director carries through.

The supporting crew the Lighting Cameraman requires for his single camera will depend on the type of mounting being used. In a small mobile operation the cameraman may himself operate a lightweight shoulder or tripod-mounted camera. He is assisted by one or two electricians who arrange his lighting. An arrangement of this kind is often used when making TV program inserts, shooting documentaries, interviews, news gathering, and similar projects. For larger camera systems, or where the camera must be precisely controlled for longer periods, it is mounted on a small wheeled dolly, or on a crane. Whether the Lighting Cameraman himself operates the camera varies with the scale of the operation. In large-scale film making he guides a *Camera Operator*, who, watching the viewfinder, controls the camera, composes the shots, and follows the action (pans/tilts).

The Camera Operator may focus by eye, using a ground-glass focusing screen in the camera viewfinder. But this method can be insufficiently accurate under certain conditions. So instead focusing is carried out by measurement alone. A *Focus-puller (Focus Assistant)* at the side of the camera adjusts the lens to maintain focus, having measured the camera-to-subject distance and calculated focus settings with a depth-of-field calculator. He also operates the zoom control.

A *Clapper Loader (2nd Assistant)* generally aids the cameraman (loading/unloading film, cue marking, logging, etc.). A *Grip (Stagehand)* assembles and pushes/guides the camera mounting, and arranges the camera rails or tracks along which the film dolly frequently travels. A team of electricians (lamp operators), generator operator) is headed by a *Gaffer (Chief Electrician)*.

Preliminary planning

Where close preliminary planning is possible, key members of the production team can assess and coordinate their individual efforts. The film's Producer, Director, Art Director (Set Designer), Lighting Cameraman (Director of Photography), and those responsible for Sound, Costume, Make-up, Special Effects, and allied crafts, all have the opportunity to prepare their specialist contributions.

The Lighting Cameraman needs to consider various aspects of the production, including:

- *Script aesthetics* – the nature of the script material, plot outline, and general interpretation.
- *Script mechanics* – a breakdown of various script details covering the types and locations of scenes, the nature of the action, etc.
- *Production treatment* – how the production is going to be shot: a storyboard of broad treatment, techniques, effects, etc.
- *Shooting arrangements* – where the production is going to be shot: studio, location, working conditions, potential problems, etc.
- *Production coordination* – interrelationships with the work of the Art Director (Set Designer), Costume Designer, Make-up, Special Effects, etc.
- *Lighting mechanics* – assessing possible lighting treatment, and equipment requirements.
- *Hire* – hire and booking arrangements, power supply availability, crewing, etc.

Lighting methods

The Gaffer (Chief Electrician) is responsible for lighting the scene to the Lighting Cameraman's requirements. Anticipating probable needs, he may rough-in the *set lighting* (i.e. rig suitable lamps for background lighting), ready for the Lighting Cameraman's final instructions. Where a team has worked together regularly they become familiar with a particular Lighting Cameraman's techniques and preferences.

Production organization varies, so that while some film Directors may hold pre-studio rehearsals in mock-up settings before shooting, many have to be content with little more than a preliminary discussion of story analysis and characterization with the actors beforehand, and a certain amount of rehearsal while the camera and sound crews are setting up their equipment.

In a typical arrangement there is a run-through in which the story-action (dialogue, business and moves) is rehearsed within the setting, as the Director checks anticipated shots through the camera viewfinder. With this knowledge of the action, the Lighting Cameraman arranges for any modifications in the background lighting that appear necessary. Now he devises the main portrait lighting. Lamps are positioned, set and balanced for the action the director has rehearsed; perhaps with the aid of stand-ins in the strategic positions.

Shooting methods

There are two basic approaches to shooting motion pictures:

- *The master scene*
- *Shot by shot*

In the *master scene* technique quite long continuous-action sequences are shot from one or two vantage points (e.g. in long shot). Particular action is then repeated, to provide a series of medium shots from further viewpoints. Finally, brief sections are repeated, to provide close-ups, detail shots, cover shots, cutaways, and inserts. Where necessary, subjects or areas can be relit or lighting adjusted to suit the next shots.

In this method there is often a great deal of duplication, and although that can assist subsequent editing, there may be poor visual continuity, due to alterations having been made between the shots. The production method allows a great deal of flexibility, but can lead to an unplanned piecemeal approach.

In the *shot-by-shot* method the director has planned each shot in great detail. There may be a pictorial *storyboard* showing exactly what is anticipated in each. Action is photographed to suit this planned concept, with sufficient overlap to allow for editing continuity.

The chief merit of this technique is that the Director thinks through the entire production, and is in a position to pass on to the rest of the team what he is seeking to do. It is an approach, however, that requires a well-developed visual sense, a practical imagination, and a workable breakdown. Otherwise there will be discontinuity, continual compromises and changes, and an uneven visual development.

Retakes

With both techniques it is regular practice to retake shots, sometimes several times, to correct, improve, or give variety of treatment, until the Director and team are satisfied. Scenery, actors' positions and camera treatment can be 'cheated' (Figure 6.1) to avoid spurious shadowing, faulty composition, etc. When each shot is completed, the lighting can be re-organized as necessary.

Fig. 6.1 Cheating
The positions of people, scenery, props, that suit one viewpoint may be less effective from another. In discontinuous shooting, positions can be slightly altered for each shot. If done carefully, the differences are not obvious. Typical reasons for *cheating* include: to improve composition; to allow the camera position to be adjusted; to allow lighting to be rearranged.

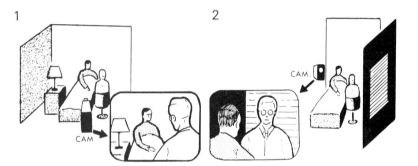

Normally, shooting is concentrated on one camera, although particular situations may require two (even three) for overall coverage. The camera dolly requires a smooth flat level floor surface to avoid picture judder during dolly moves. Especially where the ground is uneven, floor-tracks or rails are laid down and carefully leveled. As each shot is completed, these are repositioned for the next set-up.

Particular attention is paid to close-up shots, and their lighting (and make-up) is modified for the optimum effect. The more strongly defined treatment that suits the longer shots can appear too coarse under the scrutiny of the close-up shot.

Discontinuous shooting

The production is segmented into a series of *shots*, each lasting from a few seconds to a few minutes. A 90-minute film may have some 600–800 shots. With an editing cut every 3–10 seconds, some 3–15 seconds per shot have been quoted. In Hitchcock's experimental film *Rope* the exceptional shot duration of 6–8 minutes was reached, although this same master of film techniques used 1360 shots in *The Birds*. A good serial unit may achieve 120 shots a day (around 18 minutes' running time) on certain types of production, but this is untypical. A sequence may equally well take days or weeks to complete.

Checking results

The Lighting Cameraman sees an image of the scene in his camera's viewfinder. But there's many a slip between the viewfinder picture and the eventual projected picture! He can, of course, assess picture composition and framing, but he can only make educated guesses about exposure, tonal contrast, and color values in the final print.

Color film has a comparatively restricted contrast range, and the Lighting Cameraman who is aiming at a high-quality consistent product must rely heavily on careful checks before shooting:

- A *light meter* enables him to assess the lighting levels and adjust the exposure.
- A *color temperature meter* indicates the color quality of the prevailing light.
- A *color contrast viewing filter* helps the Lighting Cameraman to judge the relative tonal values and contrast within the scene; to assess the shot as the color film is likely to interpret it.

As you saw earlier, our eyes have a trick of compensating in ways that the film medium cannot, and the filter helps to overcome this dilemma. Looking through the viewing filter, you will see a darkened image in which the contrast range appears reduced, detail is lost in darker areas, and you become aware of the way strong highlights burn out; limitations that closely resemble the film's response. Within limits, the processing laboratories can improve or compensate for accidents or misjudgments, but the parameters for color film are limiting.

Continuity problems

Although individual scenes are filmed over a period, on separate occasions, often with changes of location and varying weather conditions, lighting continuity and quality need to be consistent throughout. It is an ever-present challenge, too, for costume, make-up and scenic departments. Sequences are not shot in continuity order or script order but to suit the economics and mechanics of the film. All scenes in the 'farmhouse' setting, for example, are photographed at one time, and the set is then broken up. They may show the hero in childhood and 'years later' as an ageing man. Scenes may cover all seasons of the year. Yet when joined months later into their eventual places in the completed motion picture there must be no sign of discontinuity! Against the disadvantages of these piecemeal techniques one has to balance the way they allow all facilities and work effort to be concentrated on these scenes alone, so that lighting, costume, make-up and scenic treatments can be optimized.

Restrictions and opportunities

The Lighting Cameraman has the considerable responsibility of creating the visual image that is the summation of the entire production team's efforts. He controls both the camerawork and the lighting, and is a major craftsman whose work cannot be fully assessed until a graded print can be viewed some time after shooting has ceased. In many situations there is no opportunity for later corrections.

Light-control facilities in film making are largely based on physical methods of adjustment, using scrims, flags, etc. Transportable wheeled dimmer-units and switchboards (contactor panels) with mass circuit-breakers for group switching together with individual dimmer units and switches provide electrical control.

The nature of film production allows finesse and precision. Individual lamps can be adjusted to suit the action. Lamps can be hand-held or manipulated in various ways. It is practicable to gradually cover or uncover a lamp beam during action; to use a hand, for example, to shade the light from a costume as an actor approaches and avoid its being overlit.

Basic television mechanics
The background of television production

Television is also a very diversified medium. Small units work miracles with a restricted budget and economy of equipment, and a staff of enthusiasts. At the other extreme, there are the flexible facilities of large TV networks, who regularly achieve the 'impossible', producing the highest filmic quality in an unbelievably short time.

Standards vary, of course, with the skill and the opportunities of the production group. While dramatic or spectacular lighting impresses, a large number of television productions do not call for such treatment. A high proportion of television programs depict people talking to each other in naturalistic surroundings. They require sensitive but self-effacing lighting that is immediately accepted and taken for granted.

A television Lighting Director can experience a remarkable variety of working conditions:

● From the exacting demands of precision lighting in the studio to the challenge of improvization on location;
● From a single performer in front of an open cyclorama to a studio full of scenery;
● From an off-the-cuff show which is assembled only minutes before live transmission to a meticulously planned production requiring a number of shooting days and considerable post-production work;
● From situations where the experienced Production Director has clear-cut practical ideas based on a real understanding of the system to others where a novice Director relies heavily on all the assistance the crew is willing to give.

The Lighting Director

The status and function of the Lighting Director varies considerably between organizations. There is no universal approach to television operations. Methods vary between companies and between countries. In most larger studios lighting has long been recognized as a major productional function. In small stations it may be treated as a casual routine, even as a preset installation. The status of lighting tends to reflect the organization's general attitude to program making.

In some TV companies a *Technical Supervisor* serves as a technical adviser to the production group, and organizes and coordinates a team of specialists in a technical operations crew. The *Lighting Director, Audio Engineer* (*Sound Supervisor*) and the *Senior Cameraman* head their respective teams, who operate lighting equipment, sound equipment, cameras. A separate specialist operator, the *Switcher* or *Vision Mixer*, operates the production switcher, which cuts/fades between video sources and provides certain video effects.

The Lighting Director devises and controls the complete lighting treatment for a production, and assesses the technical and artistic quality of the picture. He advises other members of the production team on the suitability of materials, techniques, etc. for the camera. He is assisted by –

● A *Vision Supervisor*, who also operates the lighting board (console);
● A *Video Operator/Shader*, who controls picture quality by remotely adjusting the camera channels;
● A crew of electricians who rig and operate all lighting equipment.

He also has a close working relationship with the studio *video engineers, telecine* and *videotape operators*, and others feeding video contributions to the production.

In another regular set-up the technical operations crew is headed by a *Technical Director*. He is a technician or engineer with many functions! As well as being a technical adviser, directing camera operations on behalf of the Production Director, carrying out video-switching and serving as supervisory engineer, he may add the job of lighting! More frequently, a specialist Lighting Director will concentrate on that aspect of the production. In a show of any size and complexity, working pressures are intense, and the work load must necessarily be spread.

Production planning

Planning in television production covers a wide spectrum: from spontaneous unrehearsed treatment to meticulous shot-by-shot discussion; from 'the same as last week' of regular formats to carefully coordinated systematic analysis. In practice, the number of TV directors who are willing or able to plan in real detail are disappointingly few. Specific planning in which the team has a clear idea of shot content, camera operations and sound treatment before rehearsals begin requires discipline, teamwork, and a firmly based understanding of the mechanics and techniques of the medium.

Many directors rely heavily on 'spontaneous inspiration', based on the shots they see on preview monitors, and feel that the pre-commitment of accurate planning stultifies artistic development. It can. Over-planning leads to soulless, uninspired mechanics where performers and crew alike find themselves unable to get into the flow of the production.

Many effective ideas arise on the spur of the moment, as you watch the action, and see the pictures before you. But freedom to leave decisions until the last minute, in order to avoid committing oneself, is a luxury that leads to ineffectual improvization, an uneconomic use of time, money and facilities, and a considerable strain on the production team. The best directors walk this tightrope with remarkable perspicacity, and create a good working framework which serves as a guide yet allows for any variations discovered during rehearsal.

What is going to happen?

In order to light any production efficiently you need the answers to a number of questions. Whether you get the answers at a full pre-production meeting with the Director, Set Designer, Audio Engineer, etc. or from a chat in the studio while looking around the set will depend on the size and nature of the production.

Sometimes you must know in advance precisely what is required. If, for instance, the Director wants the name of the show projected onto the floor this cannot be arranged at the last minute. At other times all you need is a general outline of the action; e.g. that people are going to come through a certain door.

There will be times when you have to light a production knowing all too little detail. If information is vague or unreliable it may be necessary to use a broader lighting approach than is possible with accurate planning.

Things invariably change before and during the camera rehearsal – sometimes to the point that you would have tackled the situation quite differently had you known! If you find out during rehearsals that someone is going to stand a few paces away from the planned position it may only be necessary to slightly readjust a key. But there are situations where the change might involve putting in a new key entirely, and that may not be at all simple. It depends on your facilities and the complexities of the present lighting rig.

In general terms the sort of preliminary details you need include:

■ *The subject*   The main subjects in most productions are people. You must know where people are going to be positioned, and the directions they are facing. Otherwise any key, fill and back light positions will be tentative.

■ *The cameras*   Where are the cameras to be? What lengths of shot are they taking? (In a close shot the emphasis is on portraiture; in a long shot the setting predominates.) Lighting must suit the cameras' viewpoints. If the subject is to be shot from several positions you will need to take this into account.

■ *The surroundings*   You will need to know about the general tones of the backgrounds. Are they light-toned? Then they could easily become overbright. Are they dark-toned? Then more light may be needed to prevent lower tones from merging, becoming drab and detailless. Will the subjects stand out from their background? It helps to know beforehand whether performers are to have dark or light clothes. You can, of course, alter the lighting balance on the day, but there may not be time to make major changes. If you find someone is wearing black clothing against a black background you can always add back light to separate them from it – but will the strong back light be appropriate for the situation?

■ *Atmosphere*   Are you aiming at a particular atmospheric effect, such as happy upbeat spectacle, cosy evening interior, intriguing mystery? This will influence how you distribute light and shade in the scene.

■ *Production mechanics*
● *Sound-boom.* Are sound booms being used? Where are they positioned, and what areas are they covering? (You have to avoid casting boom shadows that will be seen in shot.)
● *Lighting cues.* Are there any lighting cues – e.g. someone apparently switching room lights on?
● *Lighting effects.* Are any special lighting effects required, such as light patterns, flashing lights, follow spot?

The realities of planning

The amount of planning necessary for any show will largely depend on its complexity and the experience of those involved. A group used to working together can interrelate quickly and efficiently. But all the various requisites and paraphernalia needed to stage a production still take time and effort to prepare and gather together. Without appropriate planning, too many things can go wrong. Particularly in a live transmission, one weak link can precipitate a crisis!

Facilities, supplies, scenery, furnishings, etc. have usually to be hired or built, transported, stored . . . and brought together to the right place at the right time. In the studio, apart from the cyclorama and the floor, there is nothing to shoot until some scenic elements are positioned!

For small-scale location work you might turn up on the day with a single camera and a handful of lamps and, making instant decisions based on experience, get excellent results. But a project of any size requires systematic preparation. A large-scale public event may take months to prepare, from the preliminary 'recce' (reconnaissance) visit to the day it is taped or transmitted.

Planning approaches

'Permanent lighting'

The most basic situation is, of course, that in which the subject is lit 'permanently', and little or nothing is changed from day to day. Once the lighting is optimized it is only a matter of checking it over to ensure that all is well. Regular transmissions of newscasters, announcers, weathercasters, etc. usually fall into this category. At most, we only need to confirm that there are to be no changes – e.g. 'Today the newscaster is turning to interview someone on a nearby picture monitor'.

Flexible permanent set

Any TV studio in a local station or a network complex that is providing a regular public service has to cope with occasions where information about the program content is not available until a few hours before air time. Where this is likely to happen regularly, the best solution is unquestionably some form of all-purpose 'standard set'. We find this arrangement widely used for daily feature programs, commentary programs, current affairs analysis.

The setting is designed to provide a flexible format. It will probably include a section of open cyclorama for large-area movement (e.g. dance), one or two separate interview areas, a demonstration area, etc. With this arrangement there is usually a regular lighting rig, with several keys and back lights for each chair to allow for variations in action and camera positions. Background lighting serves from one show to the next, with minor variations.

Planning here usually comprises a brief talk-over of the probable items, and an indication of where people will be standing or sitting, which cameras will be shooting them, and the program *running order* (i.e. the order in which items are to be shot). In the event, one often finds that by the time the show is on-air, things have changed, and last-minute alterations to the lighting treatment are needed, but that is a recognized hazard.

A variant on this form of setting is the type of '*standing set*' used regularly for programs transmitted daily or several times a week. Cookery and talk shows often have this format. The scenery is left set up in the studio, while other shows use the rest of the studio staging area. Sometimes it is possible to leave all the lighting in position, too, but usually there are insufficient facilities, and lamps need to be turned or channels repatched to serve other sets in the studio.

The action in standing sets is usually confined to a series of defined areas. In a regular cookery program these typically include a display area, sink, oven, refrigerator, preparation surface, where the action follows roughly the same form from show to show.

Improvized settings

This is an off-the-cuff situation; avoided wherever possible, but inevitable at times in even the best-regulated studios. Here the team learns only a few hours before transmission that there is an item on basket weaving, quilting, folk dancing, or what have you. The Director, Set Designer, and Lighting Director talk over possible treatments. The Set Designer checks through stock settings, bric-à-brac properties, furniture . . . and devises an appropriate setting for the action! It is lit while the Director briefs the crew on what he expects to happen.

It is rare but not unknown for all this to be happening while the show is actually on-air! Any unavoidable noise is made only during film or videotape inserts into the program. But such emergencies are best kept to an absolute minimum!

Even when lighting is improvized, it is often possible to provide high-grade treatment if facilities and time permit. Where the studio has a 'saturation rig' there are usually suitable lamps available. If, however, lamps have to be re-rigged cabled and patched for the new item, opportunities can be very limited, especially if any special lighting effects (color, gobo patterns) are needed!

Regular planning

The planning process varies with organizations and with the complexity of the show. A full-scale operation generally follows a series of clearly defined steps:

1. The Director considers the general form of the production, and works out typical treatment.

2. The Director meets the Set Designer and outlines his concepts. In the case of a fully scripted production they will assess each scene or act, and perhaps master shots.

At around this time the Lighting Director receives a copy of the *rehearsal script*. If it is a drama of some kind it will outline where the action is taking place (e.g. Joe's place – a sleazy café), dialogue for all characters in each scene, the basic action, and stage instructions.

A warning! Don't be tempted at this stage to think about pictorial opportunities! There is nothing more frustrating than dreaming up treatments that turn out to be on a different wavelength from the opportunities the Director and Set Designer subsequently provide.

Fig. 6.2   The studio plan
A standard printed plan of the studio shows the staging area available, with various features and facilities. A typical metric scale of 1:50 (2 cm = 1 m) is replacing the widely used ¼ in = 1 ft scale.

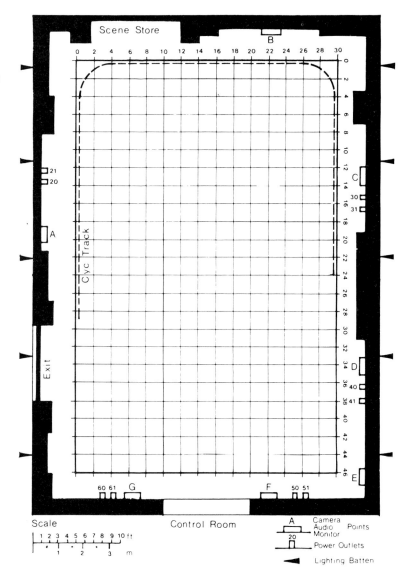

3. The Set Designer works on this material and develops a possible staging approach that promises to meet the needs of the action and locale of the production. He prepares pictorial sketches, and develops a preliminary plan. This comprises rough scale plans of individual settings drawn on tracing overlays, which are taped onto a standard print of the *studio plan*. This technique allows sets to be repositioned if necessary as a result of subsequent planning discussions.

A set of scale *elevations* shows side views of all the scenery to the same scale as the plan. These drawings include details of all vertical planes, showing walls, doors, windows, etc., and give information about their surface finish (e.g. moldings, painting, wallpaper). The elevations not only guide the construction and decoration of the setting but also help one to imagine what the actual staging will look like in three dimensions in the studio. If the elevations are stuck onto card they can then be attached to a staging plan, to provide a *model* of the complete studio layout.

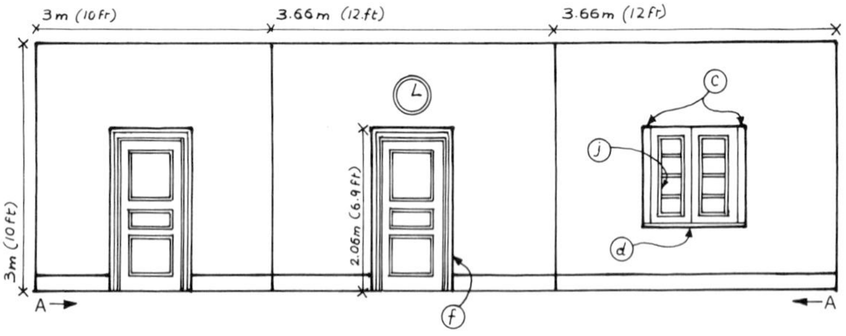

Fig. 6.3 Elevations
A scale side view of the settings, showing dimensions, detail and treatment for scenery (normally to the same scale as the studio plan). Explanatory notes are written alongside features to guide construction, painting, carpentry etc.; to identify stock units; and to indicate surface treatment, etc. (e.g. walls painted 'Dark Stone', windows void). Additional design points for carpenters, painters, and scenic artists are identified with lettered circles (e.g. ⓓ. Fix plastic molding No. 27 here).
Extra detailed information on specific features may be shown on separate large-scale elevations.

4. The Director, the Designer, and the Lighting Director may meet to assess potential lighting problems and opportunities. Any modifications are then incorporated. For less exacting treatment this stage may be omitted, but, particularly where lighting plays an important part in a production, it is as well to check out the proposed arrangements at an early stage.

5. The Director uses the revised scale plans and elevations to work out detailed production treatment: to decide on the action, the shots he wants, the development, atmosphere, etc. Concurrently, the Director will be concerned with casting, preliminary discussions with Costume and Make-up specialists, Graphics, Special Effects, and other contributors (e.g. those involved in shooting any film or video inserts for the production, model makers, etc.).

6. With a firm idea of his aims, the Director will then call a *production planning meeting* (*technical planning meeting*) to evaluate the proposed arrangements.

Fig. 6.4 Planning

1. *The overlay rough.* By drawing a scale plan of each setting on a tracing overlay you can readjust positions for optimum camerawork, lighting, and sound treatment.
2. *The sketch.* Many people have difficulty in envisioning from plans and elevations what a proposed setting is going to look like. A rough sketch can help to coordinate ideas and avoid misunderstandings.

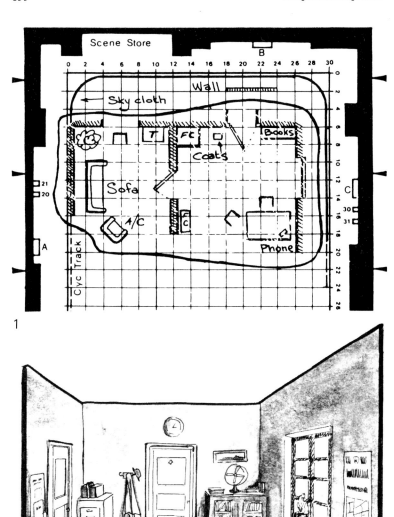

1

2

Production planning meeting

This is a crucial stage in the development of any production. It is a feasibility study of the Director's proposals, and helps all members of the team to coordinate their efforts. Without cooperative planning some very frustrating and time-wasting problems can arise. Let's look at one real-life example from a poorly planned production, for it illustrates the point well.

In a drama the situation called for a room to be lit with strong sunlight streaming through the window, casting a light pattern on the far wall. A spotlight was set up on a lighting stand (floor stand) downstage outside the window, to simulate the sun. This not only gave an excellent shadow effect but also keyed people throughout the room.

During rehearsals the Director entirely revised camera positions, and the main camera now overshot the backing outside the window. Because the backing could not be moved, the Designer decided to put sheers (nets) over the windows to prevent the camera seeing beyond the backing. These blocked off the 'sunlight' thereby killing the sun effect, the wall shadow, and the main key light for the scene. The entire set had to be relit. The result was a last-minute compromise. Effective planning avoids such situations.

Wherever possible, all the chief specialists of the team are gathered together at the same time. As well as the Director's immediate staff (Assistant, secretary, etc.) the production group will include the Designer and the Lighting Director, the Technical Director (to evaluate technical and operational problems), Audio Engineer, Make-up and Costume/Wardrobe specialists.

The *studio crew*, including the camera operators, boom operators, sound crew, video operators, scene staff, will not learn about the production until the actual camera rehearsal; apart from any preliminary information passed on by the specialists.

The Director goes through the show shot by shot, describing the action, performers' positions, moves, types of shot, any effects, etc. He may have a series of '*storyboard*' sketches that he or the Designer has prepared. He discusses the basic mechanics of the production: how many cameras he requires, their positions, when and where they will move, the order in which sequences are to be recorded, etc. Attention is concentrated on the proposed *floor plan*, on which the main furniture positions, cameras and sound booms have been drawn, as each step is checked out.

The group will also examine sketches, reference photographs, the model, samples of wall-coverings, drapes material, etc., go on to discuss proposed costume, make-up and so on, and indicate any problems or improvements they can envisage. Following the meeting, armed with this information, the specialists will start to work on their particular contributions.

For simpler productions, where the mechanics are quite straightforward, this process is considerably abbreviated. But the lighting, audio and technical specialists still need preliminary details of the production in order to check it out for possible problems and to arrange equipment, staffing, etc.

Lighting preliminaries

Following the meeting, an accurate scale *staging plan* (*floor plan, ground plan, setting plan*) is prepared by the Designer and distributed.

Fig. 6.5 Camera plan
The main working positions of each camera (1A, 1B, 1C etc.) and sound boom (A1, A2, B1, B2) are marked on the staging plan to provide the 'Camera plan'.

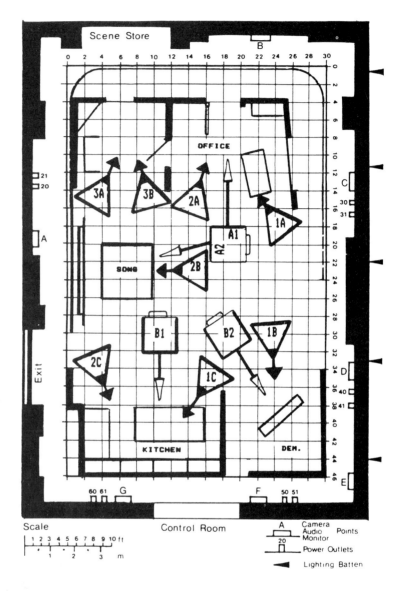

This shows all scenery and major furniture, floor painting, raised areas, seating, etc.

Using this and the scale elevations, the Lighting Director again reads through the script with notes from the planning meeting, envisaging possible treatment. Although there are times when eye-line checks with a model can help to assess potential light directions and anticipate difficulties, most of the Lighting Director's work is done with a practiced 'three-dimensional imagination', and a modicum of clairvoyance.

Pre-rehearsal/outside rehearsal

In a major television production the performers need to rehearse their lines, practice business (action), become familiar with their moves, learn the mechanics of the production (e.g. where they will be in a tight close-up), and so on. This takes some days, and studio time is too costly to allow this to be done in the final settings. Instead, most organizations carry out pre-studio rehearsals in outside halls, where set outlines are chalked or taped on the floor to provide a full-scale version of the staging plan. Vertical poles are used to mark doorways, arches, ends of walls, and substitute furniture and mock-up fittings serve to help the actors locate.

By now a *camera script* is available. This is a reprint of the *rehearsal script*, to which many operational details have been added. The script has been divided into numbered *shots*, each with an abbreviated description indicating the type of shot (close-up, mid-shot, long shot, etc.) and which camera is taking it. Transitions between the shots are marked (cut, mix/dissolve, superimposition), together with cues, effects, and other information needed to coordinate the studio crew, who will be meeting the show for the first time.

The camera script is a valuable document, particularly when scenes are shot out of order. Where the *shooting order* is considerably different from the *running order* in which the scenes will eventually be presented after editing, everyone would be quite lost without reference to a detailed script.

As the rehearsals progress, the Lighting Director visits to see a complete run-through of the production ('technical run'). Having oriented himself, interpreting the marked-out floor relative to the staging plan, the rehearsal begins.

He watches as actors go through their lines (dialogue) and moves, noting details of their positions (with 'nose lines') on a copy of the staging plan. The Director offers explanations and guidance from time to time, about shots and visual treatment. As this *action plot* develops he will probably draw in principal directions for the lighting, based on the actor and camera positions for each sequence. At least he now knows where people are standing, sitting, walking, making entrances, and has a pretty clear idea of the Director's aims. Things will inevitably alter a little later during camera rehearsal, but this is a valuable framework on which to build the lighting treatment.

This is an opportunity to get any further information on the staging, costume, make-up, etc., and to exchange ideas. Where there is cordial teamwork, the various members feel free to make constructive comment, and share problems. The Set Designer may suggest features that could be enhanced by a certain lighting treatment, while the Lighting Director may make helpful observations on a modification to the setting. They develop a mutual appreciation of each others' crafts.

As he watches, the Lighting Director notes any potential problems he can see. A difficulty that is resolved at this stage may avoid a

Fig. 6.6 Practical problems
Checking out potential problems
during planning/rehearsal avoids
dilemmas during shooting. Here the
actor could not be lit properly during
continuous action, as light was
restricted by: (1) low ceiling, (2)
covered windows, (3) actor's
position, (4) likely camera shadow.
The solution was to key the action
through a nearby door (spotlight on
lighting stand) (5).

major quandary in the studio. We need to anticipate, to see potential
snags before they arise. Let's imagine a typical situation, in which
the Lighting Director, the Production Director, Costume Designer
and Set Designer discuss a situation.

LD: 'She has her head bent down, looking at the floor (potentially
ugly portraiture). How close is your shot?' (It may not
distract in a long shot)

PD: 'A big close-up, on a narrow-angle lens, from over there.' (It's
a screen-filling portrait)

LD: 'Will she still have her hat on?'

CD: 'Yes, it has a wide brim and a veil.' (The brim shadows the
face and the veil catches the light!)

LD: 'I cannot use a camera-light, the camera is too distant (its
intensity falls off rapidly with distance). Can I put a floor
lamp down there?'

PD: 'No, Camera 3 moves over to there immediately beforehand.'

LD: 'Perhaps I can light her through the window (with sunlight).'

SD: 'There are nets over the windows (light will not penetrate).
There is a low ceiling over the bay window.'

LD: 'Can't we take the net off?'

SD: 'I don't really like the stock backcloth from that angle, the
nets soften off the exterior view.' (Painted cloths can look
artificial on angled shots)

LD: 'Can we move the actress along the seat?'

PD: 'No, it would affect the other shots. Camera 1 would not be
able to follow her over from the door.'

LD: 'Do we see the hall outside the doorway before she enters the
room?'

PD: 'No, we are watching the hero depart.'

LD: 'Then if we leave the door ajar, I could get a floor key through
from the hall onto the girl's face . . .'

A problem is solved. But it does illustrate fairly the type of
anticipation that is needed. It is not sufficient to wait to see it on
camera and then correct matters. There just is not the time!

Check out the facilities

Whether the show is large or small, whether planning has been
detailed or slight, the next stage is to check out the facilities. By now
you have a good idea about various production details. The Director
has explained about the series of lighting changes needed at the start
of the show. He would like a spotlight to follow the singer in a
darkened stage, where stars twinkle in a velvet blue sky. In the last
scene the designer has strings of lights which should 'chase' to
suggest movement. 'Oh, and by the way, it would be great to have a
rainbow effect right across the sky during the "Stormy Weather"
number . . .' *But do you have the resources to do these things?*

This is the time to check! If you have not, let the team know immediately, so that some alternative can be substituted! Instead of the 'rainbow' effect, the director might have to settle for flashes of lightning and crashes of thunder instead! Some apparently simple ideas can be difficult to put into practice. Some 'impossible' ones are quite simple.

It is always important to assess the *scale* of treatment relative to your facilities. At first sight it is quite reasonable for the Director to ask for the background lighting to change from orange . . . to blue . . . to green for three successive musical items. But what does this really involve? Supposing it requires six lamps to light the background. Then for three hues, you will need eighteen lamps and eighteen power circuits (channels) just to light the background alone. Are there any difficulties in rigging that number of lamps quite close together, at reasonably similar distances from the background? Are the power supplies available? Are there plenty of lamps and channels left to light the action? Where resources are limited, you can't take such matters for granted.

This is just an example of the way one needs to think one step ahead when planning and preparing the lighting treatment.

Cogitation!

Now is the moment to sit back and think about the ambiance, the atmosphere, the 'feel' of the visual presentation. The Set Designer has created a particular environment, which your lighting is going to interpret. Even a setting that is pretty plain and ordinary-looking, consisting of little more than an arrangement of mid-gray flats perhaps, can still be transformed through light into an attractive background for the action. Sometimes you will be presented with a plain light-toned cyc and a couple of columns, and it is up to you to use light to conjure up a magical atmosphere. It can be done. It is a matter of creative imagery and know-how.

Plans and elevations are cold impersonal drawings. They tell little of the visual opportunities that you will be able to develop on the screen. You can clothe that setting in a pattern of light that conjures the viewers' imagination. The atmospheric treatment you develop comes from latent images in your memory – from images seen in films, from places visited, from book illustrations. Such fragments accumulate and combine to form a subconscious guide to inspire your imaginings.

In the mind's eye the setting becomes a real place. In the Sherlock Holmes stories we see the claustrophobic Victorian rooms, where daylight struggles through thick lace curtains and heavy dark drapes to streak the dingy wallpaper. One can almost smell the leather chairs, touch the antimacassars, the cluttered knick-knacks, hear the steady tick of the clock . . .

Then comes the cold light of day! We have to think in practical terms exactly what we shall be doing to evoke this atmosphere. What light fixtures will we use, and where will they be placed? How does the lighting treatment for the setting tie in with that for the action, and the mood of the occasion? This is the moment to think through the accumulated information you now have about each scene, and these are summarized in Table 6.1.

Table 6.1   Points to consider when planning lighting for a studio production

Positions of settings	Check the positions of sets and action relative to the studio facilities:
	• Check potential key light positions, to ensure that lighting fixtures can be hung there. (Some lighting suspension systems are very restrictive.)
	• How do the positions of settings relate to the overhead lighting system (lighting battens, bars, barrels)? Would slight repositioning of sets improve lamps' accessibility?
	• Where settings require a number of lighting fixtures are there plenty of power outlets in that part of the studio? (Otherwise, long extension cables will be necessary to feed each lamp!)
	• Check the positions of settings relative to all scenic backgrounds, cyclorama, backdrops, scrims/gauzes, etc. Is there sufficient room to light backings/backgrounds evenly? Will they be lit from above, sides, and/or groundrows?
	• Check whether there is sufficient space around the settings for lighting fixtures on floor stands, floor lamps, concealed lighting fixtures, where these are needed.
	• If shadows or patterns are required is there sufficient room for the lamp(s) to produce the effect?
Constructional features	• Check that the way scenery is constructed and supported will not impede lighting: e.g. ropes, wire slings, support battens preventing overhead lighting fixtures being positioned.
	• Check architectural features to see whether they restrict lighting: e.g. deep windows, arches, ceilings, overhangs, etc.
	• Check for suspended items that can cast shadows and impede lighting: e.g. hung mirrors, chandeliers, tree branches, slung screens, etc.
	• Check any scenic features likely to throw distracting shadows across other parts of the setting: e.g. pillars, drapes, skeletal structures, isolated flats, etc.
	• Check for *level changes* in the settings: i.e. stairs, platforms (rostra, parallels), towers, etc. They will affect relative lamp heights.
Surface finish	• Check the tone, texture, color, and contrast of the backgrounds. These will modify the angle and intensity of the lighting needed.
Furnishings	• Check the position and form of drapes, furniture, wall pictures, main ornaments, statuary, mirrors, wall fittings, etc. They may need individual lighting, or require careful treatment to avoid reflections or shadows.
Decorative lamps	• Check the positions and types of all decorative light fittings in the settings: suspended light fittings, wall-light brackets, pole lamps/standard lamps, table lamps, concealed strip lighting, signs, decorative set lighting, etc. Are they practical (i.e. working) or non-practical (i.e. only decorative)?

Table 6.1   (cont.)

Studio 'exterior' scenes	• Consider the main light direction. ('Daylight' scenes should have a single predominant light direction.) • Is there a scenic cloth or cyclorama providing a general background? Can you light the background without spurious shadows of trees, buildings, etc. falling onto it? • Are any scenic groundrows, profiled cut-outs being used that need localized lighting (e.g. suggesting city skyline, mountains)?
Lighting effects	• Check if there are to be lighting effects, • Is colored lighting required – which colors? • Are there any projected light patterns – which patterns?
Costume and make-up treatment	• Are there any potential lighting problems – e.g. highly reflective clothing, white wigs?
Production mechanics	Check the details in each setting of: • *Performers'* positions, moves, actions. • *Cameras'* positions, moves, shots. • *Sound booms'* positions, coverage, moves.
Production treatment	Check the script for details of: • Lighting changes and cues: e.g. day, night (practicals on/off, etc.). • Follow spotlight. • Blackout. • Color changes.
Lighting facilities	• Make a guesstimate of the probable types and numbers of lighting fixtures needed relative to the staging and action. Remember that these will be influenced by the size and areas of settings. (The needs could exceed your facilities!) • Are the lighting fixtures for particular effects available in sufficient numbers (e.g. pattern projectors)?
Safety	• Anticipate fire risks, hazards.
Organization	• Will sufficient time and labor be available? • Check the time scheduled for rigging and setting lighting fixtures. • Are other concurrent studio activities (building sets, floor painting) likely to delay work? • Check times for camera rehearsal and recording, allocated breaks, etc.

Lighting approaches

So you now have a pretty clear idea of the form the production is going to take. You are ready to light it. *But where do you start?*

One thing is certain. There is little time to arrange lighting fixtures by guesswork! There is no opportunity to move lamps around until you have the most appealing effect. Rigging time is short and camera rehearsal is fully occupied, so you have to hit the target reasonably well first time. To some extent, you can redirect lamps and change their coverage during rehearsal, but there is all too little opportunity to experiment.

Occasionally you can provide alternative lighting treatments, and select the one that proves to be most effective. But that is the exception.

Although the art and craft of lighting cannot be reduced to a 'painting by numbers' routine you do need a systematic technique that works, and yet leaves you with artistic freedom. Those we shall be looking at do just that.

If you ask a dozen experienced Lighting Directors what determines exactly where they place their lighting fixtures you are likely to get a dozen answers with differing emphasis, differing priorities, but the same underlying aim. The truth is that after a while one works instinctively, not making conscious analytical judgments, but choosing from experience the most appropriate solutions for varying situations. Several experts may make the same choice, for quite different reasons.

The guiding principles that follow should help you to build up a workable treatment from the start.

Systematic lighting

Basically, there are two approaches to systematic lighting:

● In the 'look and light' method which is used in both film and TV lighting you stand within the set and, looking around, estimate where you are going to position the lighting fixtures. As you make your decisions you draw on a *rigging plot* symbols representing the various lighting fixtures on rough outlines of the settings, showing exactly where each is to be placed.
● The 'plot and light' method used in many TV studios approaches the situation differently. Here you prepare a lighting plot *beforehand*, and this overall 'map' of the lighting treatment is used both to rig the lighting fixtures in position and to aid light adjustment during the show.

Which approach is best will largely depend on the way things are organized in your particular studio, and the labor and time available. Sometimes one method is preferable to the other. It is important to appreciate the advantages and limitations of these approaches, so let's look at them in some detail.

The 'look and light' method

Here you judge the lighting fixtures needed from within the setting and the acting area. Working methodically through the action and

Fig. 6.7 Anticipating problems

1. *Lamp height.* A reasonably distant key (1) would not reach the subject (B) unless lowered into shot at (2). To place it at a more suitable height, it must be taken further from the subject (3). Now, however, it may light other intermediate subjects at (A).

2. *Varying heights of scenery and acting areas.* These must be allowed for when plotting lamp positions.

3. *Lamp shadows.* Lamps lighting Setting A may obscure those lighting Setting B. Height adjustment, beam localization, revised lamp positions may be the solution.

4. *Rigging space.* The ideal key light positions for adjacent settings may coincide. Then compromises are inevitable, and treatment must be revised.

5. *Lamp take-over points.* The take-over between the beams of successive lamps must be controlled to prevent obtrusive multi-shadows or doubling intensities. Take-over arises (*a*) vertically, (*b*) horizontally.
   Light cut-off provided by barndoors is soft-edged. The sharpness of this shadow edge depends upon the size of individual flaps relative to the lamp area. Adjacent lit areas cannot therefore be joined accurately 'edge to edge' (as is possible with hard-edged effects projectors). Wherever overlap is unacceptable, it becomes necessary to introduce shadow corridors in less critical regions. These regions may be so large that the performers move in and out of lit areas. Appropriately used, the result is effectively attractive. Overdone, the technique can appear mannered.

6. *Space restrictions.* (*a*) We may anticipate using a particularly dramatic window effect, but, (*b*) in fact, little space is available (plan). (*c*) The elevation shows the actual effect would be valueless. (*d*) If space is provided for a suitable lamp throw, a low-angle camera would shoot into the light.

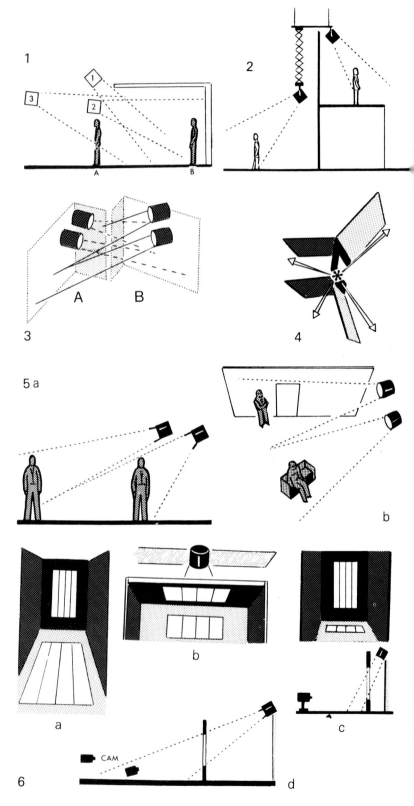

background lighting treatments, you draw lamp symbols on a plan outline of the setting to provide a *rigging plot*. This rough plot serves several useful purposes:

- It helps you to check that you have not missed anywhere while working out the treatment.
- It shows the purpose of each lighting fixture.
- It provides a record of your decisions.
- Someone else can work from the completed plot, and rig the lighting fixtures needed.

In film and location work, lamps are usually fed direct and their intensities controlled by scrims or individual dimmer units. In TV studios, the channel or dimmer number controlling each lamp will be written beside it, to form a *lighting plot*.

Whether you select lamps from those already in position, or rig each lamp in turn as you plot it, or complete the plot then rig the lamps will really depend on your circumstances. Some Lighting Cameramen and Lighting Directors prefer to have all the selected lighting fixtures in position, and then light the show lamp by lamp. Others prefer to switch on and set each lamp as it is rigged, to build up an overall effect.

Advantages

This approach is straightforward, and has a number of advantages. Above all, you are on the spot, and can see exactly what is needed. You can judge where there are problems, and do not need to estimate or imagine what the setting will look like. So there is less chance that you will misjudge.

Lighting the action

The easiest way to judge the best position for a lighting fixture is to consider the situation from *the subject's position*. So stand in the setting wherever someone is going to stand or sit during the show. (We'll assume that you are not on a raised area or a staircase.) Looking towards the direction that they will be facing, stretch an arm upwards to about 35°, to a point just to the left or right of your nose line. The simplest way to do this is to use the height of scenic flats as a guide. They are usually standardized at 3–3.6 m (10–12 ft). If a lighting fixture is hanging at 3 m (10 ft), it should be above a spot on the floor about 2.7–3 m (9–10 ft) away to provide this vertical angle. If the lighting fixtures must be higher, they must be correspondingly further away (see Figure 5.8 and 6.12).

You have now established where the key light will be located. To your other side, you can mark in a position for the filler/fill light. It should be slightly nearer, but preferably less steep than the key. Turning round, you can assess the back light position, which will be offset on the opposite side to the key, and probably a little higher. Draw the lamps on your rough plot, preferably with lines to your floor position, to remind you later of their particular purposes.

Now, standing in the next acting position, you do the same thing. But look at the key light, filler, and back light you have just set, to see whether they will still be at appropriate *horizontal* angles for this position also. In many cases, one key, filler, or back light will successfully cover a number of positions. Then you can also draw a second set of lines from the lamp symbols to these positions.

If the previous lamps are too oblique to your new position the *side* lighting may prove unsatisfactory, and you will need a further set of lamps. How many separate sets of lamps you need will depend on the area involved, how close the shots are, how varied the performers' positions are, and the variety of camera angles. By following this method you will be able to build up a rational pattern of lights to suit the action.

As you judge the positions of the key and back lights do not forget to allow for head movements. If, for instance, the person is going to look forward then sideways, split the angle, and key them from that direction. The result will probably be a good compromise.

Lighting the background

When lighting any background the main points to look out for are:

● Consider whether the background is to be lit evenly overall, shaded at the top, lit in patches or pools of light, or dappled. Are some areas to be brighter than others? Is it to be lit with color(s)?
● Avoid lighting the background from a very steep vertical angle, or an acute horizontal angle, unless you want to emphasize its shape,

Background lighting
The light intensity that is suitable for the subject itself, may be too great for the nearby background.
1. Here the subject is close to a light-toned background, and its key over-lights the surface, causing it to appear over-bright or even 'burn out'.
2. Using overall soft light may reduce the amount of light falling on the background, but the result will be flat, with poor modeling in the subject.
3. If the background is lit by an upstage key (e.g. 10H 2V), it will appear even brighter than in (1), and distorted shadows of the subject may spread across it.
4. If the subject does not move around and the key light is very closely shuttered, it may be possible to light the subject separately from the background (which can be illuminated to a lower intensity).

1

2

3

4

Background contrast
This exercise, with two different kinds of subject, demonstrates how our impressions of tonal values, contrast and modeling are influenced by the background tones.

Each subject has the same lighting treatment. Against a light background, the subjects seem to be sem-silhouetted and of lower contrast than when set up in front of a black background.

1

2

1

2

3

Confining background tones
These off-the-tube pictures show how selective wall shading can produce a feeling of confined space. Notice how the dark overhead tonal areas appear to press down onto the subjects, and the lighter areas concentrate the attention.

1

2

Foliage

1. When subjects with fine outlines (such as these white blossoms) are lit with back light alone, details may be lost as highlights merge together.
2. When frontally lit, the same blossoms are more clearly defined.
3. These laurel leaves have been lit in four different ways:
   Top left – lit from behind, creating a translucent effect.
   Bottom left – unlit leaves silhouetted against a light background.
   Top right – low intensity *back light* bounces back from the angled leaves, exaggerating their shiny surfaces.
   Bottom right – a strong localized spotlight was needed to light these leaves *frontally*. Yet they appear to be dull, lacking any shine.

3

Angled backgrounds

As you can see here, the apparent brightness of the same background tone can alter with its angle relative to the camera's viewpoint and the incident light (*cosine law*). The extent of the change depends on the surface finish and texture.

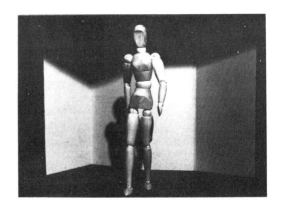

modeling and texture. As a general guide, therefore, do not have a suspended lighting fixture closer than about 2.7–3 m (9–10 ft) from the background, if it is higher than e.g. 3 m (10 ft).

- Take care not to overlight the floor near the bottom of the wall, or objects close to the wall.
- For an interior scene, make sure that the direction and shape of light falling on the background is compatible with the way the room would normally be lit (i.e. sunlight, practical lamps).
- You can often use the *same* lighting fixture to light both action and the background, if the overall effect is satisfactory.
- Where the background has been lit *separately*, check that lighting arranged for the action does not accidentally fall onto it and nullify the effect. Also take care that such background lighting does not fall on people walking near the walls, unless the effect appears quite natural.

Background treatment

How you treat the background will depend on the kind of setting you are dealing with.

- If you are lighting a neutral or stylized setting (e.g. in a talks program), backgrounds may be evenly lit, shaded, dappled, or display projected patterns. Gels may be used to colorize neutral surfaces (the cyclorama, or gray flats).
- For a musical item, dance, pop or spectacle, color may be used freely to embellish and add visual impact. These are productions in which emphasis is on *display*, and lighting plays an important and obvious part in the appeal of the presentation.
- In most forms of drama the settings are made to look as realistic as possible, to persuade the audience that they are watching action in a real place. So the lighting approach is subtle and persuasive, rather than decorative. If the Set Designer and the Lighting Director have been successful the atmosphere illusion will be so natural that the audience will simply assume that it is the real thing!

Arranging the background lighting

If you are lighting a *realistic* setting such as a living room or office, begin by standing at the planned camera positions and look carefully at the background and its furnishings. Consider the 'light sources' in the setting; i.e. the *window* if it is daylit; the *practicals* if it is night time. Decide how their light would naturally fall upon various areas of the background. Think about how sunlight would cast a light pattern on the far wall, or how pools of light would form around the practical lamps in an otherwise darkened room.

Having imagined these light patterns on the background, draperies, furniture, you have now to arrange lighting fixtures ('set lights', 'background lights') to provide them:

- You might spread the light from one lighting fixture along an entire wall – barndooring it at the top to shade the wall, and at the bottom to avoid overlighting the floor.
- You might arrange a series of fairly localized areas of light from several lighting fixtures.
- Occasionally, you might fit a sheet of aluminum foil into a diffuser frame in front of a single spotlight, and cut out holes wherever you need pools of light. This can be quicker than arranging a series of separate flags.

You have now to locate the lamp(s), and the best way is the simplest. Just stand with your back to the wall, look straight into the set ahead of you, and turn in the direction from which you want the light to come. Then look upwards until you have the appropriate *vertical angle*. That will help you to locate the spot for the lighting fixture. Using this method, you will not only avoid vague 'somewhere-up-there' kinds of selection but also will see whether there is anything likely to cast shadows onto the background.

Some Lighting Directors stand in the set beneath the lamp position they estimate will be suitable, then consider whether its angle and light-spread will be appropriate. But this method is likely to overlook obstructions, and is less reliable for the inexperienced.

Disadvantages

It is clearly a tremendous advantage to have the set already built, and to be able to *see* what you are lighting. But on the other hand, there can be considerable disadvantages too.

■ *Accessibility* Where the floor is clear beneath a lighting batten/bar you can lower it to chest level, and attach all lighting fixtures quite easily. You cable them to power outlets, and raise the batten/bar to working height, e.g. 3.5 m (12 ft).

But when a setting is already built, this will prevent many of the lighting battens/bars from being lowered below about 4.5 m (15 ft). ('Hanging space' must be left below each batten for lighting fixtures to hang.) Rigging becomes considerably more complicated; particularly if there are high walls, or ceilings, or overhanging sections.

You have to haul each lighting fixture up steps or a lighting ladder to hook it into position on each batten. This is tiring and cumbersome. Sometimes it may not be possible to place the steps close enough to rig the fixture.

In studios where there are *many* suspended light fixtures to choose from, things are easier (see *saturation rig*). But even then it may not be possible to get close enough to a selected fixture to plug it up and fit diffusers or colors. Where lamps are located on individual hangers (e.g. monopoles/telescopes) or hung separately on tackles the rigging process can take correspondingly longer.

When arranging lighting treatment in this piecemeal way you may find it difficult to alter or augment the rig later; particularly if there is insufficient space on the suspension system for subsequent changes.

■ *Time consuming* The success of this lighting method largely depends on the studio facilities. In a studio where every lighting fixture has to be taken separately from a store (or wherever it happens to be hung), and moved to its selected position one by one, rigging can be laborious and time consuming. Even if you are fortunate enough to find that some lighting fixtures are already roughly in position from a previous show, you still have to find out how they are plugged and routed in order to patch them.

Once all lighting fixtures are in position and patched and channel numbers allocated, a patching sheet can be prepared. Only then can you set up the lighting board and adjust lamps in readiness for rehearsal.

The 'plot and light' method

Advantages

The particular value of this method is that a great deal of the planning and preparation can be carried out long before the actual lighting process begins. With the information from the planning meeting, you draw out a lighting plot showing where the lamps are to be rigged, how they are to be plugged and patched. You can then plan cuing sheets and decide how to set up the lighting board. None of this has to wait until you are actually in the studio.

This method gives an overall view of the project before it gets under way. So where lighting fixtures, the available channels on the patchboard, and the power supplies are very limited it avoids the embarrassment and frustration of 'running out of lamps'. (A hazard of the piecemeal 'look and light' method.) You know the available facilities and work within them, reminded by the patching sheet.

Another plus for preparing the lighting plot in advance is that you can pass a copy to those who are going to rig the show, allowing them to see beforehand:

● The total number and types of lighting fixtures required;
● The appropriate accessories needed, such as extension cables, diffusers, sheets of color medium, flags, gobo patterns, hangers, etc., etc. (On a large show, all these separate items can take an appreciable time to organize and collect together.)
● The labor, time and work effort involved;
● Any equipment that needs to be hired, patterns made, etc.;
● Equipment that must be checked or prepared in advance.

This method is particularly practical in busy TV studios, where it enables lamps to be rigged during any available studio down time (e.g. when another show in the studio has been cleared after taping), or whenever a studio is *empty*, before any settings are erected.

The outstanding advantage of the 'plot and light' system from the Lighting Director's point of view is that it encourages you to think methodically, and to anticipate. It allows you to sit down and think through an entire production, to work out the optimum lighting treatment, together with cues, colors, effects, lighting changes, patching, etc., long before entering the studio.

You can take as much time as necessary, working through the script and developing a comprehensive plan of campaign. Unlike the off-the-cuff inspiration of the 'look and light' method, you can readily reconsider and experiment as you go.

It is not always easy when lighting from scratch in the studio to anticipate, to make allowances for the variations needed throughout the production – particularly when rigging the studio for a complex production. You may not recall, for instance, that people standing by the window are to be lit one way during a daylight scene and another during an evening scene when the practicals are on.

The plot can be photocopied to provide duplicate versions for those rigging the lamps, the operator at the lighting board, records, etc.

Disadvantages

The main limitation of this method, which is used in many TV studios, is that some people have great difficulty at first in reading plans and elevations. They cannot imagine what the staging is going to look like before seeing it in the studio. They misinterpret features, fail to anticipate, and lose the feeling of scale. They discover that what they had imagined as a large ceiling turns out to be a modest soffit. What they had dismissed as a minor feature may actually prevent them from lighting the subject on the day. When you light while looking at the setting, you can't make such mistakes. However, with a little practice, working with plans and elevations soon becomes second nature.

The form of the lighting plot

To start with, you will need a scaled *staging plan (floor plan, ground plan, setting plan)* for the production showing details of the scenery and the main furniture positions. Better still, if you have seen a pre-studio rehearsal use your *action plot*, which has the performers' and camera positions marked on it. Although rough notes may be good enough when using the 'Look and light' method, a scale plot needs to be reasonably accurate, or you'll get the distances and proportions wrong.

Many organizations use an *overlay (tracing sheet)* which is printed with details of all studio lighting facilities. It has the suspension system over the acting area marked, together with the location of all power outlets. (You can, of course, use a plain overlay and add such details where necessary.)

Model settings

By sticking elevations of the setting onto thin card, and attaching them to a studio plan, you can produce a useful scale model.

Such models help the production team to visualize camera set-ups, lighting, action, more accurately. Checking from different viewpoints, they can anticipate opportunities and problems. The model can also aid the scene crew when setting up the scenery.

1

2

If the lighting system uses a series of battens, bars, or barrels these will be numbered. So an appropriate symbol drawn on lighting bar 3 will show at a glance that a '2k' is positioned at one end and patched to dimmer channel 99.

An alternative method is to use *location marks* beside lighting fixtures. Most studio plans include numbers along two edges, which correspond to a regular series of marks (footage marks) painted along the studio walls. These are used to locate scenery (and lighting fixtures) accurately on the studio floor during setting. Where lighting fixtures are hung from individual telescopes or pantographs they can be located by quoting their positions relative to these marks – e.g. a '2k' at 15/12.

Drawing the lighting plot

Although a lighting plot may be unnecessary when working beside the camera on location or in the studio it is essential when assessing pictures remotely from a lighting control desk in a TV Production Control Room. It enables you to identify each lamp in a moment.

Without an accurate plot a TV studio rehearsal can become a nightmare. You are left wondering which lamp is causing the boom shadow, or why there is a hotspot on the cyc. Unless you are using only a handful of lighting fixtures and can remember their associated positions and channel dimmer numbers, you are rudderless! Even a 'back-of-envelope' sketch is better than nothing.

It is easier to draw the various symbols representing different kinds of lighting fixtures if you use a stencil of some kind, with scale-size outlines (see Figure 6.8). There are official standards, but many organizations use their own symbols. Do not use tiny lamp symbols on a scale plot, for they will mislead you about the amount of space needed or rigging. (You may draw six lamp symbols on a bar and find that there is really only room to rig three lighting fixtures!)

Fig. 6.8 Lamp stencil
A stencil with patterns for the main types of lamps used, to aid neat lighting plots.
Left column: Fresnel spotlight, lensless reflector spot, HMI arc.
Center column: Scoop, Softlight (internally reflected), small broad.
Right column: Cyc light/groundrow, effects spot/profile spot, follow spot.

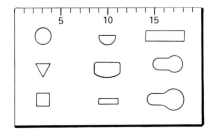

You will find it helpful to use clear outlines to represent all suspended lighting fixtures, and shaded symbols for lamps on lighting stands or on the ground. Symbols can be colored to indicate color medium to be fitted. Diffusers, barndoors, snoots, flags can be indicated with adjacent symbols.

We'll assume that you have in front of you a copy of the scale *staging plan* showing where people are standing and moving around within the set outlines (action plot), together with main camera positions. (You may be working on an overlay, or directly onto the printed plan.)

Let's take a close look at a practical method of plotting. And remember, these are indications to guide you, *not* rigid rules!

Whenever you draw a scale symbol for a lighting fixture allow a reasonable amount of space around it to allow for turning the lamp, and attached accessories (open barndoors, color frames). If you crowd lamps together on your plot you will not be able to rig them that way in the studio.

Lighting people

1. Start with the first action position you are going to light. A circle on the plan represents a person, with a *nose line* showing the direction in which they are looking.

Are they likely to be turning their head to look in a different direction – to talk to another person, to look at a picture monitor, to face another camera? If they are, then you must decide whether a single key light will suit *both* positions or whether you must use two separate key lights; e.g. one for when they are full face to camera and another when they have turned towards the monitor.

2. Assuming the simplest situation, in which a person is facing the camera, draw a *key light* axis (i.e. the light beam) up to about 30° to this nose line. Preferably, try to key the nose on the side further from the sound boom. If you are lighting a profile or an angled head, remember that the key may be on the far side of the head from the nose line.

3. About 3–3.6 m (10–12 ft) away along the key axis, draw the spotlight symbol (Figure 6.9).

4. Extend the axis *through* the head and beyond, and at about two-thirds of the key light distance (say, 3 m (10 ft) away) draw a spotlight symbol for the *back light*.

Fig. 6.9 Drawing the plot
1. The keylight direction is offset from the nose line, about 3 m (10 ft) away minimum.
2. Draw a symbol to represent the person (with nose direction), draw in the keylight. Fill light illuminates the shaded area of the subject.
3. A slightly offset back light and background lighting complete the basic lighting.

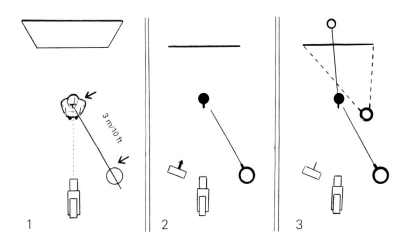

5. Assuming that fill light is necessary (as it usually is), place a soft light source at about 3–4 m (10–12 ft) away, on the opposite side from the key light. In most TV studios a soft light source on a floor stand will probably get in the way of cameras and be seen in long shots, so you will need to hang the filler at about 3 m (10 ft) high.

Do not overlook the way the light intensity from soft light sources falls off with distance. They all too easily overlight someone who is near, yet are insufficient for someone e.g. 6 m (20 ft) away. In a large set you may need additional overhead fill light mid-stage to be faded in for close shots when action is near the back of the set.

It is a good idea to include quite a generous amount of filler in the plot, for it is easier to rig it initially than to add extra lighting fixtures during rehearsals. But start with fill light kept to a *minimum*, only increasing it where necessary. Do not be tempted to flood the place with soft light, or you will achieve flat, characterless illumination. Modern TV camera systems and film emulsions do not need a high-intensity 'base light' ('foundation light') to minimize the contrast range. If shadowy areas are too dense, add appropriate fresnel spots to light them rather than boost the overall intensity of soft light. (Don't use fill light to 'correct' the poor modeling from badly angled key lights!)

6. The power ratings needed for these lamps will depend on –

● The light levels your cameras need for their working stops;
● How close the lamps are to the subject;
● The efficiency of the light sources.

If the key light is fairly close a lensless spot (e.g. 1 kW) may prove suitable. But where lamps are further away, higher powers will be needed. Typically, a key light may be 2 kW, a back light anything from 1 kW to 500 W, and soft light filler around 1.5 to 2 kW.

Your choice of lamp power really depends on the area you want to light, and how far away the lighting fixture is from the subject – particularly if it is a soft-light source. The closer a lamp is, the less its light spread. The coverage of a close fresnel spot on a floor stand can be quite restrictive. In larger studios, with lamps further away or cameras working at smaller lens stops, 5 kW keys will probably be necessary.

 Where you want an impressive effect, such as a strong window shadow, a powerful lamp may be essential. ('Powerful' is a relative term, depending on your local light levels. It may be a 10 kW quartz light, a 6 kW HMI arc, a 150 amp carbon arc.) A low-power lamp will be ineffectual. If you do not have an appropriate lighting fixture, try to achieve the effect another way (e.g. a projected pattern). Where you *have* to use a lamp that is really too powerful for the purpose, use a scrim, diffuser, or neutral density medium in front of it. (In some fixtures you can switch lamp filaments.) Dimming the lamp will lower its color temperature appreciably – fine for sunsets or firelight effects, but otherwise inappropriate.

Fig. 6.10   First assess the situation
The scene to be lit, showing the main light directions (compatible with daylight).

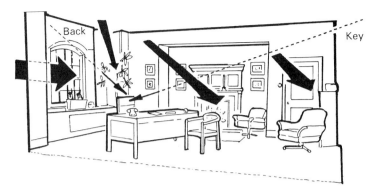

7. Check that the positions you have chosen for each of the lighting fixtures is practicable. Are there any hazards?

● Think three-dimensionally. Remember that there are walls and other obstructions to be taken into account when positioning your lamps. They are visible on your plans and elevations.
Is there *room* for the lighting fixture? If there is insufficient space, you may find that, for example, a fixture is far too close to the backing it is lighting. Instead of even lighting it produces just a localized hotspot.
You may inadvertently choose a lamp position that *cannot be reached* when the set is erected (e.g. over a large ceiling). When the set is in place, the lamp may be *too high to be used* (e.g. a batten over a high wall).
Something may *obstruct* the light beam (e.g. a pillar, hanging scenery).
Will the lamp be in shot? Will it cause lens flares, or boom shadows, or camera shadows, or throw one person's shadow on another? Will the light fitting itself cast shadows (e.g. onto a nearby wall)?

● A soft light pointing towards someone near a wall will certainly fill them, but will it flood over the entire wall too?

● A key at right angles to a wall may overlight the wall. One that is shining along a wall may cast long shadows across its surface. Such potential problems can be anticipated at this stage.

Fig. 6.11   The lighting plot
After lighting the action areas (chairs), the set is systematically lit to simulate a sunny day. Lamps are patched into supply channels (here 1–13) and each lamp's channel number written within or beside it.

Keys 4, 10. Back lights 11, 13. Fill light 6, 8, 9. Sunlight through window 1. Background lighting 12, 5, 7. Backdrop lighting 3, 2.

○ 2 kW fresnel spot – hung
● 2 kW fresnel spot – on floor stand
▭ 1500 W scoop
▬ 1500 W striplight on floor
⌂ 4 kW softlight (internally reflected)

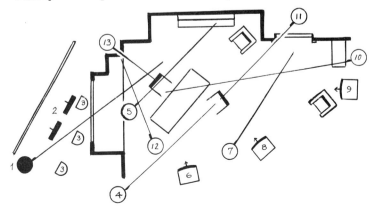

This completes the lighting plot for that person. The result may not be art, but it will be along the right lines. With experience, you will get the feel of your particular lighting fixtures, and how they behave at various distances.

8. Next, consider whether the person is going to move a short distance. Or perhaps another person is going to join him.

Will you be lighting the two places separately, or are they close enough together to share the same key, back light and filler? Will these lamps *suit* both positions?

The problem with the *single key* is that its intensity and angle may not really be right for both positions. (A half vertical scrim or diffuser could reduce the light level at either position.)

On the other hand, if you use *two separate keys* you may have problems in the takeover between them; a fall-off or overlap between their adjacent light beams.

If the two positions are very close together you may be unable to light them separately, anyway. Perhaps the key for one position will serve as back light for the other.

9. If a person moves from this position to another several paces away how will they be lit? Will they still be within the spread of that key, or will they now be in another position's key? Do you need an intermediate key to take over, and maintain a similar light level throughout the walk? Occasionally, you can create a more interesting atmosphere by letting them move out of the light into shade.

10. Having lit that position, move on to the next, and apply the same principles there. Check whether the light from one area falls onto the next and creates any ill-effects (e.g. one position's back light may side light another nearby). Can you clear this by careful barndooring or a flag?

11. Where the action is too spread out for localized treatment, or there are too many separate positions, or you have too few lighting fixtures, you may need to consider one of the *area lighting* methods instead. They can still provide very satisfactory results for broader forms of production. You should check rehearsal pictures carefully, and see whether extra local treatment is needed to improve certain shots.

Lighting the setting

Having lit the action, you can start to light the setting itself. Your decisions will be based on the principles we discussed earlier. The overall atmosphere you want to create in the setting (type of environment, mood, high key/low key, time of day, etc.) will influence your choices.

1. *The walls*
Divide the setting into a series of sections (e.g. a side wall), and
decide on the most suitable treatment for each:

- A *single spotlight* to provide:
  An even tone overall.
  A shaded background.
  Localized blobs or a dappled effect.
- A *series of blended spotlights* providing an even brightness overall.
- A *series of localized spotlights* providing controlled patches of light.
  Avoid lighting walls with suspended soft light (e.g. scoops,
  space-lights) unless bright tops to walls are appropriate.

2. *Set structure*
Look out for features of the set that need carefully angled lighting:

- Deep recesses, arches, beams, soffits, ceilings. Steep lighting may
  not reach subjects beneath an overhanging area. Oblique lighting
  may cast distorting shadows.
- Check for situations where a lamp intended for one purpose can
  inadvertently overlight something else en route. For example, a
  lamp intended to key someone entering a room may overlight the
  door as it opens. In this case barndooring is no solution, for it will
  cut light off the subject itself. Instead you may need to re-angle the
  lamp.
- Where people are standing above the normal floor level (on
  staircases, raised areas, parallels/rostra) they will usually require
  correspondingly higher keys. Hangers (drop arms), pantographs,
  or telescopes/monopoles are used to adjust lamps to various
  heights, and to prevent one lamp getting in the way of another
  close by.

3. *Overshoot/shoot-off*
Check that lamps will not be visible through openings in the sets
(windows, arches) from the planned camera positions.

4. *Set decorations*
Some aspects of the set dressing may need special attention.
Dark drapes or dense foliage may require extra illumination.
Mirrors or shiny surfaces may need carefully angled light to avoid
reflections.

5. *Backings (outside windows, doors, etc.)*
Will the backings be lit evenly overall, lit from below, lit from
above?

6. *Cyclorama, cloths, scrims/scenic gauze, backdrops*
Are there large areas of plain surface to be lit? Do they require even
overall lighting, bottom lighting, decorative lighting (e.g. shadows,
light patterns, color)?

7. *Offstage lighting*
Lighting from fixtures outside the acting area:
● Shining through windows, doors, etc. (e.g. sunlight, moonlight).
● Back lighting scenic units such as pillars, arches, translucent screens.
● Providing safety lighting for offstage areas in shadow (unlit safety lane behind drapes or cloths).
● Audience lighting.

8. *Lighting effects*
Is any auxiliary lighting needed?

● Practicals – Table lamps, chandeliers, etc.
● Decorative inbuilt lighting on sets – Concealed strip lights, translucent panels, chaser lights, signs, etc.
● 'Natural effects' – Firelight, lightning, etc.
● Moving effects – Passing lights or shadows (e.g. in a vehicle); torchlight, etc.

9. *Lighting changes*
Organize which lamp channels you are going to switch/dim/cross-fade for lighting changes (i.e. day-to-night changes; room-light switching; color changes).

Summing up

This has been a generalized overview of the plotting process, but it contains the essentials. At every step you will soon find yourself instinctively looking ahead, comparing, considering alternative approaches. How long it takes to prepare a lighting plot must vary, of course, with the complexity of the show and the experience of the Lighting Director. But, typically, it should not take more than a couple of hours overall to produce a full plot that *works on the day*.

You see the real benefit of preliminary plotting when you arrive in the studio to light the show, and find every lamp in the right position with its accessories fitted, patched and ready to set. You are free to concentrate on creative lighting, rather than waiting around and worrying about hardware and rigging problems.

Rigging the lighting fixtures

You have already met various aspects of *lamp rigging*. You have seen some of the problems of positioning and plugging/patching the lamps for the production. Whether it is a minor rig requiring only a few lighting fixtures or a major effort, the secret lies in good organization.

In most studios you will find a number of suspended lighting fixtures left from previous shows, as well as units in an adjacent store. If it is an off-the-cuff production and time is short, you may be able to use a certain number of lamps that are reasonably near to the positions you want. Where the same sections of a staging area are used day after day for similar kinds of action, some of the key, filler and back light positions may suit one show after another. Rigging in these circumstances may simply involve adding a few extra lamps taken from other parts of the studio.

When rigging from scratch, first check the lighting plot and list the number and types of units, hangers, accessories, colors, etc. required for the new production. Collect them together ready for use. If you are working in an empty studio (where scenery is to be erected after lighting), work along the suspension system battens/bars one at a time, moving or adding lighting fixtures to correspond with the lighting plot.

If there are already fixtures on the batten, consider whether they can be left in position although unused, with their barndoor flaps tucked in or their accessories removed (e.g. flags). Always leave plenty of space around each lighting fixture to permit it to be tilted and turned in any direction, and ensure that its cable is free to allow movement.

Do not leave gobos (patterns) in projector lamps, or they will have become 'lost' when next needed. Do not actually remove unused lamps from the suspension system unless you are going to need them, or you may finish up with a pile of lighting fixtures that you then have to store!

If the settings have already been built in the studio it is usually preferable to rig one set at a time. There may be problems in getting access to some suspended barrels/bars, and if some unexpected difficulties arise (e.g. an obstruction) try to provide an alternative lighting fixture in the vicinity.

Saturation/blanket rig

Some studios use a *saturation rig* or *blanket rig* system, in which a large number of lighting fixtures are left permanently in position on overhead battens/bars ready for rapid selection. In one studio design a series of individually hoisted bars or barrels form a regular pattern over the staging area. Each is provided with two dual-purpose lighting fixtures, which can be converted from a fresnel spot to an internally reflected soft-light source by a control on the housing. Attached to a pantograph suspended from a rolling trolley, the fixture can be slid to any position along the lighting barrel. Additional fixtures can be attached to a barrel, using clamps or adjustable hangers (drop-arms). To rig lamps *between* the regular barrels, extra clip-on cross-barrels are used.

When a system of this kind is used the rigging time and effort are considerably reduced. Whenever possible, all lighting barrels are rigged in an empty studio, then raised to the ceiling to provide complete freedom for the scenic crew to bring in and set up scenery. The barrels are later lowered to position the lamps at their respective working heights.

Table 6.2   The techniques of setting lamps

The subject	• Scrutinize the subject, and consider the particular features you want to emphasize or suppress.
	• Will there by any obvious problems with that subject (e.g. a person with deep-set eyes or a peaked cap, who needs a lower key light)?
	• Anticipate troublesome light reflections (e.g. on an oil painting). Will a carefully positioned key avoid bad speculars, or is dulling spray needed?
	• How critical will the shots be? If revealingly close, will there be sufficient depth of field? Possible camera shadow problems?
	• Is the subject stationary, or is it moving around?
	• Is it being shot from several angles? If so, from which directions?
	• Are shadows likely to fall on the subject from people nearby, or parts of the scenery (hanging chandeliers, arches, tree branches)? Lighting angles may need to be adjusted to suit the situation.
The key light	Arrange an offset key light, suitable for the camera position(s), subject direction(s), and action.
	• Does the key light cover the subject? If the subject moves, will it still be unsuitable?
	• Does the key need to cover more than one subject; an area perhaps?
	• Does the light coverage need restricting? Is there any unwanted spill light onto nearby areas (people, setting)? Use barndoors, flags.
	• Does that key suit the various camera angles? Will it be necessary to cross-fade to a second key for a different camera viewpoint?
	• Check and adjust the key light intensity. (Adjust light level by slightly spotting/flooding; switching lamps; adding diffuser.)
	• Is the key light likely to cause boom shadows?
	• Is the key light position – *Too steep* (i.e. too high)? Look for dark eye sockets; long nose and neck shadows.
	• *Too shallow* (i.e. too low)? It can dazzle talent; reflect in glasses; cause background hotspots; throw subject shadows onto the background; cast camera shadows onto people.
	• *Too offset* (i.e. too far round to one side)? Causes a head profile on the person's shoulder; produces a half-lit face.
	• Try to avoid placing people closer to the background than e.g. 1.8–2.7 m (6–9 ft) so that they can be lit properly, separately from the background.
The fill light	• Except for high-key or 'open-air' scenes, avoid high-intensity overall fill light (*base light*). It flattens modeling, reduces contrast unduly, and overlights backgrounds.
	• Do not use high-intensity fill light to disguise the effects of badly positioned keys.
	• Position the fill to illuminate the shadows; not to add illumination to the key light level.
	• Avoid steep or widely angled fill light.

Table 6.2   (cont.)

- Fill light must be diffused. If necessary, place diffuser over soft-light sources.
- Check that the fill light does not over-illuminate items nearer the camera than the main subject. Are the back and side walls of the setting too bright, due to excess fill light?
- Typical fill light intensity is around half key-light level.
- The more frontal the key light (i.e. the nearer to the camera lens axis), the less the need for filler.
- *Exceptionally*, in dark surroundings (e.g. a singer on a darkened stage) localized fill light can be provided by a dim, very diffused spotlight (despite the extra shadow it casts).
- Do not rely on light bounced from the floor to provide fill light. It is uncontrollable, and of low intensity.
- A lamp fixed to a camera can provide fill light, but its effectiveness varies considerably with camera distance. It is liable to be reflected in shiny background surfaces, glasses, etc.

The back light
- *Is the back light necessary?* It may not be needed for some subjects, or where edge lighting is used.
- Avoid *steep back light*. It becomes ugly 'top light' – flattening the head, and hitting the nose tip. If someone is close to a background and cannot be backlit properly, omit it. Do *not* use top light or side light instead.
- Avoid *very shallow back light*. Lamps come into shot. Lens flares may develop.
- Check that back light for one camera position is not creating u*gly side light* for another viewpoint.
- *Avoid excess back light*. Intensity is typically 1–1½ the key light level. (Excess creates unnatural hot borders to subjects, and overlights hair.)

Background lighting
- Is the background associated with a particular *style, atmosphere, mood*, that needs to be carried through in the lighting treatment?
- Will the background *predominate*, or is it an *incidental* behind mainly close shots of the subject? This may decide how much detail is put into background lighting.
- Is there any danger that the subject might *blend* into the background tones?
- Do any areas of background tone need *compensatory* lighting (i.e. too dark-toned, too light-toned, excessive contrast)?
- Does the background require *specific* lighting (e.g. a series of spotlit sections) or *broad* lighting (e.g. a flatly lit cyc)?
- Are there are practical lamps (e.g. wall lamps) that will influence the lighting treatment?
- Aim to *light the subject before lighting the background*. Any subject lighting falling on the background may make extra background lighting there unnecessary. Where possible, however, keep the subject and background lighting separate, for they usually require quite different treatment.
- Wherever possible, relate the light direction to the environment; i.e. visible windows, light fittings.
- Generally *shade off walls* above shoulder height to improve the subject's prominence.
- Check background lighting for distracting shadows, hotspots, patches, specular reflections.
- Avoid *bright* areas near the top of the picture.
- Take care not to exaggerate *tonal contrasts*.
- Avoid light *scraping along surfaces*, revealing and emphasizing surface blemishes, bumps, joins, etc. Look for spurious shadows from wall fittings.

Setting lighting fixtures

Methods of adjustment

Although roughly positioned ('rough-trimmed') by the rigging crew, each lighting fixture will need to be turned and tilted, and its height, coverage and intensity adjusted to suit its particular purpose. Many lighting fixtures can only be adjusted by hand, so altering the tilt and pan controls, barndoors, etc. of a suspended lamp, involves climbing a step-ladder or lighting ladder (see *Safety*).

Television lighting fixtures today are widely available with special controls that enable them to be adjusted either by hand or by a hook-on *lighting pole*. The tubular aluminum lighting pole is sectionalized or telescopic, and enables you to adjust lamps from e.g. 2 to 4.5 m (6.5 to 15 ft) above the ground: pan, tilt, focus, switch filaments convert from hard to soft mode, and barndoor adjustment (rotate and flap positions).

On some lighting fixtures such adjustments can be made remotely by electronic control systems. Although necessarily more expensive, they are particularly suitable for isolated or inaccessible positions.

Building up the lighting treatment

There are situations where one cannot avoid having *all* the lighting fixtures lit at the same time; e.g. in a directly powered rig where lamps are not fed from a switch panel or dimmer board. If you have to light under these conditions, avoid looking into lamps, and have an assistant wave a hand or lighting pole over the lamp he is setting while you watch its moving shadow in the setting to check its coverage. Where there is a lighting switchboard in the studio that allows any channel to be switched on independently of the lighting control board this can be used both to test and set lamps.

When setting a number of lamps, you can either

● Switch a lamp on, set it, then switch it off and go onto the next, or
● Switch a lamp on, set it, then switch on the next as well, to build up an overall effect.

There are advantages and drawbacks in either system, but on balance, the second is probably better, for it avoids accidental 'doubling-up' (i.e. two or more lamps inadvertently lighting the same area).

Setting the lamps

There are really only two places to stand when adjusting ('setting') a lighting fixture: either at the subject's or the camera's position. It is generally best to stand within the set while someone else adjusts the fixture. Admittedly there are odd occasions when it might be quicker

185   *The production process*

Fig. 6.12 Lamp height and distance

Having decided on the lamp's direction its *vertical angle* will depend on its height and distance.

1. For a given height, the further away it is hung, the shallower its vertical angle. Height is not normally indicated on plots, but lamps are usually rigged at around 3.5 m (12 ft), and adjusted during setting.

2. The same vertical angle can result from a closer low lamp (A) and a more distant higher lamp (B).

   (a) *The closer lamp:* requires less output (lower power, less heat). Its coverage is more easily restricted. Fixtures are more accessible. *But* the light spread may be limited. More fixtures may be needed to cover the acting area. Light intensity varies noticeably as a person's distance from the lamp alters.
   Lower lamps may obstruct cameras, sound booms, scenery. They may come into shot, cause lens flares.

   (b) *The high distant lamp:* provides a more even light spread. Fewer fixtures are needed to cover a given area (so fewer to adjust).
   Light intensity is more constant as the subject distance alters.
   But because the distance ('throw') is greater, lighting fixtures must be more powerful. It is more difficult to restrict light to localized areas (spill problems; doubling). Lighting fixtures may be less accessible.

3. This elevation shows the vertical angles at various distances and lamp heights. As a person moves closer to a lamp the vertical lighting angle becomes steeper. Someone lit by a lamp 3 m (10 ft) above ground level should not move closer to it than e.g. 1.5 m (5 ft), or the steep angle will degrade portraiture. If the head is tilting downwards, 3 m (10 ft) may be a minimum distance.

4. This scale shows the vertical lighting angle for typical standing and sitting positions, with a lamp hung at 3 m (10 ft). A downward tilting head will increase these angles by up to 20°.

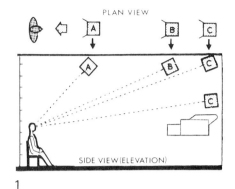

1

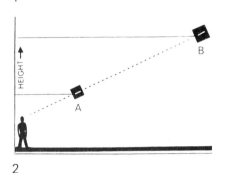

2

3

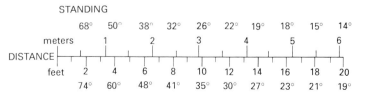

4   SITTING

to tap a barndoor flap yourself than laboriously explain, but that is another matter. If your assistant is standing on a step-ladder beside the fixture, he may be able to see odd problems that are not so obvious from your viewpoint, such as hanging cables or obstructing lamps that need to be cleared. Whether you rely on shouted instructions, hand signals, or radio intercom to guide an assistant varies with local custom!

When setting a *fresnel spotlight* many people prefer to begin by fully spotting the lamp. It helps to identify that lamp among others in a lit area, and makes it easier to center it on the subject. The fixture is then fully flooded as a general routine before fine setting.

As you guide the person setting the lighting fixture (fine-trimming), concentrate on *what is happening in the set* and avoid looking into the lamp yourself. If you look at the scene through half-closed eyes it will help you to see tonal balance more clearly.

There are times, though, when you do need to look directly at a lit spotlight:

- To assess its degree of flood or spot;
- To check if it is centered on your position by observing the light pattern in the fresnel lens (see Figure 6.13).
- To check the barndoor position.

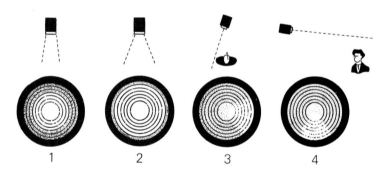

Fig. 6.13   Setting a fresnel spotlight
Looking through a viewing filter at a fresnel spot, from the subject position, you can tell from the lens pattern that the lamp is:
1. Fully spotted;
2. Fully flooded;
3. Turned slightly to the left;
4. Tilted upwards.
Always use a filter.

When you must look into a lamp, **always** use a *viewing filter* or a piece of dark color medium (neutral density). Otherwise you will temporarily blind yourself, and strong after-images will spoil your tonal judgment for a while.

Checking lamp coverage

There are several quick methods of confirming the coverage of each lighting fixture:

Fig. 6.14   Tracking the light source
The line joining the shadow to the subject points to the light source.

- To trace which lamp is causing a particular shadow or hotspot, hold your hand out and project an imaginary line from the shadow to your hand. This line points to the light source (Figure 6.14).

- As you look around, you can continually diagnose light directions by extending a line from a shadow to the feature producing it. The positions of shadows immediately indicate light direction: the shadow spread below (or above) a feature reveals the lamp's vertical angle. Shadows stretch to the side of the feature as the horizontal angle increases.
- When standing in an actor's position checking the action lighting you can immediately identify the number of light sources illuminating you by examining your own shadows on the floor. At a glance you can not only check their directions but their lengths are a reminder of the lamps' heights. Their hardness shows the degree of diffusion.
- To check how even the illumination is over an area, hold a hand with the palm towards you and a finger curled at right angles to it. As you move your hand around, watch the density of the finger shadow. Even slight light fall-off is immediately obvious.
- 'Flashing' lamps (i.e. switching them on and off) is a useful tactic for identifying light sources – you see if the shadow or specular disappears. (Not practicable, of course, for discharge lamps.) It is mostly used during rehearsal while watching a preview picture to check which lamp is causing a particular problem.

Checking the lighting treatment

As you set a lamp, carefully check what it is doing:

Coverage
- Is the lamp's beam covering the plotted area? Wave the lighting pole or a hand over the lamp.
- Does the lamp need 'flooding' or 'spotting' a little to adjust its coverage?
- Does it require barndooring to restrict the light spread?
- Does the lamp require a snoot or a flag? (These are much less used in TV lighting than in Film, because of the lack of time to adjust them.)
- When setting a soft light source, does it need to be tipped down to avoid spreading over distant areas?
- Is the light accidentally spilling onto nearby walls, passing through windows onto the backing beyond, diluting colored lighting, shadows or light patterns?

Brightness
- Is the lamp brightness insufficient or excessive? (Spot or flood the lamp, switch to a double or single filament; reduce or add scrims; adjust the dimmer; move the lamp closer or further away; change the lighting fixture.)
- Is the *light intensity* even? (Any hotspot, shading, light fall-off, dark center?)

Unwanted effects

- Are there any *spurious shadows or reflections*? Check for unattractive shadows, multi-shadows, light overlaps, specular reflections. (Any lamp shadows on walls or people? Check shadows of wall and hung practicals; check for 'dirty light'.) Check back light spread in anticipation of any lens flares. (Use top barndoor flaps to shield light off the camera positions.)

Even lighting

- If you are using a series of lamps to flatly light a surface check it from a distance through half-closed eyes or a filter for overlaps or unevenness. Waving over each lamp can be a very useful guide.

Light patterns

- Check any cast shadows or projected patterns. (Are the intensities appropriate? Are they suitably sharp and undistorted?)

Colors

- If you are using colored light check that it is consistent; e.g. that all gels in a groundrow are producing a similar result. (Old and new gels can vary in hue and density; there may be hue or shade errors.)

Lighting measurements

Now we come to an important difference between motion-picture lighting and typical television lighting.

In film making, frequent measurements of light intensity and color temperature are essential if exposure and color quality are to be consistent. There are times when an experienced guess will have to suffice, but, as we have seen, the eye is all too easily fooled. Admittedly, within reason, film laboratories can compensate for inaccuracies during processing, but no lighting cameraman treats such matters casually.

TV cameras are generally aligned to suit a 'standard' light level and color temperature (e.g. 150 fc/1615 lux at 2950 K). When unfamiliar with local studio conditions, or when lighting location interiors, it is good practice to ensure that your general light levels are around those values by making preliminary checks with a light meter and color temperature meter. With experience, as you become accustomed to working with the same lamps in the same studio, you will find that you can approximate within about half a stop, and only need to make occasional light checks.

The point is, that the TV camera itself becomes your light meter. Relative intensities are adjusted by eye while setting the lamps, and are finely balanced by watching preview monitor pictures during rehearsals. Although light intensities may be altered by filament switching or fitting scrims (wires), most fine trimming is done by varying channel dimmers. Admittedly, color temperature variations can develop when using this technique but where unacceptable they are corrected (usually by filament switching).

Working conditions

In an ideal world the TV Lighting Director would find himself in a quiet studio with his crew, working at a leisurely pace, calmly setting the lamps, experimenting, perfecting, until, satisfied with results, he is ready for the camera rehearsal. It does happen – but very rarely!

Most *small* studios have only a modest amount of equipment, and one needs to be very circumspect. However, with ingenuity and careful treatment, you can often make a little go a long way. Even turning or replugging lamps during the show can help you out of an embarrassing shortage of lighting fixtures. You might, for example, turn the filler on the right of a set in another direction to provide fill light on a neighboring setting when that scene is over. After the song in which a follow spot was used, you might use the channel supplying the spot, to power the 'sunlight' in another set.

In a large TV studio (e.g. $24 \times 30\,\text{m}/80 \times 100\,\text{ft}$) equipment and facilities are often generous, but the schedule is demanding. There may, for instance, be as many as ten substantial settings, requiring some 200–250 lamps in all, together with their various accessories. As you have seen, these all need careful checking and adjustment. The scheduled time available to set all lighting fixtures from scratch and to iron out problems may be around four hours or less!

Studio design and staffing will influence details, but let's see how one network studio organizes on this scale.

Although lighting fixtures may be rigged and scenery erected during the day these jobs are often carried out overnight, to maximize studio throughput. The next day, on arrival the lighting crew lower the lighting bars to potential working heights, check any unresolved overnight problems, and place floor lamps in position (i.e. groundrows, lighting stands, lamps attached to scenery). All practical lamps are wired up and checked.

To speed up the process in a show of this complexity the Lighting Director will probably be 'split working' with an assistant. As they each guide the electricians who are setting the lamps they use radio intercom to the operator at a *lighting hoist panel* on a studio wall, who is switching channels on as required and adjusting the heights of the various lighting bars. The operator at the *lighting board* (*lighting console*) in the production control area is also in contact, and is setting up the files, cues, etc. ready for rehearsal, and noting any amendments to the lighting plot and patching. Problem lamps are investigated and any rigging difficulties overcome.

But they are not alone! While they organize the lighting treatment many others are scurrying around dressing the settings, attaching bric-à-brac to walls, fixing drapes. Carpenters are hammering home some additional bracing for a set. Painters are refurbishing damaged scenery. Scenic artists are simulating paving on the floor. Joins between flats are being taped over. Bursts of 'mist' waft across the studio as the special effects are prepared. Cobwebs are being blown around in the attic set. A hansom cab is trundled into position . . . All these last-minute tasks add to the hurly-burly of the occasion.

Then the operations crew arrives, to set up their cameras and sound booms, position picture monitors and loudspeaker units, all with their respective floor cables . . .

The clock ticks relentlessly on, and after a meal break it is time for camera rehearsal.

Working problems

Although, on a show of this size, the Lighting Director will have made a point of checking out the basic settings while they were being built in a scenic workshop and advised the Set Designer of any potential difficulties, there may still be unexpected last-minute complications as the sets are completed and dressed in the studio. Let's take a look at the kinds of problems that arise during this lighting process and the solutions one applies.

We are setting the lights on a *living room* set. The action lighting has worked out well, although over by the dark desk, which is a main action point, the key is tending to overlight the 'white' wall. (A point to check later on camera.) The scenic cloth outside the window is rather wrinkled. Diffusing and lowering the lamps lighting it should improve matters.

Over here we shall have to compromise. The spotlight lighting the dark wall panels is casting long, ugly shadows below the decorative wall-lights. Lowering the lamp to a shallower angle will reduce these shadows. But now a bright specular reflection appears on the wall. We could change the lamp's direction, but then it would probably hit a person standing nearby. The best option is probably to leave the lamp in this lower position, and strongly diffuse it to make the reflection less obvious. Dulling spray will reduce the sheen on the wall surface.

The velour drapes over here are very dark, and will need extra light to give them modeling. We increase the intensity of the lamp by switching in another filament (it is a dual-filament bulb). But now it overlights the white statuette just in front of the drapes. Closing a barndoor flap to cut the light off the statuette will rob the drapes of light too. The lamp is too far away to shield the light off with a flag. Will moving the statuette out from the wall help? No, light still spills onto it. The Designer may need to spray the statuette down (darken it) or replace it, rather than allow the drapes to be underexposed.

Moving into the 'bedroom' set, we have a predicament. There is a fresnel spotlight on a lighting stand outside the window, providing 'moonlight'. It was to have been at a lower intensity when the girl stands at the window (so as not to burn out the face) and brighter for the close shots when she is asleep. However, a hired cupboard by the window has proved to be much larger than planned. It prevents the moonlight reaching the bed at all. So we have to find an alternative method. Supposing we rig an ellipsoidal spot, projecting a window pattern onto the bed, and barndoor the 'moonlight' to confine it to the window position? On-camera it will look as if the effect is due to a single source, the moon, illuminating the entire room. A compromise, but it works.

And so the process goes on. But accurate planning and forethought bring their rewards and problems are comparatively few.

Camera rehearsal

The nature of camera rehearsal

Now the Director, the actors, and the various specialists who have been organizing and rehearsing the production are seeing the settings that have hitherto existed only in their imaginations. However, for the *entire* studio crew, including the cameramen, boom operators, lighting crew, stage hands, etc., the show is quite new and unfamiliar. They have to discover both the nature of the production and their role in it as they go through the camera rehearsal. Learning as they go, they refer to various written data (running order, cue sheets, script, camera card, etc.) and listen on earphones to *intercom* instructions (*production talkback*) from the Director and other members of the production team.

The most widely used rehearsal arrangement is for the Director to sit in the production control room watching a series of *preview picture monitors*. Each of these monitors shows the continuous output of its respective camera (or other picture source). The Director guides the team through the *intercom system* (*production talkback*). The program sound is heard over a nearby loudspeaker.

On a separate *master monitor* we can see whichever camera the Director has selected on the production switcher. This picture, when on *line* (*main channel, studio output, transmission*), will be recorded on videotape during the taping period or transmitted live at the scheduled time.

Watching the picture monitors, the Director rehearses the action. In a long shot the actor opens the door, and the camera tightens to a close-up as he sits at the table. The Director decides that if the actor were to come round the table the other way the result would be more dynamic. He explains the point over intercom to the *Floor Manager* (*FM*), who is the Director's chief assistant on the studio floor. The FM guides the actor, who, on cue, goes through the revised action. The Lighting Director, with an eagle eye on his own picture monitor, checks whether any changes are needed. Better reduce the key during the walk perhaps, or the sunlight might now burn out the face at one point.

This, in essence, is how the camera rehearsal develops. Whatever the scale of the production, camera rehearsal involves finding out, revising, polishing. If the Director is satisfied with the results on camera he goes on to the next sequence. Otherwise he revises and improves. It is up to the Lighting Director to let the Director know if he cannot accommodate these changes. So, too, the Sound Engineer, Switcher, and others will seek to adjust to these improvizations. However well planned and pre-rehearsed, there are always alterations during camera rehearsal, as the Director revises his ideas about shot development or modifies the action to improve the shot.

These are the conditions under which TV lighting is carried out. The *film* Lighting Cameraman continually liaises with the Director and is responsible for all his single camera's shots. In *television* the Lighting Director is faced with a succession of pictures coming from several cameras that are controlled by the production Director, and has only a tenuous influence on the shots.

Working pressures are considerable. Time is precious. It is possible for a crew to begin rehearsals for a 30-minute drama in the morning and go home that evening after the entire show has been taped or transmitted *live!* Thanks to videotape, fewer productions are transmitted as they happen. Instead, they are recorded in sections or shot by shot with corrective retakes and subsequently edited together, with added music and effects (audio sweetening). Even 'straight-through' performances will usually have retakes, cutaways, detail shots, 'nod shots' edited into place afterwards.

Rehearsal methods

Traditionally, several regular methods have emerged for tackling this marathon.

■ *Dry run* or *walk-through*   Occasionally, rehearsals may begin with a *dry run* or *walk-thro'* in which the actors go through action for each scene while camera and sound crews watch and the Director explains the treatment and continuity. Although this method does help to familiarize the studio crew with the production it is a luxury where time is limited.

■ *Camera blocking* (*blocking out, stopping run-through, stagger-through*)   The first (and sometimes the only!) camera rehearsal is normally *camera blocking*. Most Directors do all their camera blocking from the production control room in the way we have just seen, continually assessing preview pictures, and aided by their script. They will not usually leave the control room to go into the studio during rehearsal unless there are serious problems.

On the other hand, there are Directors who prefer *floor blocking*, and conduct this rehearsal from the studio floor while standing near the action and watching a movable picture monitor. Each scene is 'blocked and run' in turn. Having gone through the action slowly and corrected inaccuracies as they arise, the sequence is then repeated at normal pace to confirm coordination and timing. The Director may watch the revised version in the production control room.

Having completed this camera rehearsal, there is usually a short break during which notes are given to the crew and performers to clarify any problems, and the Lighting Director, Set Designer, Sound Engineer, etc. have an opportunity to make any adjustments or changes that are needed.

Where time permits, there may then be a *final run-through* (*dress rehearsal, dress run*) which, hopefully, will be a fault-free 'on-air' quality performance.

■ *Rehearse–record* This method is really a form of floor blocking, in which the Director blocks and runs each shot or sequence and then tapes it *immediately afterwards*. Any retakes are recorded before going on to rehearse the next section. Whether shooting involves one camera or several depends on the nature of the scene. Each sequence may need to be relit, cameras repositioned, and scenery reset or cheated.

The separate videotaped sections are later edited together with appropriate transitions (cuts, mixes) and bridging sounds. Hypothetically, this pseudo-filmic technique could provide optimum results for each shot. But in practice, the process can be extremely frustrating and time consuming, with quite unnecessary compromises. Indeed it is not unknown for a Director to 'save time' by recording the *first* camera rehearsal in case it happens to be successful – although everyone is seeing the shot for the first time and has had no opportunity to adjust any aspect of the scene (lighting, scenery, make-up, costume, etc.)! Things have to work first time!

There is no question that this method aids the Director who cannot plan his shots or visualize them accurately beforehand and has difficulties in staging action for multi-camera production. But progress is slow and often tedious compared with multi-camera methods.

It also has many major weaknesses. Where the scenes are shot out of order for any reason there can be diverse opinions about continuity – whether the drapes were pulled by this shot . . . whether the rain had stopped . . . or he had taken off his coat . . . or the practical lamps werc off. Visual continuity can be poor as things are changed or corrected in the course of the recording period. And, of course, in the end an extensive amount of post-production editing is required; often quite needlessly.

From the Lighting Director's viewpoint there is too little time available for any of the production team to correct imperfections or shortcomings, and that is a serious limitation. The traditional method of working through the entire show during the rehearsal period, then recording it in a separate session, allows one to snatch odd moments to put things right, to adjust a barndoor, to put in a diffuser, to rig a new lamp if need be, while the Director is revising shots or action, or rehearsing the next scene. There will be odd breaks, when the designer is able to replace some drapes, or an actor's overlight shirt can be replaced with a slightly darker one, or the Make-up Artist can darken-down white hair that is picking up the light. Video engineers can check a camera that appears slightly soft focus.

The *rehearse–record* method allows little time for other than brief changes or improvements. If a wall is too bright, or an ugly shadow appears, or a key angle needs altering, action stops entirely until it is put right . . . or the imperfect version is recorded nevertheless.

When faced with such an arrangement the Lighting Director has to be philosophical! He provides a plot and lighting treatment based on what the Director *says* he is going to do when planning the show, then continually modifies as time permits, to suit what actually happens. At best, results may be filmic, and tailor-made for each shot. At worst . . . !

Lighting the rehearsal

Correcting lighting problems

There are two approaches to correcting lighting problems. The Lighting Director or his assistant can note any alterations needed during the camera rehearsal and make the changes during a break; or deal with them as they arise, while the rehearsal continues. Although some changes can be made without disturbing the performers unduly (e.g. adjusting barndoors, turning/tilting lamps), others would be too distruptive (e.g. rigging new lamps), and must normally wait until rehearsal is over.

There are some lighting defects you can diagnose simply by looking at the picture. It is easy to see that a light pattern needs adjustment or some background shading is uneven. Other defects may need on-the-spot investigation – e.g. How can we avoid that boom shadow?).

There is always the dilemma that, while the Lighting Director is working on the studio floor during rehearsals, he is missing rehearsal pictures and unaware of any further problems that are developing. On the other hand, if he leaves faults temporarily uncorrected he may not appreciate the seriousness of a particular problem until checking the situation later. A reliable assistant who can carry out running corrections during rehearsals is a boon indeed!

While some alterations, such as the background shading, only directly concern the Lighting Director himself, other changes may affect the rest the production team. Let's take the boom shadow, for example. We need the solution here immediately:

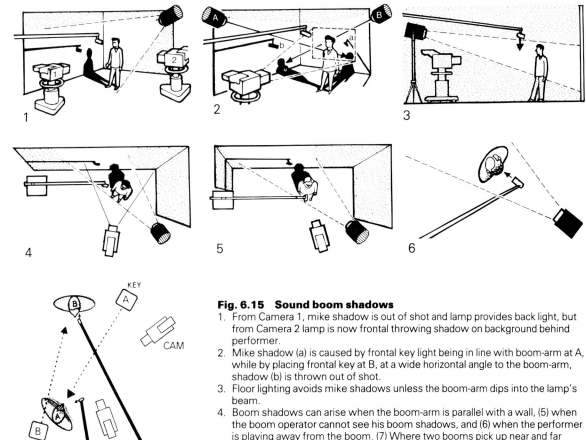

Fig. 6.15 Sound boom shadows
1. From Camera 1, mike shadow is out of shot and lamp provides back light, but from Camera 2 lamp is now frontal throwing shadow on background behind performer.
2. Mike shadow (a) is caused by frontal key light being in line with boom-arm at A, while by placing frontal key at B, at a wide horizontal angle to the boom-arm, shadow (b) is thrown out of shot.
3. Floor lighting avoids mike shadows unless the boom-arm dips into the lamp's beam.
4. Boom shadows can arise when the boom-arm is parallel with a wall, (5) when the boom operator cannot see his boom shadows, and (6) when the performer is playing away from the boom. (7) Where two booms pick up near and far action, the latter can shadow the closer source.

- Can the boom operator avoid the shadow now that he is aware of it? It may only require a slightly different microphone position.
- Will adjusting the top barndoor on the key light remove the shadow? It may not.
- Will we have to relight from another angle? What effect will that have on other shots?
- Should the boom operator move his boom position? This may then affect camera positions.
- Will a different method of sound pick-up be necessary (e.g. a personal microphone)?
- Will the Director need to adjust the action or the position of the actor? That could affect other shots.
- Would it help the boom swing if the actor stopped talking while moving?

Until the problem is resolved no-one will know, and it is certainly not a matter to bring up when the rehearsal has ended! It is the kind of situation that requires urgent discussion with the Sound Engineer and the Director, and may even stop the rehearsal.

Table 6.3   Methods of eliminating boom shadows

	Method	Result
By removing the shadow altogether	Switching off the offending light.	This interferes with lighting treatment.
	Shading off that area with a barndoor or gobo.	Normally satisfactory, providing the subject remains lit.
		Shadowing walls above shoulder height is customary, to make the subject more prominent.
By throwing the shadow out of shot	By placing the key light at a large horizontal angle relative to the boom arm.	A good working principle in all set lighting.
	By throwing the shadow onto a surface not seen on camera when the microphone is in position.	The normal lighting procedure.
By hiding the shadow	Arranging for it to coincide with a dark, broken up, or unseen angle of background.	Effective where possible, to augment other methods.
	By keeping the shadow still, and hoping that it will be overlooked.	Only suitable when inconspicuous, and when sound source is static.
By diluting the shadow with more light		Liable to overlight the surface, reduce surface modeling, or create multi-shadows. Inadvisable.
By using soft light instead of hard in that area of setting		Occasionally successful, but liable to lead to flat, characterless lighting.
By using floor stand lamps instead of suspended lamps		Light creeps under the boom arm, avoiding shadows. But floor lamps occupy floor space; can impede camera and boom movements; can cast shadows of performers flat-on to walls.
By altering the position of the sound boom relative to the lighting treatment		May interfere with continuous sound pick-up or impede camera moves.
By changing method of sound pick-up	By using a low fishpole, hand-held shotgun (rifle) mike, slung, concealed, personal mikes, prerecording, etc.	These solutions may result in less flexible sound pick-up.

Teamwork

Where a production team works closely together they will continually interrelate. The Costume Designer will check that the Lighting Director can cope with a white shirt; Make-up will explain that they intend using a darker foundation on an actor; the Set Designer agrees to change a table covering; a cameraman indicates that he has insufficient depth of field, etc. Teamwork is essential, and there is continual discussion.

Guided by the Lighting Director, the Video Operator has been examining film inserts and color-matching them to the studio output. Decisions have to be made to compensate these stock shots. 'If the color balance is adjusted so that the horses appear natural, the foliage looks strangely reddish. Adjust for green foliage, and the horses turn blue!'

During the first camera rehearsal the Video Operator works 'hands off' until the lighting balance has been adjusted. Otherwise we could find that while the person at the lighting control board was deliberately dimming a lamp to reduce overall picture brightness the Video Operator was opening up the camera lens aperture to compensate!

In creative lighting, balance is an integral part of the entire visual concept. Once the relative intensities of all the lamps have been adjusted, the Video Operator can subtly readjust the camera channels to improve the matching of pictures, and enhance the pictorial effect.

The lighting control board (lighting console)

The *lighting control board (lighting console)* today is a very sophisticated piece of equipment, and in a more complex show may require continual operation. It has several functions during a production:

1. Switching
   Ensuring that the appropriate lighting fixtures are switched on (and at appropriate intensities).
2. Balance
   Static balance: arranging the relative intensities of lighting fixtures, to achieve a particular effect, atmosphere, time of day, etc.
   Dynamic balance: changing the relative intensities of lamps during action for atmospheric effect.
   Corrective balance: adjustments made to compensate for production dynamics; e.g. someone moving out of their key light.
3. Changes
   Immediate changes: altering the lighting treatment at the end of an item (e.g. killing all frontal lighting to create a silhouette effect).
   Progressive changes: gradually dimming channels, altering contrast, varying colors.
4. Action cues
   Performer actuated: coinciding with a performer's action (e.g. switching off room lights).
   Timed effects: coinciding with dialogue or other cues (e.g. explosions, fire, lightning).

During camera rehearsals the Lighting Director sits by the lighting control board, watching preview and line picture monitors, assessing the lighting treatment and picture quality.

You may need to alter the lighting arrangements for a number of reasons:

- *Inaccurate lighting* – badly positioned lamps, poorly set lamps, unsuccessful effect.
- *Productional changes* – revised action, revised shots, different camera positions, extra shorts, changed treatment.
- *System failures* – lamps blown or accidentally moved.
- *Operational problems* – boom shadow, camera shadow, lens flare.
- *Scenic changes* – set dressings or scenic pieces added since the set was lit.
- *Scenic defects* – poor backdrop, unacceptable tonal values, shiny surfaces.

The dynamics of rehearsal

During rehearsal the Lighting Director develops the trick of splitting his attention, continually monitoring:

- Listening to intercom instructions.
- Listening to program sound.
- Assessing and modifying lighting balance.
- Looking out for any visual defects.
- Noting any changes in the action, performers' positions, camera shots.
- Checking on any scenic alterations.
- Preparing for upcoming cues, etc., etc.

The points that arise are diverse:

- 'We are over-shooting (shooting-off) the edge of the set.' If the Designer is going to add some extra flattage, we must check that it does not cut off the key light.
- 'The actor has been repositioned at the table.' He is now in the other person's key. It will need trimming.
- 'We must let the Make-up Artist know that we cannot reduce the back-light on that bald head.' They will reduce perspiration and darken it.
- 'The actor tends to tilt his head down and get black eyes, we must get the Director to have a word with him.'
- 'I wonder if the Set Designer has seen how the photo-backing outside that window is bulging, causing it to be unevenly lit.'
- 'The practical table lamps are far too dim. Is it possible to fit Photofloods?'
- 'The cables for the lamp hidden behind the chair are visible. They need tidying away.'

- 'Have the stage hands enough room to move that wild wall without entangling with lamps?'
- 'Camera 2 seems to have drifted – a red color bias.'
- 'The fire flicker effect looks very mechanical.'

Such miscellaneous topics are noted and dealt with in due course.

Away from the studio the preview pictures are your only clues to what is happening. Only by watching them closely can you interpret studio activities. Some control rooms have observation windows into the studio, but the view is usually very limited.

As people move around from one area to another within a setting they will be lit by a series of lamps. You will often need to decide which they are in at any moment. The trick is to use furniture and parts of the setting as reference points. If they are near the window, the key is 14, while if they are beside the table it is 31. (This idea works well, except for large open areas, where reference points are sparse.)

So when a person enters a room and is badly overlit we glance at the lighting plot to check that Channel 47 is the key confined to the door area . . . flash the lamp to confirm identification . . . and reduce its intensity by fading it down a little. The lighting balance is now fine.

Well, that is how it is most of the time, but there are those other moments too! We see a shot of someone walking, and he passes through a blazing hotspot . . . but he has walked on and the camera has panned with him . . . We can no longer seen the problem patch. We know that there *is* a hot lamp but we cannot see it to identify and correct. The mystery can only be solved by going into the studio and investigating on the spot.

Careful balance checks with a light-meter before rehearsal could have anticipated this kind of fault . . . but was there time to measure and balance light intensities?

Close-up lighting of people can very exacting, and under typical TV production conditions a challenge, to say the least:

- Where cameras move around to any extent, lamps on floor stands (and their cables) can be a nuisance.
- Directors frequently use narrow lens angles (long-focus/telephoto lens) to take close shots from a distance, notwithstanding the perspective distortion, even where it is impracticable to arrange good lighting.
- Frontal fill light has usually to be suspended at around 3–3.5 m (10–12 ft) to avoid impeding cameras or sound booms; and where camera cranes are used, even higher.
- Key lights are in many cases located above the side walls of a setting at similar heights.

These are the prices one must pay for high productivity. Nevertheless, the final product frequently achieves remarkably high artistic standards.

Lighting balance

As we discussed earlier, *balance* is the technique of adjusting the relative intensities of the lamps after they have been *set*; i.e. their angles and coverage adjusted. Balance is the most elusive facet of the art of lighting. You can produce dramatic differences just by altering the brightnesses of a few lamps. A studio exterior, for example, may be transformed from 'tropical sunshine' to 'moonlight', from 'approaching storm' to 'treacherous night' by re-balancing the same lamps. But even quite small changes in proportions will often have a considerable influence on the feel of the picture.

Relative brightness

An effective lighting balance usually involves several distinct stages:

- *Action lighting/portraiture.* Adjusting the brightness of the key lights, and the relative intensities of the fill and backlights, to create a particular atmospheric effect: high key, low key, etc.
- *Subject/background contrast.* Adjusting the background brightness to provide a good tonal contrast between people and their backgrounds.
- *Scenic contrasts.* Adjusting the relative brightness of parts of the scene to create a three-dimensional illusion.

Interior scenes
- *Walls.* Adjusting the relative brightnesses of the walls. Walls containing windows are generally proportionally darker during daylight.
- *Ceilings.* The brightness of any ceiling varies considerably, according to the window's position; the time of day; light reflected from fallen snow or sand; room lighting (practical lamps).
- *Adjoining areas.* Adjusting the brightness of adjoining areas (e.g. hallway, other rooms seen beyond) to that of the main room.
- *Interior/exterior balance.* Adjusting the subjective differences in brightness between an interior scene and the exterior seen beyond a window.
- *Exterior light intensity.* The strength of light entering windows (sunlight, moonlight, street lamps) compared with the apparent brightness of the exterior visible through windows. (A large dark area outside a window can make sunlight appear overbright or incongruous.)
- *Practical lamps.* Relating the brightness of practical lamps in a night scene to their apparent effect on the immediate surroundings. Adjusting the relative lightness of various parts of a room – considering their distances from the light source(s).

Exterior scenes
- *General light level.* Adjusting the effective strength and contrast of the 'natural light'.

● *Relative brightnesses.* Adjusting the relative brightness of various
  parts of the scene:
  The brightness of the sky.
  The density of shadows.
  The apparent brightness of interiors seen within buildings.
  At night, the brightness of any exterior lighting, and its
  effectiveness.

Effective levels

When balancing the lighting for any picture the important thing to
bear in mind is that you are not seeking to copy reality but to achieve
a *subjective effect.* You have always to consider –

● The light levels should not fall below the minimum required by
  the camera at the working aperture you have selected. Insufficient
  light will simply under-expose the picture.
● Conversely, the maximum levels must be limited, unless you want
  the lightest tones to burn out.
● The camera can only handle a limited contrast range.
● The TV receiver can only reproduce limited tonal subtlety and
  contrast (particularly under home-viewing conditions).

Let's look at a practical example:

> We are in darkened room. It is siesta. The window is tightly
> shuttered against the hot sun that beats down onto the
> whitewashed walls of the deserted street. The sleeper awakens and
> throws open the shutters.

Here we have the combined problems –

● Too little light in the room;
● Too much light in the exterior;
● Excessive contrast between the interior and the exterior.

The room is virtually in darkness. We have to find some excuse to
introduce light. Suppose a few rays of sunlight streak through the
shuttered windows. That would give a visual 'explanation' for
selective rim lighting – enough light for us to discern what is within
the room.

On a true location it is unlikely that we would be able to expose for
detail in both the exterior and the interior. The lighting contrast is
too great.

● Adjust the lens aperture to suit the interior, and the scene outside
  the window would be entirely over-exposed.
● Expose for the exterior, and the interior would be barely visible.

To lessen the difference it would probably be necessary to fit a filter
over the window opening (Wratten 85 plus neutral density) to
color-balance and reduce (hold back) the natural light. The interior

would be lit to a very low intensity; sufficient to allow it to appear just visible (e.g. low-intensity bounce light with local back light). In the studio the 'exterior' lighting can be balanced so that it appears very bright yet still remains within the exposure range.

Tonal illusions

Considering that a 'black' surface may actually reflect more light under intense illumination than a 'white' surface under low illumination, 'brightness' and 'darkness' are clearly very subjective judgments. By *reducing* the amount of light on parts of a scene you can actually make other parts look brighter! A *key light* will often appear stronger if you reduce the amount of fill light or if you lower the brightness of the background behind the subject.

Where a *background* is overlit it will often create the impression that foreground subjects are much darker than they really are.

- If you stop the camera lens down a little to *reduce* the effective brightness of the background it will reduce all tones in the picture similarly, and the subject will look darker still.
- If you open the camera lens aperture a little to *increase* the effective brightness of the subject it will improve, but the background will become even brighter!
- Add light to the subjects to compensate, and you will find that:
  You have increased the overall light level.
  Another area nearby may now seem insufficiently lit by comparison.
  The extra light will probably spill onto the background and brighten it further.

Overall contrast

By adjusting the overall contrast in a scene you can change its entire ambiance.

- On the one hand, if there is little or no fill light, and shadows are unrelieved, the effect can be: dynamic, strong, coarse, stark, ugly, restrictive, confined . . .
- On the other hand, if you increase the fill light until the fill-to-key ratio is similar, shadows are illuminated, and the effect may appear delicate, fresh, open-air, flat, dull, undynamic, drab, weak . . .

The apparent tonal balance can appear to change with camera positions. If, for example, lighting is predominantly frontal, little or no fill light is needed. The contrast ratio can be low.

Cut to another angle, where the same lighting is now behaving as side light, or even back light, and we now have little frontal lighting and a high contrast. The effective lighting balance is quite different.

Table 6.4   Changing the light intensity

When *light levels* are too low or too high they can be changed in several ways.
(INC = increase light intensity; DEC = decrease light intensity.)

Dimmers	Alter dimmer settings to INC/DEC but check the picture for changes in color quality. Simple and immediate control, but can involve extensive equipment.
Diffuser (scrim)	Diffusers have the advantage that you can treat just part of a light beam if necessary. DEC: place a diffuser over lamp; add further diffuser material. INC: reduce/remove diffuser.
Spotlights	INC: spot up the lamp slightly. DEC: flood the lamp slightly. Simple, but must be done manually, and affects the light spread.
Softlight sources	Switch on/off sections of a multi-lamp softlight source. This does not affect color temperature. But the change in light level may be too great.
Lamp distance	INC: move the lamp closer to the subject (now brighter, but its coverage will be reduced). DEC: move the lamp further from subject (light intensity falls, but its coverage will increase). Simple, but may not be practicable.
Lamp power	Replace the lamp with another of higher or lower power. Except when lamp filaments can be switched (dual-filament lamps), this involves rigging a new lamp.
Extra lamps	INC: add more lamps. (Beware more shadows or overlighting other areas.) DEC: remove lamps (e.g. switch off some of the soft light).

Lighting quandaries

In lighting, as with all crafts, there are many problems that one encounters regularly. Some have arbitrary solutions. Others can only be improved by modifying the situation itself.

Clearly, one can't expect other people to rearrange scenes every time there are lighting difficulties. It is often a matter of priorities. If the Director alters a shot, will the new version have less dramatic impact? If he uses the shot in spite of the problems, will the shortcomings of the lighting detract from its value?

Much depends, of course, on whether

● You are able to light shot by shot;
● There are brief breaks in which to improve the lighting;
● The action is continuous.

Solutions that are obvious when you can tailor the lighting to a specific shot are not always practicable during continuous production.

Sometimes, as we saw earlier, lighting may be degraded throughout a scene by some design feature such as a ceiling that is only visible in the odd shot. The question, then, is whether to redesign the set to suit those particular shots (e.g. use a partial ceiling piece) or accept the lighting limitations.

In previous chapters we have already seen a number of examples of situations that pose difficulties for lighting. Further quandaries are shown in Figure 6.16.

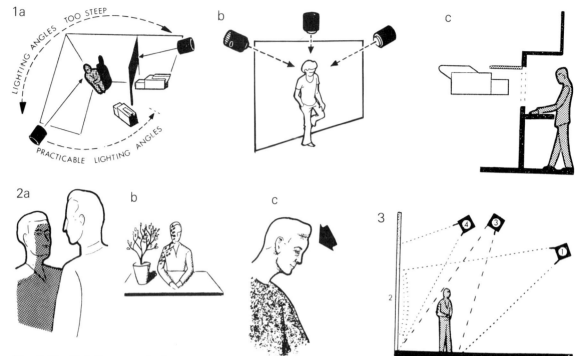

Fig. 6.16 Lighting quandaries

Here are a number of common difficulties where solutions are not immediately obvious, or compromise is unavoidable.

1. *Restricted lighting treatment.*
   (a) When a person is in the corner of a room effective lighting angles are limited; particularly if scenery blocks lighting.
   (b) When someone stands against a wall angles are limited. (Avoid top lighting.)
   (c) The camera may shoot where it is difficult to light appropriately (e.g. camera traps, in open doorways, closets).

2. *Shadowing.*
   (a) When two people stand close together one may shadow the other. Repositioning is usually more practical than relighting.
   (b) Nearby objects may cast shadows onto your subject. Relighting to avoid the shadow may degrade the lighting treatment.
   (c) Tilting the head downwards exaggerates modeling. In closer shots a camera light may relieve the unattractive effect.

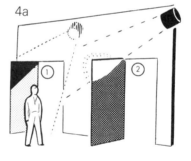

3. *Background lighting.*
   You can use a key light to illuminate both the action and the background (1). But this may overlight light-toned surfaces (2). Restricting light to the action alone (3) overcomes this problem, but may involve steep lighting angles (poor portraiture). If the background lighting is steep, in order to keep it off the action area (4), it may create harsh set modeling.

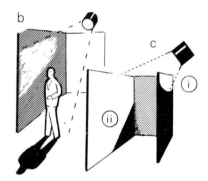

4. *Spurious shadows and hotspots.*
   (a) Light through a doorway can produce a distracting shadow on the open door (1). A distant key can catch the top of a door as it is opened, causing a localized hotspot.
   (b) Light can inadvertently scrape along adjacent walls, causing a visible light streak, casting elongated shadows, and emphasizing texture.
   (c) Where a lamp is positioned behind flats, part of its beam may be cut off (i) and shadows cast on the lit surface (ii).

1. Restricted lighting treatment

A. *Position.* When people are positioned near walls, or in the corner of a room, lighting angles are limited. Wall shadows are unavoidable and back light is impracticable.

   Scenery. Scenery may allow the camera to see the subject but prevent a key being suitably positioned (e.g. when shooting through a door or window).

B. *Peep-holes.* When the camera shoots through a very restricted opening such as a small pull-away panel in a wall or closet (camera-trap) it may be able to see subjects that are unlit by normal set-lighting or a camera-light. Additional local lighting may be needed.

C. *Furniture.* Some seating can limit lighting angles; e.g. a wing chair cuts off back light, side fill or key, allowing only a frontal key light.

2. Shadowing

A. *Mutual shadowing.* When two people stand in line with the key light one may shadow the other (often the smaller). Wherever possible, it is simpler and quicker to re-angle the people, than to re-light.

B. *Props.* When displaying models, foliage, etc. these may cast shadows onto the demonstrator.

C. *Exaggerated modeling.* When a head is tilted downwards its modeling is exaggerated, eye sockets and neck are thrown into shadow.

3. Background lighting

● A key light may be used to illuminate both the performer and the background.

● But if the background is light-toned it may become overlit.

● To avoid lighting the background and casting a shadow onto it you could raise the key, but this would probably result in steep, ugly portrait lighting.

● Any attempt to light the background separately would require a steep, localized spotlight that would edge-light its surface, causing unattractive modeling.

4. Spurious shadows and hotspots

A. *Door openings*

   1. When lighting through a door opening, check that there is not an ugly shadow on the door as it opens.

   2. A key for a distant position may accidentally light a nearer door when it is opened, producing a spurious patch of light on it.

B. *Spurious wall streaks.* A key light or back light can inadvertently streak along an adjacent wall, creating a spurious patch of light, exaggerating texture, and casting elongated shadows.

Fig. 6.16 (cont.)

5. *Scenic obstructions.*
   (a) *Horizontal obstructions* such as ceilings, arches, borders, beams, can obstruct lighting. So instead of the preferred angle (1), the lamp has to be too steep (2), or too low (3). It may not be possible to cover all the action at a suitable lighting angle.
   (b) *Vertical obstructions* (such as walls) can limit possible lighting angles.
   It is sometimes possible to overcome these problems by set design (false returns in flattage), concealed lamps, or cheating the action or scenery for specific shots.

6. *Camera shadow.*
   A close camera may cast a shadow onto the subject; especially if it is high. The lamp may be barndoored (but this cuts light off the subject itself). Lowering the camera or moving it further away may be the quickest remedy.

7. *Reverse-angle shots.*
   The light intensity needed for a dramatic shot (1) may be excessive from another viewpoint (2). Typical situations include light streaming through windows, doorways. (Usually only a problem during continuous action.)

8. *Light penetration.*
   When light comes through restrictive surroundings it can severely limit the amount of illumination reaching the subject – e.g. leafy branches, tracery, window blinds, scrim, nets, sheers, screens, etc. The surroundings may become overbright before sufficient light has penetrated. Shadows may obscure or camouflage the subject.

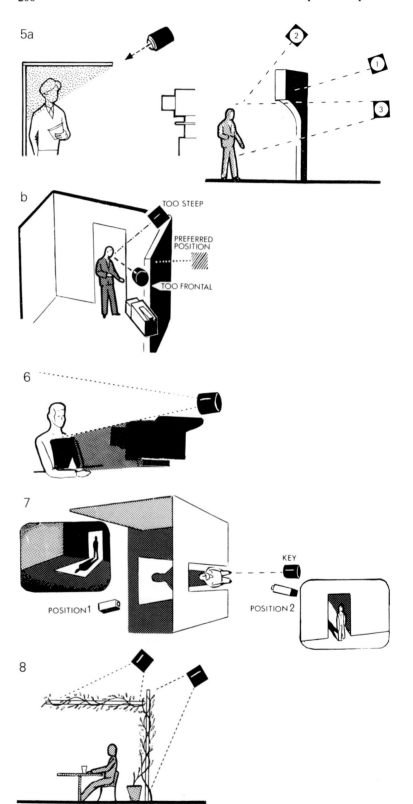

C. *Corner-cutting shadows.* Where a key light is positioned just *behind* a flat the lower part of its beam will be cut off when tipped down and the top rear of that flat accidentally lit (i). The result will be a large wall shadow and an unlit area within the setting (ii).

5. Scenic obstructions

A. *Horizontal obstructions.* Some scenic features such as arches, drapes, beams, borders, fascia boards, overhangs may prevent a key light from being positioned at the optimum angle (1). Instead, it has either to be located at a steeper angle (2) in order to get over the obstruction or underneath it at a low angle (3), with the risk of a camera shadow.

B. *Vertical obstructions.* Where a camera is positioned close to a side wall it may be difficult to key the subject from an appropriate angle. It may be possible to cheat by using a concealed key, modify the scenery, or adjust the performer's position to suit the shot.

6. Camera shadows

Whenever a camera works close to a subject, especially when looking downwards, it may cast a shadow onto it. Barndooring (bottom flap) is no remedy, for this leaves the subject unlit. The regular solutions are:

● To lower the camera (the shadow is lower on the subject);
● To raise the key light (this lowers the shadow, but increases modeling);
● To move the camera further away (using a narrower lens-angle);
● To move the camera round the subject out of the light-beam.

7. Reverse-angle shots

The high light level required to produce a strongly dramatic effect from position 1 may be excessive from position 2. This dilemma frequently arises during continuous production, when shots are intercut between either side of a door or window. The lighting balance can be cheated when the separate viewpoints are shot separately. (See Figure 6.1.)

8. Light penetration

It is often assumed that if a subject is placed near an illuminated background (such as a rear-lit panel), or a surface that is sufficiently perforated to allow light through, it will be effectively lit. With certain exceptions, this is a myth!

Lit surfaces (lit ceiling, wall panels, translucent canopy, overhead lighting). If the light from a nearby surface is sufficiently strong to fully illuminate the subject it will be too bright on camera. The

Fig. 6.16   (cont.)

9. *Ground lamps.*
Where action is at or near ground level typical problems include:
   (1) Light streaking along the floor.
   (2) Intermediate objects being overlit.
   (3) Shadow cast onto the subject; or
   (4) Onto the surroundings.

10. *Lighting a piano.*
   (a) Shadow problems include:
   (1) Performer's shadow on keys;
   (2) Shadowed fingers;
   (3) Lid (or stick) shadow on pianist.
   Illumination problems:
   (4) Steep lighting;
   (5) Dazzling low key light, caused by trying to avoid shadow problems (2) and (3);
   (6) Inside of piano dark;
   (7) Hotspots on piano body.
   (b) Typical lighting approach.

11. *Background pattern position.*
A centralized background pattern changes in frame position as the camera viewpoint alters. This frustrates pictorial balance and continuity whenever cameras move.

12. *Camera-lights.*
   (a) While providing low-level filler camera-lights may over-illuminate parts of the subject.
   (b) One person's camera-light may over-illuminate another person positioned nearer the camera. Its lighting-angle for the nearer subject may be unsuitable; or may provide a strong kicker, edge-lighting his face.
   (c) Camera lights can produce spurious light reflections in spectacles, and shiny surfaces.

9

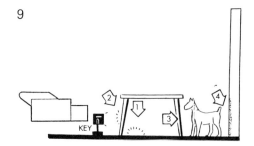

10a

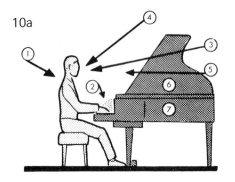

b

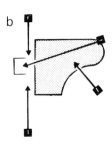

11

12

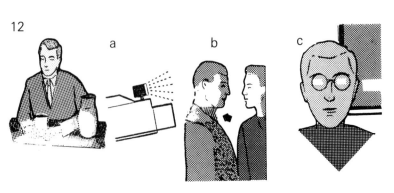

subject appears dark relative to its illuminated background. The trick is to keep the intensity of the illuminated surfaces down sufficiently to still appear fully lit, yet add strong subject lighting from a compatible direction.

Broken-up light. When incident light is restricted by leaves, blinds, decorative grilles, lattice, etc. too little passes through it to illuminate the subject properly.

● Extensive cast shadows may obscure or camouflage the subject.
● The edges of the material creating the shadows are liable to be over-bright.

Rather than rely on infiltrating light alone to illuminate someone seated beneath a canopy of foliage, it will probably be necessary to introduce an additional closely dappled spotlight or localized reflected light to build up the levels on the subject.

9. Ground lamps

Sometimes, when shooting a subject around ground level, the only way it can be lit successfully is by using a floor lamp. Things to avoid here, include:

● Light streaking along the floor;
● Intermediate subjects becoming overlit, and
● Casting shadows onto the subject;
● Distracting shadows on the background.

10. Lighting a grand piano

A. *Regular problems.* When lighting a pianist, typical points check out include:

● The pianist's shadow falling onto the keyboard.
● The piano shadow falling onto the pianist's hands.
● The shadow of the raised lid or stick falling onto the pianist.
● Steep lighting intended to avoid 2 and 3, by lighting *over* the lid, causing unattractive portraiture.
● Light from a key on a lighting stand intended to avoid 2 and 3, by lighting *under* the lid, dazzling the pianist.
● Dark inner lid to piano. Lid shadow on strings and mechanism.
● Hot spots reflected from body of piano (espcially camera-light).

B. *Typical lighting approach.* Regular methods of lighting a piano include;

● Key/back lights at each end of the keyboard.
● An angled key light from the far side of the piano (see Figure 6.16, part 10).
● A local spotlight to illuminate the interior/front of the piano.

Fig. 6.17 Lighting continuity

Technical and artistic continuity problems arise in several forms.

1. As a camera moves around a subject what was a rimmed, decorative, or night effect becomes progressively flatter and unattractive.
2. Cross shooting between people facing to and from the light source produces intercut bright and dark pictures. (a) Exterior, (b) interior–exterior shooting. From different intercut viewpoints, any contrasty lighting balance may present varying brightness and atmosphere. More even lighting may produce a poorer pictorial effect.
3. Backed by a light wall (1) a face appears dark. Against a dark wall (2) it appears light. (Intercut viewpoints are disturbing.)
4. From a sunlit window to within a room, light intensity would naturally fall. Interior light levels must be maintained, however, for wide-area portraiture and pictorial quality. But evenly distributed lighting may lessen the illusion of an interior.

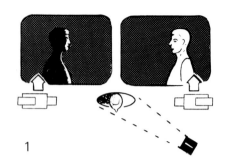

1

2a

b

LAMP

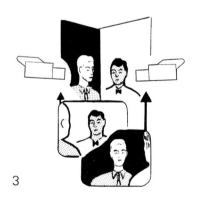

3

4

11. Background pattern position

A centralized pattern on the far background will change position relative to the subject as the camera viewpoint alters. This frustrates pictorial balance.

12. Camera-light problems

A. *Hot foreground.* While providing low-level filler, a camera-light may overlight parts of the subject.

B. *'Side-swiping'.* One person's camera-light may over-illuminate another person who is nearer the camera. Its lighting angle may be quite unsuitable; or may provide a strong kicker, edge-lighting his face.

C. *Reflections.* Camera-lights can produce spurious light reflections in spectacles and shiny surfaces in the background.

7 Lighting on location

As the costs of program making in the studio rise, an increasing number of productions are shot on location. Instead of using a built studio setting, drama is shot in a real-life situation, modified perhaps for the occasion.

Location shooting has great advantages as far as realism is concerned, but it poses major problems too for the program maker.

Natural light

Natural light may be convenient and free but it is deceptively unreliable. The sun's direction and altitude continually change, varying with the hour, the time of year and the latitude (Figure 7.1). Its quality also alters considerably.

The light illuminating subjects out in the open is of three kinds:

					AM												PM				
	3	4	5	6	7	8	9	10	11	12	13	14	15	16	17	18	19	20	21
JAN						R	8	13	16	18	16	13	8		S				
FEB					R	10	17	22	26	28	26	22	17	10		S			
MAR				R	10	18	26	33	37	38	37	33	26	18	10		S		
APR			R	10	18	27	35	42	47	49	47	42	35	27	18	10	S		
MAY		R	10	16	25	35	43	51	57	59	57	51	43	35	25	16	10	S	
JUN	R		9	18	27	37	46	53	59	61	59	53	46	37	27	18	9		S
JUL		R	10	16	25	35	43	51	57	59	57	51	43	35	25	16	10	S	
AUG			R	10	18	27	35	42	47	49	47	42	35	27	18	10	S		
SEP				R	10	18	26	33	37	38	37	33	26	18	10	S			
OCT					R	10	17	22	26	28	26	22	17	10	S				
NOV						R	8	13	16	18	16	13	8	S					
DEC						R		10	13	15	13	10		S					

Figures show the sun's vertical angle (altitude):
e.g. at midday in May 59°, in December 15°
This example is for Latitude 52° North
For every 5°E, add 20 minutes
For every 5°W subtract 20 minutes

Fig. 7.1 The apparent movement of the sun
1. Typical sunrise (R) and sunset (S) times, and hours of daylight.
2. *The sun's vertical angle (altitude).* The maximum angle the sun reaches during the day varies throughout the year.
3. *Rising and setting direction.* Although the sun rises 'in the east', and sets 'in the west', the actual direction alters throughout the year. (Note: The times for the southern hemisphere are transposed by 6 months.)

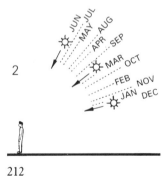

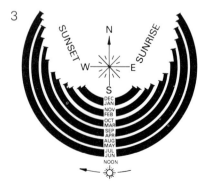

212

- Direct *sunlight*;
- The high color temperature *sky light* from the total sky around us; and
- Light that is *reflected* (and tinted) by nearby surfaces.

Full sunlight in a clear sky creates harsh modeling and strong shadows. These shadows may be colored by blue skylight (as in snow scenes) or by reflected light. As the sun becomes veiled by atmospheric changes, cloud, mist, haze, rain, or fog, the lighting contrast falls and modeling is reduced. Under an overcast sky light levels fall and illumination becomes dull, flat, and shadowless.

Even a thin layer of cloud passing over the sun can modify the intensity and sharpness of the light from one minute to the next. You have only to watch from a hilltop as passing clouds shadow the landscape to see how rapidly light changes.

1

2

Natural light (1)
This series of shots was taken at noon on a summer's day. With the steep sun behind the camera, all features of the landscape are crisp and well modeled.
1. There is a considerable depth of field, and even distant subjects are clearly seen.
2. In the closer shot, details of the bridge, the surface texture (and the decay!) are sharply defined.

If you are shooting in natural light there is always the prospect that picture quality will alter between scenes shot at different times. The *color quality* of natural light also changes quite noticeably. At sunrise and sunset a combination of the yellowish-red sunlight and the bluish skylight illuminating shadows can upset color values in the picture. Regular checks on the *color temperature* of the light may reveal that you need to introduce a compensatory lens filter, or white balance the video camera to keep the color balance within acceptable limits. Although compensation can be introduced during processing or videotape editing (TARIF adjustment), it is preferable to provide a consistent recording.

So while accepting the benefits of natural daylight you must recognize its limitations, and be ready to adapt or compensate for them in some way:

- The direction of the light changes, over an arc from east to south to west (Figure 7.2).
- The light quality is inconstant. It can range from flat to strongly contrasted.

1

2

3

Natural light (2)
Seen from the reverse side of the bridge (i.e. facing in
the opposite direction), the effect is quite different.
1. In the distant view, shadows merge. All features
   are less distinct (they are lit by soft sky-light).
   Foliage is back-lit and translucent.
2. As we get closer, some detials become clearer, but
   most of the scene is silhouetted against the light.
3. Even in a closer view, the shot is still
   disappointingly featureless, and texture minimal.
   Although color would help us to distinguish
   between various parts of the scene, the overall
   effect would still be dull, flat and lifeless.

- The color temperature alters with time, direction and weather conditions.
- The light intensity varies both in a short term and throughout the day. It may be too strong or too weak for your working lens aperture (*f*-stop). It may fluctuate during a shot.
- The distribution of light is random. One subject may be in shadow while others are brilliantly illuminated. The subject you want may be dimly lit or unmodeled, while another is prominently lit.
- Natural shadows may be distracting or inappropriate, and will vary throughout the day.
- The light may cause spurious lens flares or glare that degrade the picture.
- Scenic contrast is frequently far too high for the system, while at other times it is disappointingly even-toned.

Sometimes Nature provides just the conditions you want. There are actually times when the heart gladdens to encounter wreathing mists, dull leaden skies, snow, and other variants from Nature's repertoire.

Fig. 7.2   Natural lighting and location

The effect of sunlight changes throughout the day, altering the appearance of the location subject. Features on north-facing walls are never effectively illuminated by natural light (absence of modeling).

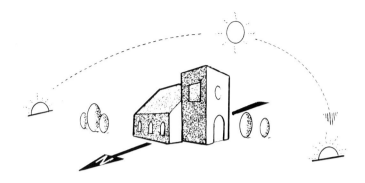

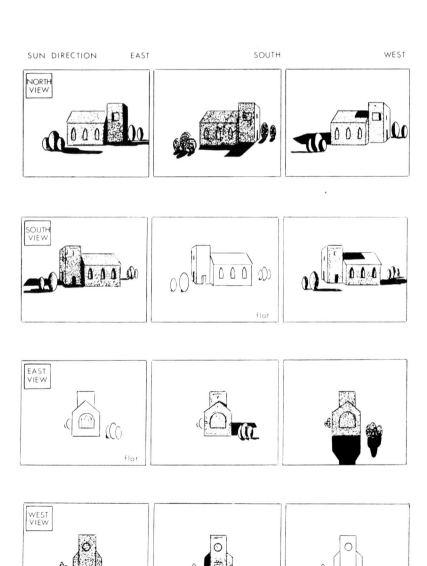

When the light is not right you have several options. Sometimes altering your viewpoint to suit the prevailing light direction could be the solution. It may be a matter of waiting until the clouds pass over or the rain stops. Shooting at a different time of day may be the answer.

Within limits, you can often adapt to the prevailing conditions. In a high-contrast situation, for example, you may expose for shadows and let the highlights take care of themselves, or vice versa. Gamma can be reduced during development or by circuit adjustments (video). Conversely, when scenic contrast is low you can use a higher gamma to exaggerate tonal values.

1

2

Natural lighting contrast
1. When we are preoccupied with the subject, lighting treatment may be incidental, as long as we can see clearly.
2. Sometimes the harsh lighting that emphasizes form, structure and texture creates the best effect.
3. The delicate half-tones of this high-key scene come from the diffuse light of an overcast sky.

3

Wherever possible it is advisable to check out typical lighting conditions in advance of shooting. Where, for instance, the outside of a building figures prominently in a shot it is as well to be forewarned if it faces north. Sunlight will *never* throw its texture and structural details into sharp relief. It will always be flatly lit by sky-light only, and can fall into deep shadow.

Even where the sun lights a building from the right direction it may be so steep that the modeling is coarse and crude. Remember,

1

2

3

Textural daylight
When daylight is at the right angle, it can strongly
emphasize texture.
1. Here edge-lighting from a steep sun throws the
   rough stonework into strongly-contrasted relief.
2. At sunset, even a lonely fledgling casts long
   shadows. The stone's texture is emphasized, but
   see how it reveals the shallow depth of field.
3. *Contre-jour* shots can transform even the most
   mundane subject. Here the light shines through the
   leaves, and emphasizes the irregular cobbled path.

the sun's vertical angle varies considerably throughout the day,
depending on the time of year. And, of course, the sun may not be
shining anyway!

On many occasions the appearance of the buildings themselves is
often quite incidental, but we are concerned about the shadows they
cast. It is as well to know beforehand that the area where the action
takes place is likely to be in deep shadow at the time you will be
shooting!

Exterior shooting

Day exteriors

Sunlight direction
If there is a free choice, try to position the camera to suit the sun.
Frontal sunlight (i.e. from behind the camera onto the subject)
produces a flat, harsh, unattractive effect, as well as being likely to
dazzle talent. Colored surfaces may appear rather too saturated
under such direct light. When you are shooting towards the sun, and
subjects are silhouetted, you will probably need strong fill light to
illuminate shadow areas; probably from reflectors.

Supplementary lighting

Unless daylight is waning or the weather overcast, most supplementary lighting you use in the open air is likely to prove disappointingly ineffectual.

■ *Booster lamps* If you are shooting in strong sunlight, and introduce *booster* lights to illuminate the shadow areas, even powerful 225 amp arcs ('brutes') or large HMI sources may be barely bright enough to have any real effect. They do have the advantage, though, that their uncorrected color temperatures can conveniently match that of daylight.

For closer shots, tungsten-halogen lamps (quartz lights) may serve as booster keys or shadow fillters. But blue or dichroic compensatory filters will be necessary to adjust their color quality to the prevailing light; so reducing their output to some extent.

The main drawback of light sources on location is the problem of supplying power. Even when suitable high-current supplies are available, there is still the hazard of cabling over long distances (voltage drops, cable bulk, vulnerability). In many situations a large mobile generator becomes necessary; and that can introduce noise problems.

■ *Diffusers* Where sunlight is excessive you can use large-area diffusers for localized shots but cut down the amount of light falling on the subject. Sizes of these scrims (nets, silks, butterflies) vary from around 1.22 m (4 ft) square to as much as 3–4 m (12–15 ft). They may be suspended as canopies or supported in gobo stands. Frequently used when filming in sunny locations, these large diffusers can be very effective. But they are vulnerable in even a slight wind.

To avoid the cruder modeling of the natural sunlight you may use several layers of diffusing material to reduce its effective strength, and light the subjects with reflected or incandescent light. Occasionally, you may cut off the sunlight altogether with a dense black velvety material such as duvetyne held in a light frame ('blackout', 'solid'), and rely on skylight alone.

■ *Reflectors* It is very easy to underestimate the value of reflectors. In the natural world around us reflected light enables us to see details in shadowy areas; it reduces overall contrast in the scene. You have only to watch the face tones of someone walking near a light-toned wall or reading a newspaper in strong sunlight to see the effects of reflected light. Using specially prepared reflector boards, you can divert sunlight to serve as key lights, filler, and background lighting without consuming a single amp!

The brightness and hardness of the reflected light is influenced by the type of surface used for the reflector board. So you choose the material to suit the job; depending on whether you want the hard light from a metallic surface or the more diffuse light from a white surface. While silver reflectors are most valuable for dark walls and

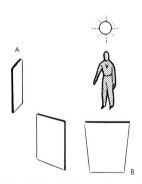

Fig. 7.3 Reflector boards
A reflector's effectiveness depends on its angle relative to the sun. It will not function when placed (A) at right angles to the light. When faced towards the sun (B), the maximum amount of light is reflected

foliage, the warmer light from white or gold surfaces may be more suitable for portraiture. You can also improvise reflectors, using various materials such as sheets of expanded polystyrene (Styrofoam; Jabolite), plastic sheeting, projection screens, white card or paper.

Because the effectiveness of a reflector is so dependent on the sun's direction and intensity its success will vary with local light conditions. The angle of the board and its flatness will affect its coverage. If you flex a reflector to form a convex or concave surface the spread of the reflected light will change. The closer the subject is to the camera, the smaller the reflector board you will need. If your subjects are distant or spread around, you will need a larger reflective surface or a number of boards.

Night exteriors

Fundamentally, there are two approaches when shooting night scenes:

1. 'Day-for-night';
2. 'Night-for-night'.

'Day-for-night'

This is a convenient and economical subterfuge in which you shoot the scene by daylight in such a way that the final picture appears to be a night shot. This approach is usually the only reasonable recourse where large areas are involved. Any attempt to create an overall moonlight effect through extensive lighting would not only look pretty unconvincing but would require a considerable amount of power.

Basically, the idea is to darken the shot overall yet leave sufficient distinguishable features for the audience to be able to discern the subjects. Because color is less evident at night we want to reduce the shot's color impact. In all cases, the effect is highly empirical, and depends on the degree of darkening you are aiming at; and that is where video cameras have a particular advantage, for you can see results immediately.

The usual methods are to shoot in daylight, reducing the exposure by about 1½ to 3 stops (by stopping down, or using neutral density filters), and to use

● A heavy *night filter* on the lens, giving a blue cast to the entire scene. The effect is stylized, almost a visual cliché, but easy to apply.
● A *graded sky filter* on the lens, which darkens the upper part of the shot, leaving the lower section unaffected. The result is an arbitrary tonal demarcation, but where the horizon coincides with the gradation the effect can be quite successful.

● A *polarizing filter* on the lens. Given the right conditions, this selectively darkens the blue sky. But it may result in increased color saturation for some subjects, rather than the low luminance you are seeking.

In addition to filtering, you can improve the night illusion by 'printing down' and/or emphasizing the blue cast during film processing, and by increasing gamma. Sometimes even just using tungsten-balanced film in daylight with greatly reduced exposure may do the trick. The color channels of video cameras can be rebalanced.

How well these methods work will vary with the tonal contrasts in the original scene. In general, important subjects need to be in the mid to upper tonal range, while darker values merge together. At worst, pictures become dense, 'muddy', and unconvincing.

You can often enhance the night effect by shooting towards the sun (*contre-jour*), the strong back light giving definition to the scene, and simulating moonlight. Where there are visible features that we normally associate with night scenes, such as street lamps, illuminated signs, car headlamps, and illuminated windows in houses, it is best to shoot the location under failing light, for then these sources can be lit (perhaps emphasized with added lighting) to create a more convincing illusion.

Wherever possible, avoid having the sky in shot. Unless that part of the shot is very heavily filtered it will be a giveaway.

'Night-for-night'

This method, so called to avoid confusion with 'night shooting' (i.e. studio work during the night hours), does involve shooting after nightfall. It uses selective illumination, together with any existing luminants such as street lamps, moonlight, camp-fire, light spilling from windows, etc.

You will probably find that the direction of lighting is critical, and should be confined to side and back positions. If the camera viewpoint moves, you may need to adjust the lighting angles to keep light in the 9H . . . 12H . . . 3H region. Frontal light should be kept to an absolute minimum (e.g. to reveal an expression in close-up), and then considerably localized and broken up with shadow. Filler is best avoided altogether.

Each kind of exterior location poses its own typical problems. If you are shooting in a forest at night, for instance, you may find that by the time sufficient light is reflected from foliage to reveal detail the leaves appear unnaturally glossy and obtrusive. Although you will normally be using lighting fixtures on stands, low directional light is seldom convincing unless there is some logical reason for it (e.g. light streaming from the open door of a nearby dwelling).

Where shooting is confined to a fairly limited area the problems of insufficient light may not be too great; although the situation is not helped by the fact that you will probably be using light blue or pale green filters over lamps, to simulate moonlight, and this will reduce their light output. In wider shots you may find that the light from your sources is disproportionately illuminating the surroundings near the lamps, so that, for example, although light levels are satisfactory on the main subject, tree trunks and foliage either side of it are overlit.

Night scenes shot beside the sea or a lake are especially difficult. Lighted areas on a sandy beach look phoney, even if they are broken up with a cookie/cucoloris. The exception is when the story line justifies the effect (e.g. a car's headlamps illuminating the action). A person or a boat is likely to appear as a brightly lit subject in black detailless surroundings. (It is often difficult to light them from preferred angles.) Any reflections from the water stand out as strong speculars.

When large structures are seen in night exteriors (buildings, sheds, walls) they quickly take on a floodlit appearance unless the lighting is well broken up. If dappled too extensively, though, the light patches may camouflage the subject so that it is difficult to interpret! A street scene lit by concealed equipment on stands alone tends to appear quite unrealistic. The old trick of freshly hosing the streets to gain dramatic value has become a visual cliché. Where the location includes a fair amount of environmental lighting, little more may be needed, except that 'cheated in' for close shots.

Location interiors

A wide variety

'Interiors' shot on location range from private homes to factories, from public events to coal mines. The variety is considerable, but the underlying principles are the same. Each type of locale offers its own characteristic problems and opportunities. What appears at first sight to be a difficult locale may in practice only require an odd fixture or two to supplement the existing lighting, while a 'simple' job can call for a great deal of ingenuity.

The kind of illumination you find at the location often determines your opportunities. Some interiors are flooded with daylight, others have extensive in-built fluorescent lighting. In some locations decorative lighting predominates, while in others there is little or no light at all.

Production requirements

The location itself is only one aspect of the project. Equally important is what the director is going to shoot there! Whether you are faced with a three-lamp improvised set-up or an extensive precision rig largely depends on the Director's production treatment. Even if the interior itself is vast, it may still be practicable to use a modest rig if you only need to light *a small part of it* at a time. If the

Director insists on shooting all round the walls and the ceiling in one continuous take, your opportunities are likely to be limited indeed.

Sometimes you can restrict the action area, and so cut down on lighting needs. If, for instance, you are shooting in a large hangar, instead of lighting the entire enclosure (or allowing it to fall into darkness), a few scenic flats, screens, cardboard box 'walls' or whatever is appropriate, may limit the amount of background seen and simplify the project.

The nature and the purpose of the production will usually decide the part the location plays.

Visible background

We may be mostly concerned with the action in a large arena while the environment itself remains purely incidental, requiring little or no additional lighting.

Structural

If the lighting is aiming to show the architectural beauty of an ancient building you are into an entirely different ball game. Not only must the lighting fixtures be carefully positioned to reveal form and texture but the greatest care has to be taken to avoid any damage. Lights must not appear in shot.

Unseen spotlights model various features, each lamp blending with its neighbor. You have to avoid either 'floodlighting' or an unnaturally 'spotty' effect. If the light is disproportionate or too evenly distributed it could produce an indefinable 'cheapness' and artificiality, which one wants to avoid at all costs.

The additional lighting has to be *compatible* with the environment. Although, for instance, lighting fixtures on the ground might illuminate some carved stonework very effectively it is arguable whether one should use that approach if the light direction or the effect look at all unusual for the location. Unfamiliar lighting treatment can produce a false, theatrical impression.

Naturalistic

Where a location interior is to look competely natural you may occasionally find that the existing illumination achieves just that. You need do little more than augment it with the odd local key to improve portraiture.

However, there are many situations where detailed additional lighting is a must. If the day is dull and overcast, 'sunlight through the window' may be provided by an HMI arc outside. On the other hand, if natural light is embarrassingly strong, you may need to fit a neutral density filter over the window opening to reduce it, or an orange-colored conversion filter which will color-balance its light with your own incandescent lighting inside the room.

Sometimes the existing illumination is an embarrassment, for it is in the wrong direction for your purposes, or too strong, or creating hotspots on walls, or distracting the attention in some other way.

You may be able to overcome difficulties by avoiding shooting anywhere near the window. Then, although it remains out of shot, daylight contributes to the general light level in the room. Perhaps you can rearrange the position of the subject so that the natural light falls upon it from a better angle.

Would it be better to pull window shades/blinds/drapes to cut off or considerably reduce the daylight? Then, of course, you will be relying almost entirely on your own lighting.

Many Lighting Directors make extensive use of *bounce light* to illuminate interiors. The scattered light provided by spotlights pointing at an unseen part of the ceiling or a white surface provides a useful overall fill light of indeterminate direction. This is a quick, simple solution to interior lighting which can produce quite satisfactory pictures.

But do not overlook the limitations of bounce light:

● It is a wasteful process. Only a fraction of the illumination from the lighting fixture is reflected as soft light.
● The light scatters everywhere, illuminating everything.
● The overall tonal contrast is considerably reduced.
● The reduction or even absence of shadows and shading results in flat overall illumination.
● The effect of light from above can be quite incompatible with the character of the interior.
● Walls and ceiling will generally appear lighter than the lower walls; an effect that destroys any feeling of enclosure.
● Where you use powerful lamps to obtain sufficient bounce light you may encounter ventilation problems in confined spaces.

Although it may occasionally be preferable to use a strongly diffused broad source instead, remember that, unlike bounce light, this will illuminate nearer subjects more strongly than others further away.

Atmospheric
Here the lighting treatment is deliberately arranged to engender a particular atmosphere: a room lit by the warm glow of a log fire, a smoke-filled tavern, the chill sense of desolation in a ruin . . . The lighting rig will be concealed, making use of high wall-to-wall telescopic poles, hidden lighting stands, and similar supports.

Decorative
The location interior may be transformed to create a decorative illusion: colored light, hanging decorative practicals, swirling reflections from a mirror-ball perhaps, to add an exciting magic to the dance routines in an old barn. Visible lamps add to the occasion.

Table 7.1   Lighting on location

Typical equipment

Lamps
- Camera light.
- Hand-held lamp.
- Lensless spot (e.g. 250, 650, 800, 1000, 2000 W) with barndoor.
- Small broad (e.g. 600–1000 W).
- Floodlight bank/Minibrute (e.g. with four to nine 650 W PAR lamps).
- HMI arc (e.g. 200, 575 W; 1.2, 2.5, to 6 kW).

Accessories
- Lightweight lamp stands.
- Medium/heavy-duty lamp stands.
- Spring-loaded support poles ('barricuda', 'polecat').
- Gaffer grips/alligator grips.
- Wall plates (for PAR lamps).
- Gaffer tape.
- Sandbags.
- Diffuser/scrim/wires.
- 'Daylight' color medium (gels).
- Dichroic filters.
- Window filter material (rolls and sheets) – Wratten 85/light orange gel.
- Neutral density filter material (rolls or sheets). May be combined with color correction.
- Reflector boards or sheets.
- Mobile power supplies (e.g. battery-belt, 30 V DC battery pack).
- AC power adapter (DC from household mains).
- Power cables, multi-outlet cables/spider boxes.

Typical household supplies
110–120 V, 15 A wall outlets (1650–1800 W max.).
220–240 V, 13 A wall outlets (2860–3120 W max.).

Treatment

Exterior shots – DAY
Sunlight
- Avoid shooting into the sun, or having talent looking into the sun.
- Reflect sunlight with a hard reflector (metallic faced) as key, or soft reflector (white coated) for fill light.
- In close shots color-corrected lamp can fill shadows.
- Bright backgrounds are best avoided, especially when using auto-iris.

Dull day; failing light
- Color-corrected light can provide key light or fill shadows.

Exterior shots – NIGHT
- Quartz lighting can provide key, fill, and back lighting for most situations. For larger areas, higher-powered sources are necessary (e.g. HMI arcs).
- Where necessary, either rebalance the camera to suit any local light sources (and filter your own lamps to match) or overpower them locally with your quartz light, and ignore background color inaccuracies. Extensive background lighting for night shots can require high-power lamps.
- Avoid moving the camera around where possible (lag effects visible; comet tails from highlights). Large lens apertures are often unavoidable (producing shallow depth of field). Extra video gain may emphasize picture noise.

Table 7.1 (cont.)

Interior shots – DAY	● The camera has to be balanced to the color temperature of *either* the daylight or the interior lighting.
	● Where strong sunlight enters windows: pull shades or blinds, put filter material over windows, perhaps plus neutral density media to reduce sunlight strength. Or avoid shooting windows. Alternatively, filter your lamps to match daylight.
	● Camera or hand lights (e.g. held 30–40° off lens axis) can serve as keys to fill light for nearby subjects, but:
	The light is very localized. (Very apparent in longer shots.)
	The very frontal lighting can easily dazzle people.
	The lamp casts adjacent shadows on a close background.
	A camera light flattens modeling.
	The light reflects in shiny backgrounds (e.g. glass, polished paneling).
	The light may be unsteady.
	The lamp may overlight close subjects, while leaving more distant ones underlit.
Interior shots – NIGHT	● Either supplement any existing lighting or replace it with more suitably angled and balanced lighting.
	● Where backgrounds are overlight, restrict main lighting to the subject. Overdark backgrounds may need extra lighting.
	Lighting arrangements may range from camera or hand lamps to three-point lighting. The most suitable form of lamp support depends on local conditions (on stands, clipped to structures or furniture, on telescopic poles from floor to ceiling, wall plates taped to structures, etc.).

Safety
On location it is important to take extra safety precautions in all aspects of lighting. For example:

Equipment condition	No frayed wires, loose parts.
Secure lamp fittings	Lamps should be firmly fixed and supported, and safety bonds fitted. All lamp stands should be secured or bottom-weighted to prevent overbalancing.
Power overloads	Keep within the power ratings of available sources. Remember, other nearby equipment may be sharing available power with your lamps.
Grounding/earthing	Have all lamp fittings individually fused and switched, and properly grounded/earthed.
Hot lamp fittings	Lamps can all too easily scorch or burn up nearby cables, drapes, paper, plastic, etc. In confined spaces, ventilation may be insufficient. Color medium can fume and smell. Overrun lamps can overheat fittings and nearby surfaces.
Floor cables	People can trip over cables, or inadvertently tug them and upset lamps. Tuck cables away, under rugs, near walls, hang them (gaffer tape or wire loops).
Water	Take great care when water is about! It can cause short circuits, or electrocute without warning. Water spray or rain on lamps will cause them to explode unless they are suitably designed for these conditions.

Public events

Events of this sort range from public meetings to orchestral concerts, from ballet to circus, from ice hockey to rodeo, from church services to parades. Each presents its own special lighting problems. One factor is common to them all where lighting is concerned, and that is the need for stringent *safety precautions* when working in any public place. Lighting fixtures should not be positioned anywhere near public access areas. Extra security bonds should be fitted to any overhead units. All cables should be tucked away out of reach. Particular care is needed when fitting any accessories (color medium, scrims, barndoors) to ensure that there is no chance of their falling.

It would seem reasonable to assume that, as these events invariably have their own integral lighting, little extra would be needed. But in practice, the existing lighting may not be suitable for the camera for a number of reasons:

- There may be insufficient light.
- It may be of unsuitable quality.
- Its contrast may be excessive.
- Light directions may be inappropriate for the cameras' positions.
- Although local illumination may be fine for the audience who are seeing action from a distance, the camera may reveal ugly modeling in closer shots.

As an example, let's look at the kinds of difficulties one can have when taking the camera into a theater.

Stage lighting that looks so excellent when sitting in the audience will often be insufficiently bright or excessively contrasty for good photographic quality. You may need to scale up or augment the existing lighting for the camera. To create a three-dimensional impression on the screen, light directions may need to be changed. Decorative colored light on the stage may appear garish or unattractive under the camera's scrutiny.

The traditional theater proscenium arch and audience layout generally involves frontal lighting from the circle or balcony, which often produces multi-shadows that are very obvious on the screen. Lighting over the stage area is invariably from high battens interposed between cloths, drapes and hung scenery. This keeps lamps out of the audience's eye-line but produces steep, 'toppy' illumination that looks quite ugly on camera. Additional lamps on 'perches' (vertical rails either side of the proscenium arch) and any floor stands provide predominantly side lighting.

At any location, you have to make the decision whether you are going to adjust the camera system to suit the local illumination (by lens filters, or rebalancing the video camera) or filter your lamps to match the local conditions. In a location where fluorescent lighting predominantes it might well be best to use a portable fluorescent fill light unit (120–160 W) to match the available light in an office, factory, store or schoolroom, and filter to that color quality. Alternatively, you might decide to swamp local lighting with tungsten-halogen lamps (quartz light), and disregard any color inaccuracies in the background.

The mechanics of the locale can certainly inhibit your lighting treatment. How far you can augment or change any existing lighting

for the camera will often depend on whether the event is being
specially provided for the medium or whether the camera is
'eavesdropping' on a regular presentation.

For many public events, such as indoor sports (boxing, table
tennis, swimming, skating), lighting installations often follow
patterns agreed as standard with the respective organizations.
Experience has shown how some lighting arrangements are liable to
dazzle or confuse competitors' judgment.

Similarly, there are situations where you need to consider the
nature of the action and arrange the lighting for safety's sake. With
high wire and balancing acts in circuses, for instance, lamps may
need to be steeper or moved to one side in order not to inconvenience
performers.

Insufficient light

When shooting in particularly dim surroundings there are several
methods of ensuring that you get at least a useable image:

Film
- Use the largest available lens aperture. (But this considerably
  reduces the depth of field.)
- Use a fast film stock.
- Shoot with monochrome (black and white) which has the highest
  sensitivity and tint or colorize the print.
- Use forced processing, pre-flashing, latensification (post-fogging)
  to increase the effective speed of the film. (Liable to affect picture
  grain, tonal values, and color rendition.)
- Use a slower shutter speed (e.g. 6 frames per second effectively
  provides four times as much light). There must be no movement
  in the shot or it will be correspondingly speeded up.
- Use the maximum shutter opening angle. (It is likely to be fully
  open anyway.)
- Take a flash still photograph and shoot an enlargement of this.
- Fit an electronic image intensifier to the camera. (The greenish
  monochrome image is adequate for some purposes.)

TV/Video camera
- Use the largest available lens aperture. (But this considerably
  reduces the depth of field.)
- Increase video gain. This increases video noise ('snow').
  Background irregularities become more clearly visible. Spurious
  video effects are more evident when the camera has insufficient
  light (e.g. streaking, lag, smear).
- Take a flash still photograph and shoot an enlargement of this.

Failing all else, there are always the more extreme solutions of

- Using a camera with a more sensitive type of pick-up tube;
- Fitting an electronic image intensifier to the camera.

Equipment

In the end, how you can tackle any project must be determined by such factors as budget, time, manpower, and available facilities. It is one thing to be able to hire any necessary equipment and quite another to have to cope with only a handful of lamps.

Choice of equipment is directly influenced by the scale of the operation. When working in a small room, for instance, you will need few lamps and relatively little power. But those lamps need to be appropriate for the job. There is little space to place light fittings.

Use a bulky lighting stand and fresnel spotlight, and you will not only take up a lot of valuable floor space but the lamp's spread may well be insufficient to cover the action. Here the lensless spotlight ('Redhead') and the small *broad* with their wide light-spread have advantages. Barndoors and flags can be used to confine the light. In a large room or hall such fixtures would have a more limited value, and instead, a fresnel spotlight could be ideal.

A lamp's distance from the subject will affect its performance to some extent. Generally speaking, the further a lamp is from the subject, the more powerful it will need to be to maintain the same light level, so whereas a 500 W fresnel spotlight may provide a good key light at e.g. 3 m (10 ft) a 5000 W lamp (5 kW) may be needed at three times that distance.

However, it is very important that you bear in mind that lighting fixtures of similar power but *different design* will provide noticeably different light intensities at any given distance.

When a fresnel spotlight is some distance away (a long 'throw') it becomes more difficult to restrict its light to specific areas with barndoors. Close barndoor flaps tend to affect the lamp's light output rather than its coverage.

Installations

A high proportion of location interior shooting takes place not in vast arenas or echoing cathedrals but in domestic-scale surroundings – offices, living rooms, schoolrooms, laboratories, etc. The types of the lighting equipment used matches the scale of such projects. Typical lightweight units include (see Chapter 10):

- Hand-held lamps;
- Attached portable lamps – in clamps, clip lights, hung-on or taped-on lamp fittings;
- Telescopic lighting stands;
- Support poles.

Sometimes facilities can be *improvised* for a brief take: a Photoflood in an existing light fitting in a room, a clip-light attached to a pole that can reach an out-of-the-way position, or a lighting fixture hanging from a rope. A hand-held lamp, or a lighting stand that is 'tracked out' as the camera moves, may overcome an 'insoluble problem'. A camera-light may supplement a regular set-up.

It may be tempting at times to rely on soft light (bounced or from units on lighting stands) to overcome local difficulties, but this simple solution produces very limited results. It is far better, wherever possible, to attempt to light the subject 'properly' with a three-point set-up.

The maximum lamp height in many interiors is very restricted. Although ceilings may be at 2.5–3 m (8–10 ft) the lamps will usually be correspondingly lower, so good vertical angles are generally possible.

Lighting fixtures may be fitted into tall lighting stands or clamped to support poles (ceiling-to-floor or wall-to-wall), but always pause to consider whether lamps are likely to damage or overheat nearby surfaces. Any closer than about 1 m (3 ft) may be unwise, particularly where rising heat can accumulate.

Lamps clipped, clamped, or taped to doors, window fittings, furniture, etc. can be very convenient ways of squeezing light into awkward places. But again, you will need to check safety carefully. Such arrangements leave the floor clear but trailing cables need to be tucked away to avoid becoming a hazard. The base of even a small lighting stand will need bottom-weighting with sand or water bags, especially if it is fully extended. It is worth taking the precaution of taping the cable to the column to prevent its weight pulling the lamp out of position or unbalancing it.

Unlike studio lighting, where lamps tend to be steeper than one would like, location lighting for small to medium interiors can bring its own problems:

● Reflections of lamps in windows, wall pictures, shiny surfaces, eye-glasses.
● Shadows of people and room furnishings, falling on the background directly behind them. (Where people are talking to an unseen interviewer their key can usually be angled to throw their shadow out of shot.)
● Walls cannot be shaded to any degree.
● The action lighting cannot usually be separated from the light falling on background.

In some situations it is unimportant whether the camera happens to see your lamps and equipment. In others it would destroy the entire ambiance of the occasion. When shooting in someone's home, for instance, you may need to disguise a stand used for back lighting by placing a piece of furniture or even a low drape or board in front of it. On the other hand, if only the vertical tube of the lighting stand is visible, and the base hidden, it may not be recognizable, so can be left visible.

Frugal lighting

Lighting on location is a challenge at the best of times. But when your total lighting equipment comprises a few lamps in the back of the car, as it does for so many mobile film and TV units, every lamp counts! So let's look at possible solutions when lighting on this scale:

Shooting without lights – in lit surroundings

You will usually have selected a preferred viewpoint, checked the light intensity relative to the lens aperture (*f*-stop) you are using, and confirmed that the background is satisfactory (not intrusive, not too bright, without distractions). If the light *direction* shows the subject effectively and the exposure is right there are no problems. You can go right ahead.

If there is insufficient light
● Then, as you saw earlier, a larger lens aperture may be the answer, or increased video gain.
● If there are drapes or a roller blind at a window even a slight adjustment often improves conditions to a surprising extent.
● Is there any extra lighting available where you are shooting, such as ceiling lights, that will lift the levels sufficiently?
● Sometimes when shooting fairly small items such as *objets d'art* one finds that the subject itself is sufficiently lit but its background is too dark. Perhaps you can take a closer shot, excluding much of the surroundings. Alternatively, you might insert a dummy background (e.g. a sheet of cardboard or a portable cine screen) to isolate the subject and show it more clearly.
● There are times when the best solution is to move the subject; e.g. take a statuette from a dark corner of the room outside into the daylight.

If the light direction is not really satisfactory
● Perhaps repositioning the subject relative to the light or changing the camera viewpoint will improve matters.
● Don't forget the value of a reflector when lighting interiors. Reflected light may considerably improve the shot by redirecting available light. A matte white surface can provide excellent fill light when reflecting sunlight. A metallic surface will reflect either sunlight to provide a key light or soft daylight to provide filler. For a very close shot even a sheet of white paper, card or plastic, or a metal tray can be pressed into service. Sometimes a relay-reflector will do the trick if the angle of the natural light is not suitable; light from a mirror or metallic surface placed in the sunlight is bounded onto a second reflective surface, which illuminates the subject.

● When shooting a daylit interior consider whether it might be worth waiting for the light to change. It might be better to delay shooting until clouds pass over the sun rather than shoot in bright sunlight with harsh shadows and high contrast.

Camera lights

By attaching a small portable quartz-light to the top of the camera ('spot bar') you can ensure that, wherever the camera shoots, some light is going to fall on the subject! But a camera light is a mixed blessing, and it is as well to appreciate its limitations. How evident these are will often depend on whether your single lamp is the only luminant or whether it is supplementing other illumination.

The camera light's main advantage is that it does not require a second pair of hands. It can provide a modeling light for close exterior shots on a dull day, or fill light for hard shadows when shooting someone in sunlight.

It is invaluable when shooting under difficult conditions. If, for example, you have to move around while following someone it might be the only way of ensuring that the person remains lit. If your camera has *auto-iris* and is self-adjusting to variations in picture content the camera lamp may help to stabilize the subject brightness under these conditions, when it would otherwise vary with a changing background.

However, the camera light does have its drawbacks. It adds to the camera's overall weight. Its light is extremely frontal and so tends to flatten out subject modeling. Its light will reflect in spectacles and any glass or shiny surfaces behind the subject as an intense white blob. People facing the camera may find the light dazzling. Its light is localized, and when you are using a wide lens-angle on the camera (zoomed out) the limits of the lamp's beam-spread may be clearly visible.

Finally, the illumination from the camera light does not carry far and, particularly in longer shots, things near the camera can appear overlit while anything further away remains virtually unlit.

Hand-held lamps (sun gun)

A hand-held lamp separate from the camera does require the assistance of a second person, but this can have several advantages. It is less weight for the cameraman to have to carry. The assistant can anticipate where to point the lamp and so avoid the 'searchlight effect' that can occur as a camera light pans around. The assistant can probably hold the lamp high (not easy for any length of time) at a better angle to light the subject than the camera light would, and avoid direct 'kickback' reflections.

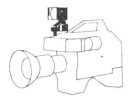

Fig. 7.4 Lightweight camera lamp
A small spotlight clamped to the top of a lightweight camera (or camcorder/combo) is a regular utility lamp for mobile cameras.

Fig. 7.5 Hand-held lamp
A small hand-held spotlight provides a useful flexible light source on location.

The hand-held lamp can be more powerful than a camera light, e.g. from 250 W to 1000 W. It may be battery or AC powered. Some designs have a cooling fan; those without tend to overheat. When shooting in daylight you can fit a dichroic filter or color-correction filter medium to raise the lamp's color temperature to match.

Single stand lamp

The only light available

Whatever you are lighting, the priorities are: the subject's *key light*, then *filler* (*fill light*) to illuminate its shadows, and finally a back light and/or background lighting, according to the situation. If your single lamp is providing most of the illumination in very dim or unlit surroundings you will have to use it relatively frontally; perhaps slightly to the side and above the camera.

A *single soft light* such as a 'broad' will provide a more attractive effect than a hard light source; particularly if you use a diffuser (scrim) of tracing paper or frosted plastic medium in front of the lamp. This will cut down the light but improve the pictorial effect. Although this soft light will scatter around, at least it will prevent the surroundings falling into deep shadow.

If you use only a *single hard light source*, pictures may be over-contrasty unless you fix a sheet of diffuser over it (notwithstanding light loss). Otherwise your subject may be left isolated in a pool of light within dark surroundings, or casting a strong obtrusive shadow. The further the lamp is off the lens axis, the more pronounced will be the light fall-off on the other side of the subject. Sometimes you can fill this to some extent with specular light from a metal-foil reflector.

In lit surroundings

How you use the lamp will largely depend on how satisfactory the existing lighting is:

■ *Key light*   The single lamp can

● Bring up the light level to improve exposure;
● Allow you to stop down (for greater depth of field);
● Ensure clearer modeling in the subject;
● Provide a low-intensity highlight (catchlight) that adds sparkle, glisten or glitter to it. Eyelights, you will remember, add vitality to portrait lighting.

The key light might be allowed to illuminate the background, to improve its general visibility.

If you are not careful you may inadvertently find yourself overlighting a close-up subject, then having to stop the lens down to compensate exposure. If you do this, the overall brightness of the rest of the scene will appear to fall off.

When shooting under sunny conditions you can use the sun as an attractive back light, and key the subject with your single lamp. Alternatively, you can use the sun as a key light, and introduce your lamp (and/or a reflector) as filler.

■ *Fill light*   The lamp can be used to illuminate dense shadows cast by other lighting. Soft light is preferable as it does not cast further shadows, but its intensity may fall off too rapidly with distance to be useful. Instead you may find it better to use a hard light (diffused perhaps) to relieve shadows.

■ *Back light*   Sometimes you will find a subject that is perfectly well illuminated but the picture lacks that extra quality that back light can add. Even when shooting in the open air on an overcast day, the addition of a single back light can often enhance the effect.

Positioning a back light may not be easy. If you cannot use a tall lighting stand hidden behind the subject it may be necessary to clamp it to a horizontal support pole or even suspend the lamp.

■ *Background light*   Even when the main subject and its surroundings are well lit, additional background lighting may improve its appearance, bringing out the texture and shape of items there, giving the whole scene a more three-dimensional look. And, of course, where the background is dark and unrelieved, extra lighting may be necessary to reveal any detail.

Using two lamps

The extra fixture improves your opportunities considerably. You can use the lamps separately or combine them in any of the permutations we have been looking at. They can become the shared key/back lights for two people. Carefully positioned, they will often light the background too.

Using three lamps

By now, you will have realized that just because you have three light sources you do not necessarily have to use them in a 'three-point' pattern; especially when the surroundings are already lit to some extent. Your priorities will depend on which aspects are most lacking in the shot. As always, it is the effect on the screen that counts.

Lighting density

If you are shooting in well-lit surroundings, or shots are fairly restricted, the extra lighting needed may be minimal. But there are situations that immediately ring a warning bell in the mind of an experienced Lighting Director. They are 'lamp hungry'. Here are typical situations where features in the scene or the production treatment will usually call for a large number of lighting fixtures. If you haven't got them, then you will have to think of another way of tackling the problem, or advise the Director of the difficulties:

Technical
- When you need considerable depth of field and the lens must be stopped well down.
- When your lamps are of low power or have a very restricted spread.

Area
- When you have to light a large area evenly – e.g. an audience where the Director is taking close reaction shots.
- When you have to provide the entire lighting for a wide shot; particularly if the far background is a considerable distance away.
- Where action moves over a wide area (e.g. as in a dance routine) and people and the surroundings need to be lit throughout.

Multi-subjects
- Where a number of subjects are spread around, each needing individual lighting. (This includes situations where groups of dancers are widespread in an arena to an interior where architectural features are separately lit.)
- Where each subject in a display group is to be lit individually.

Scenic
- Where there is a large area of background, such as a cyclorama, which has to be evenly lit overall.
- Where there are a series of planes in the scene that need individual lighting treatment; e.g. one set of screens or arches behind another.
- Where you want to show detail or modeling in a large dark-toned background.
- Where a location is awkwardly shaped, each lamp may only be able to reach a limited area.

Operational
- When the camera moves around showing a series of areas that require separate lighting.
- When a subject is shot from several directions, the respective backgrounds will be visible from each direction, and can add up to a surprisingly large overall area, particularly if distant.

Effects
● Where a series of projected patterns are used to build up an overall effect.
● Where several color changes are required on a cyclorama and/or foreground scenic units.
● Where there are atmospheric lighting changes (e.g. morning to night).

8 Atmospheric lighting

The Lighting Director, like the rest of the production team, follows an ephemeral art of make-believe. At the touch of a button he may transform a plain, dull, ordinary-looking background into an attractive highly evocative image. On the other hand, unsympathetic lighting can reduce a sensitively designed setting to crude artificiality on camera.

There are no 'rules' for persuasive picture making, but there are certainly many contributory factors, and we shall be studying these here.

The intangibility of lighting

'Visual appeal' is an elusive quality and correspondingly difficult to define. In one situation you can create an arresting picture with a single carefully placed light source. In another it may be necessary to build up an effect patiently, lamp by lamp.

When an audience comments on 'good lighting' you will usually find that they are responding to a particularly impressive dramatic effect. But most skilled lighting is self-effacing. The more subtle the treatment, the more natural or 'obvious' it appears to be. As the audience concentrates on performance they do not 'notice' the lighting, but they are influenced by its effect.

Looking at a *setting* in the studio, we can appreciate the skill with which the Designer has re-created a living room, kitchen, or whatever. But when we look around at the *lighting* for that scene we simply see a lot of bright lights. A carefully balanced lighting treatment just appears as so much illumination.

There is a wide gulf between what the camera sees and what appears on the screen. When we look at a *picture* of the scene we are influenced by its *pictorial composition*; the interplay of line, form, tone and color. We see an evocative atmospheric illusion; a convincing impression in which patterns of light and shade guide our attention, induce emotions.

But none of that is at all obvious when we go into the studio and look around. The eye sees no interrelationships. The casual glance cannot evaluate how it will be interpreted by the camera, nor assess how the appearance of the scene will be changed by slight adjustments of exposure, gamma, color balance, etc. Part of the Lighting Director's art lies in making allowances for this transmutation between the appearance of the scene in the studio and the final picture on the screen. He learns to anticipate and interpret the various disparities.

The gulf between

The creative artist in motion-picture and television production has one outstanding difficulty that inevitably divides him from his audience. He can never see his work from a fresh, unbiased viewpoint. He can only guess at how his audience will interpret and react to what they are seeing.

Familiarity with each move, each shot, each setting in a production prevents one from seeing with the innocent eye of the audience. There are no first impressions. Something that seems obvious to him may pass them by. Something he feels needs to be emphasized may have been understood first time by the audience, and not need underlining.

You can demonstrate this effect easily enough by rerunning a section from any videotape over and over. Each time you see it, various aspects of the scene will take on subtly different values. You will begin to notice little details that had not struck you at first viewing. Something that interested you at first sight you later disregard.

When working on a production you can lose a sense of proportion through growing familiarity. You need continually to think in terms of what is there on the screen rather than what you know is there in the studio. It may be a great effect in the studio but does it have *screen* value? A large decorative pattern projected onto the background may have taken some time and ingenuity to produce. But how well does the *camera* see the effect? To the audience it may be little more than a defocused blur, and seldom appear in shot except as odd blobs of light! If there is a backing beyond a doorway that is only seen now and again when the door is opened, is it worth devoting time and equipment to an elaborate window pattern there rather than a simpler treatment? Or will it add an extra touch of realism to the setting?

One often finds that it is attention to detail of that sort that makes a setting come alive. If a photo-backdrop outside a window is lit with care the entire interior scene becomes totally realistic. But light it badly, revealing its bulges and creases, and incredibly, the room itself can appear unconvincing and false.

The camera is the audience's eye. They rely on what the camera shows and the microphone reveals, to form an impression of the surroundings. They are reacting not to what is there but what they *think* is there. We can deliberately fool them; we may accidentally mislead them. But we must not forget that their interpretation is in our hands.

We rely heavily on an audience's credulity. As someone looks towards the camera and a shadow falls across their face, the audience assumes that it comes from a real window in the 'fourth wall' where the camera is located. In fact it may come from a projected gobo or a window flat held by stage hands.

On the other hand, the audience is sometimes distracted and puzzled by an effect that we accept unquestioningly. They may wonder about that strange shadow falling on the background. We disregard it, for we know that in the studio there is an overhanging tree-branch hung just out of shot that they cannot see.

Lighting styles

Pictorial style

Fundamentally, there are three broad approaches through which we can depict the three-dimensional world in the flat picture:

- *Silhouette*
- *Notan*
- *Chiaroscuro*

Although we accept these pictorial styles unquestioningly they are, strictly speaking, quite different graphical concepts.

■ *Silhouette* In this treatment one concentrates on the subject's *outline* and suppresses all surface details. In the true silhouette the subject appears black against a light background – a treatment used for dramatic, mysterious, and decorative effects. We usually see silhouettes as unlit ornamental forms (tracery, skeletal screens), outlined shapes against a bright sky, unlit objects backed by an illuminated panel.

Semi-silhouettes arise when shooting subjects against the sun (*contre-jour*) where back light predominates and frontal lighting levels are low. Again the treatment is decorative, and little or no detail is usually visible in the subject.

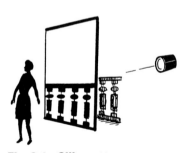

Fig. 8.1 Silhouettes
Silhouettes can be provided either by shooting an unlit subject against a lit background or shadows cast onto a rear-illuminated area.

■ *Notan* Here pattern rather than form predominates. Emphasis is on the *surface* of the subject; on decorative details, ornamentation, surface tones and color. The impression is essentially flat, two-dimensional. Shadows and shading are eliminated. You can achieve this effect with widespread diffused low-contrast lighting; back light being reduced to a minimum or excluded altogether.

Because texture, modeling and form are reduced or suppressed, attention is concentrated on tonal and color relationships. The high-key result is mostly used for fashion shows and large-area dance routines.

■ *Chiaroscuro* This is such a familiar style that we are apt to overlook that it *is* a deliberate technique. Here emphasis is on *dimension*. One's aim is to convey an impression of solidity and depth. This effect is achieved by carefully controlled tonal gradation, tonal separation between planes, progressive tonal values, shadow formations, and textural control.

When effectively applied, the subject and the scene appear totally three-dimensional. The picture has arresting, persuasive vitality.

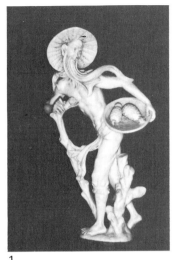

1

2

3

Pictorial styles
Here we have the three basic pictorial styles:
1. *Notan* – where emphasis is on the subject's surface tones, color and outline; a flat two-dimensional effect.
2. *Silhouette* – where all we see of the subject is its outline shape.
3. *Chiaroscuro* – where light and shade reveal form and texture and create a three-dimensional effect.

Many painters, including Rembrandt, Caravaggio and Dali, have achieved remarkable illusions through chiaroscuro techniques.

Pictorial effect

Dominant tones influence both the picture's mood and its aesthetic appeal.

■ *Low-key* Where mid-gray to black tones predominate, the picture can have a somber, sober air about it. The darker tones may give it a comforting, cozy quality suggesting restfulness. Or it can have a heavy, tragic quality.

■ *Very low-key* Where black areas predominate in the picture, relieved with smaller areas of lighter values, the effect is highly dramatic. It may simply convey an impression of 'darkness' and 'night', but usually there are emotional overtones of mystery, and the sinister.

■ *High-key* When mid-gray to white tones predominate in the picture there is invariably a feeling of lightness, and cheerfulness.

■ *Very high-key* Where the picture is full of light-gray to white tones it can convey impressions of delicacy, simplicity, gaiety, openness, spaciousness, triviality, or the etherial.

When subjects are presented in front of a totally white background, usually with shadowless notan lighting treatment, it is often termed *limbo lighting*. But, strictly speaking, this is a staging effect. Conversely, *cameo lighting* is a scenic style in which subjects are lit against a black background.

Tonal distribution

The way tones are arranged within a frame has a considerable influence on picture impact.

- If the tones are broken up into many small areas the picture will tend to lack unity, and appear disjointed.
- When the picture has large, well-marked areas of similar tone they give it significance and strength, lending it a certain vigor.
- Where dark-toned areas are relieved by smaller distinct light areas the effect tends to be grave, solemn, and mysterious.
- Conversely, where a light tone is relieved by smaller distinct dark areas the overall effect is of liveliness, brightness, cheerfulness, delicacy.
- Unrelieved dark tone containing few contrasting areas tends to be dull and depressing.

Pictorial quality

Pictorial quality is highly subjective. Our impressions depend so much on the range of tonal values in the picture, tonal contrast, the sharpness of demarcations between tones, where they are located in the picture frame. Background tones that look prominent, arresting, dramatic when sharply focused can merge to become vague and featureless when defocused. Crisply well-defined contrasts so appropriate to one situation may be entirely wrong for another, where delicately graded tones would be more suitable.

Tonal values are only a broad guide to pictorial quality. A shot in which there are no dark tonal areas may look thin and lacking in body. But then again, it may look delicate and refined. Without lighter tones, a picture may lack vigor and sparkle; but it would be exactly right if we want to convey the wan light of fading dusk, or a foggy locale. A picture that has few half-tones and strong contrasts can seem harsh or coarse, or it may have vitality and dramatic force.

Pictorial treatment

Lighting offers a wide range of pictorial styles. Some are direct and obvious, others have subtle nuances. We can analyze them under the three broad headings: *illumination – realism – atmospheric*.

Table 8.1   Approaches to pictorial treatment

Illumination – overall (flat)	Main consideration is visibility. Lighting is almost entirely frontal and flat. Tends to *notan*. At its best, subjects distinguishable from each other, and from their background. At worst: flat, characterless pictures; ambiguously merging planes; low pictorial appeal.
Illumination – solid	By a careful balance of frontal and back light subjects are made to appear solid. This and clarity are the principal considerations of this chiaroscuro approach. The lighting suggests no particular atmosphere.
Realism – direct imitation	*Direct* imitation of effects seen in real life, e.g. sunlight through a window imitated by a lamp shining through it at a similar angle.
Realism – indirect imitation	Imitation of a natural effect, but achieved by a *contrived* method, e.g. simulating sunlight by lighting a backing beyond a window and projecting a window shadow onto an adjacent wall, from a more convenient position.
Realism – simulated realism	An imitation of a natural effect, where there is no direct justification for it from within the visible scene, e.g. a window shadow on a far wall, that comes from an unseen (probably non-existent) window. (This may be a projected slide, cast from a cut-out stencil or a real off-stage window.)
Atmospheric – 'natural'	A lighting treatment in which natural effects are not accurately reproduced, but *suggested* by discreet lighting. A pattern of light that highlights and suppresses pictorial detail selectively, to create an appealing effect. Suggesting realism in most instances, but seldom strictly accurate for that particular environment. A typical motion-picture approach.
Atmospheric – decorative	Associative light patterns, e.g. of leafy branches on a plain background.
Atmospheric – abstract	Light patterns that have no direct imitative associations, but create a visual appeal; e.g. a silhouetted unknown person, lit only by a rectangular slit of light across his eyes. A flickerwheel pattern cast over an exciting dance sequence.

Illumination

■ *Overall illumination*   Here the subject is evenly flooded with light. Visibility and clarity are the main aims. Clearly, this is the way we usually light backdrops, pictorial backgrounds, sky-cloths, graphics, title cards. When we light three-dimensional subjects in this fashion the result is a *notan* effect. Subjects are distinguished from their surroundings through tonal and color differences. Ineptly handled, flat lighting of this sort can cause planes to merge ambiguously.

■ *Solid illumination*   This is a typical three-point lighting approach, in which subjects are well modeled and outlined in light to create a three-dimensional illusion. Again the overall pictorial effect is neutral and non-representational.

Realism

■ *Direct imitation*   The main aim of this approach is to replicate in the studio the form of illumination one finds in a particular type of location. In many cases it *substitutes* a lighting fixture for the actual source. If, for instance, the scene shows a room that would normally be lit solely from above by a skylight this effect is simulated by an overhead light shining through a void in the ceiling. In a scene lit by a single candle or a lantern a miniature lamp ('peanut') is hidden in the fitting held by the actor to illuminate the action.

■ *Indirect imitation*   In this approach we imitate the *effect* of natural lighting, but by a contrived method. Although a room appears to be lit by local light sources such as a window or practical lamp, illumination from that direction is augmented by extra lighting to provide the optimum visual effect.

We might project 'sunlight patterns' onto a wall, that supposedly come from a nearby window. A pool of light around a practical lamp seems to be coming from the lamp itself, but is really from a localized spotlight.

Where some natural source of light appears in the picture you need to take care that any additional lights do not seem to conflict with it (e.g. a person with a candle on their left, lit from a lamp on their right). Where a light source is not *visible* (e.g. a window in the 'fourth wall') the light direction is less critical.

■ *Simulated realism*   Here we see effects that appear to be quite natural – yet are quite artificially created.

● Outside a bedroom door, the shadow of a balustrade falls across the wall, presumably from the hall below. In fact it is a cut-out framework in front of a lamp on the ground.
● A person sits warming in front of the fire. There is no fire or fireplace; just a lamp on the ground and a fire flicker stick.
● Light from a stained glass window falls onto a distant wall. It is from a projected slide.

Exactly how the effect is created is immaterial, but it must be convincingly realistic. Typically it will be from a projected slide, a cut-out stencil, or an off-stage window.

Atmospheric

■ *Naturalistic*   Here, although the setting appears to be naturally lit, detail is deliberately highlighted or suppressed to develop an atmospheric impression, a particular mood. While specific patches of light and shade in the picture may not be truly justified they *seem* natural enough:

● One might, for example, *similate* an attractive shadow of a window blind on the wall of an office. It appears quite natural, and adds to the illusion of realism – yet the effect is improbable in a real location.
● A piece of furniture or statuary is discreetly lit by restricted lighting, to reveal its form and texture. In reality, in such a location it would fall unnoticed into dull shadow.

Impressions of light direction
We interpret light direction from (1) the way light is distributed over the subject, (2) the position and shape of light on the background, and (3) the direction of shadows. In (4) the gesture seems to reinforce these clues.

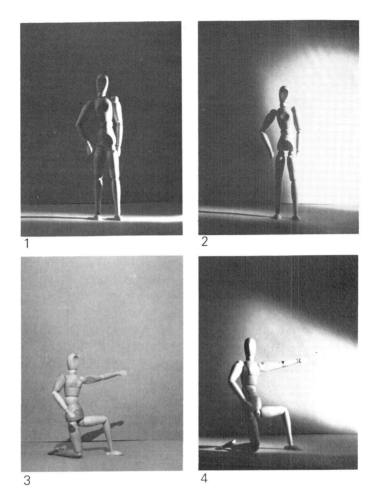

1

2

3

4

● A section of ceiling is dimly lit . . . by a lamp on the ground, hidden by a coat 'casually' draped on a chair. Because the ceiling is visible, it adds to the feeling of enclosure in low shots. Normally the ceiling would be in shadow, and make no contribution to the 'feel' of the room.

This is a typical motion-picture approach, which is also followed in progressive television lighting.

■ *Symbolic* Sometimes a particular environment can be suggested simply by patterns of light and shade alone. There may be little or no built setting. Regular examples are shadows of prison bars to imply a prison environment, leafy patterns suggesting countryside or woodland, rippling reflections denoting nearby water.

Decorative effects can stimulate the imagination. They are also a very economical and convenient way of suggesting an environment. But if over-emphasized, effects of this kind can appear mannered and eye-catching.

Fig. 8.2 Pictorial effect
1. Light and shade can produce a natural, environmental effect.
2. Over-fussy manipulation of lighting can create a mannered, self-conscious pictorialism that draws undue attention to itself.

■ *Stylized* This is a form of pictorialism which makes extensive use of decorative treatment. It is a self-conscious approach that is preoccupied with visual effects, even where they are not justified. The pictorial impact is often striking, but usually quite artificial and theatrical.

It is characterized by exaggeration. Interiors are strongly lit, sunshine predominates, walls are normally light-toned and un-shaded. There is usually excessive fill light, often from camera height or below. The lighting for low-key scenes is highly dramatic and strongly contrasted.

This pictorial treatment is generally accompanied by separately shot extremely stylized high-key portraiture: no nose-shadows in close-ups, generous frontal fill light, and obligatory back light in all circumstances. For many years this style was widely used in feature film making.

■ *Abstract* Here light patterns are used for visual effect alone. They may have no direct imitative associations. The effects may be purely decorative, or they may be for fun, to shock, to arrest attention, to intrigue, to stimulate . . . but above all, let them be appropriate! They can take several forms:

● *Decorative* – e.g. random color reflections from metal foil. Projected gobo patterns. Multi-color dappling.
● *Fantasy* – e.g. light from a whirling color wheel illuminating an exciting dance sequence. A projected flame effect using green/magenta light. Laser backlight with smoke effects. Clouds of multi-colored fog.
● *Bizarre* – e.g. a page of newsprint projected across a face. Stroboscopic light causing movement to be seen as a series of brief 'stills'.

'Natural' lighting

This is probably one of the most controversial topics in the field of lighting – the 'natural approach'.

Tiring of the predictable sameness of certain pictorial treatments, the ever-perfect portrait and the ever-present back light, some critics assert the need for a lighting treatment that is more naturalistic. They contend that, however variable available light may be (natural or artificial), it *is* the kind of illumination we have come to expect in the real world. Any form of stylized lighting approach is unrealistic and contrived, and diminishes the realism of the scene.

Against that it can be argued that film and television are, by their very nature, *artifices*. They can only *simulate* reality, and are necessarily *un*natural. The screen is inherently restrictive. The variations in scale and viewpoint as shot-size and camera set-up change, the conventions of editing . . . are all part of the disciplines and conventions of the media. It is because available light produces such diverse and often unattractive effects on the screen that many lighting techniques evolved.

If we attempt to directly imitate natural lighting the results will be equally unpredictable and *inconsistent*. At best, they can be excellent, but, on balance, they tend to be mediocre if we do nothing more than mimic ('*Direct imitation*').

As we discussed earlier, looking around the real world, eye and brain continually interpret and normalize what we see. The screen, on the other hand, concentrates our attention so that many unattractive effects we would normally disregard in life become prominent and distracting.

The realities of natural light

Look critically at some pictures shot on location with only 'available light'. Is the portraiture flattering, attractive, unkind, ugly? You will probably find hot tops to heads, black eye sockets, half-illuminated faces. While one person squints towards the sun with nearly closed eyes his companion may be strongly lit from behind, with glowing ears and nose and frizzy hair. In the background behind them some items are overbright while others are barely discernible and merge with their surroundings.

Make a point of looking around critically in a shop, a bus, your living room, in an elevator, in a corridor, in a store, out in the street. How do people and places *really* look under natural lighting? Existing interior lighting may be effective yet produce ghastly portraiture. All those unattractive effects are there to see. But we usually ignore them. Under the camera's scrutiny, the closer the shot, the more obvious they are.

Ask yourself this leading question. If these various effects on the screen were the results of *your studio lighting* would you be content to leave things this way, or would you want to move the key, or add (or remove) some fill light? Would you really want to imitate the natural effects you see here? Simply substituting a studio lighting fixture for the normal light source is no reliable solution, and it is worth looking at examples in some detail to see why.

Sunlight

Let us assume that we are shooting a setting representing a room lit by daylight. Presumably, to light this situation naturally we would need a powerful light source outside the window of the room setting to provide a 'substitute sun'. But does it provide us with 'natural sunlight'?

When a real interior is lit by sunlight alone, random light is reflected in all directions from the walls, floor, and ceiling. A studio setting has no ceiling or fourth wall to reflect light; so there will be less scatter and the interior will be darker.

Light in the studio setting would also be more contrasty. Shot from inside the room, people would be silhouetted against the bright window, with little reflected light. Standing side-on to the window, a face will be half lit, bisected by light.

Light levels would vary considerably throughout the room, increasing rapidly near the window. So when intercutting between varying viewpoints, exposure would vary. Is this really acceptable? Reducing the intensity of the studio 'sunlight' so that the people near the window are less strongly lit will also reduce the amount of light penetrating the room.

So in this example the Lighting Director probably finds himself having to augment this 'natural effect' by introducing extra illumination for those angles where the subject looks less attractive, or is less visible. Is this additional light compatible with the real effect on location? Or does it create a false and unconvincing atmosphere? If inappropriately treated, even a real location can look phoney!

As we introduce extra lighting to produce a correctly exposed picture, or to suit various camera positions, the treatment can take us further and further from the natural effect. We may well end up with a profusion of shadows and incompatible light directions.

Overhead soft light

One further example of 'naturalistic' lighting that has its followers is the use of *overhead soft light sources* (e.g. 'space lights, chicken coops, troughs, five-lights') to provide overall scattered light for interior scenes. Excellent for 'exterior' settings where they simulate sky light, or large-area lighting, its use for most interiors is questionable. Diffused overhead lighting is only appropriate in certain situations (e.g. illuminated ceilings, skylights, glass roofs). In smaller interiors it produces unshaded walls that are overlit at the top, and even an 'open-air' effect.

Influential lighting

Light intensities

Man's reactions to the lighting of his surroundings are of a very fundamental nature, so that lighting conditions and mood are often closely interlinked:

- *Sunlight* epitomizes light and openness, and can be powerfully oppressive, invigorating, exhausting.
- *Darkness* is concealing, and has overtones of mystery or treachery.

Following such associative thoughts, we find that even the way an empty room is lit will set a mood, create an atmosphere. Light walls can look stark, unwelcoming, crude. Or they may give a room an open airy appearance. Dark walls may look dull and somber, or rich and sophisticated. A low general lighting level punctuated with pools of light has snug comfortable associations (a folk-memory of the camp-fire perhaps?).

Fig. 8.3 Directing attention
Tonal relationships and clarity of detail enable one to direct attention to particular parts of the picture.
1. Isolated tone.
2. Graded tone.
3. Compositional line.
4. Clarity of detail (depth of field adjustment).

1

2   3   4

Directing attention

You can use light to guide the eye; to induce people to look in certain directions, to concentrate on particular features.

- The eye is attracted by isolated tone; e.g. a small light area within a large darker area.
- Adjacent areas of strongly contrasting tones (e.g. black and white) draw the attention.
- Saturated colors, particularly those of lighter hue, attract the eye more than pastel (desaturated) hues.
- Our attention goes to areas of sharp detail, paying less attention to unsharp areas (unless movement or saturated color distracts us).
- The eye is guided by graded tones; usually moving from darker to lighter areas.
- The attention is moved by converging compositional lines.

1

2

3

Concentrating attention through light
These off-screen shots show how light can be used to
direct attention:
1. Here the light distribution creates an impression of
   depth and mystery.
2. Our attention is concentrated downstage, but there
   is an air of expectancy. Is someone about to come
   down the stairs?
3. We concentrate on the central character's
   reactions to the semi-silhouetted action in the
   foreground.

Background lighting can play an important part in concentrating
attention on a subject. You can, by lightening or darkening its
background, make a subject more or less prominent. Shading or
localized light behind a subject can be used to draw attention to it.

Revealing facts

You can use light to reveal information *selectively*:

- You may *emphasize chosen features* of a subject or a scene – e.g. by
  picking out certain details with light while leaving the rest in
  shadow.
- You can deliberately *create misleading or ambiguous information* –
  e.g. by lighting two different planes to the same tone, they may
  appear continuous, even when at different distances from the
  camera.
- As you have seen, light can be angled to *exaggerate or suppress*
  texture and form.
- Light can *suddenly reveal information*: e.g. the night prowler lights
  a cigarette and we see his face in the match flame. A lightning flash
  or a passing car's headlamps can momentarily illuminate a scene,
  to create a dramatic climax.

Concealing facts

You can arrange light to *conceal information* from an audience. The imagination is stimulated when the eye has only tantalizingly restricted visual clues:

● A person is silhouetted against the sun, so cannot be identified.
● Shadows conceal identifying scars on the watcher's face.
● The room lights fail, so we cannot see who is winning the fight.
● Heavy shadow patterns camouflage a subject, preventing our discerning its real form.

Associations of light

You can use light to evoke a particular mood, a time of day, a certain environment:

● Upwards lighting creating an uncanny or horrific effect.
● The steep light suggesting a midday sun, or the low-angled light of dawn or sunset.
● The rhythmically passing lights of an unseen train flick across the face of a waiting passenger.
● The fluctuating light on the faces of an audience watching a movie.
● Tree shadows passing over the faces of people in a car.

Associations of shade

As you have seen, tonal values can have a major influence on one's reactions to a picture. Where tonal contrasts are *sharply emphasized* this tends to isolate areas and to make them more definite. High contrast gives subjects a harsher, perhaps more angular appearance. The overall effect is of crispness, hardness, definiteness, vitality and dynamism.

On the other hand, where one tone gradually grades into another, and contrast is low, areas are blended together, and there is a feeling of continuity. The resultant effect is more rounded and subtle, suggestive of beauty, softness, restfulness, vagueness, mystery and lack of vigor.

Shadows

Without the shadows that provide such valuable visual clues we would often have difficulty in interpreting the world about us. Looking at pictures of the three-dimensional scene in front of the camera, we rely on shadows in order to rationalize what we are seeing. If you arrange lighting so as to obliterate all shadows you will introduce an element of ambiguity. It will become much more difficult to judge dimension and distance accurately in the flat

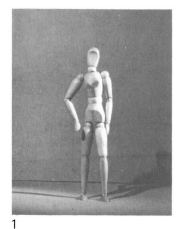

1

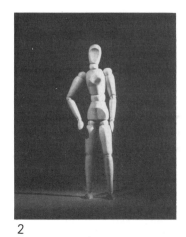

2

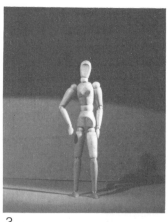

3

Background shading
Background tones not only affect a subject's prominence, but our interpretation of mood and space. While *lighter* tones create an impression of spaciousness, *darker* tones give a closed-in effect.

1. Mid tones convey neutrality; but medium-toned subjects tend to merge with their background, unless separated through lighting.
2. Darker backgrounds throw mid-toned to light-toned subjects into relief.
3. Background shading introduces a sense of restriction. If the tonal demarcation is *sharp*, the effect is coarse, firm, definite, crude.
4. Where there is *gradual* tonal gradation, the result is more subtle; delicate, rich, vague.
5. The lower the shading point on the background, the greater the feeling of restriction. But where the subject seems to 'break through' the shaded region, it can appear to be stronger and more forceful.

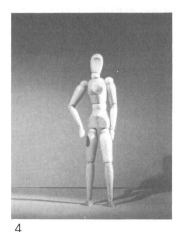

4

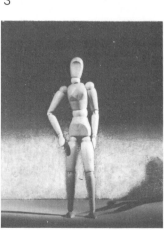

5

screened image, for we have to rely on perspective and proportions alone to figure it out.

If the camera shoots someone standing in front of a photographic backdrop, then, provided foreground and background are reasonably matched, he will appear to be within the background scene. But if his shadow or that of a foreground item falls onto the backdrop we will immediately reinterpret the picture, and realize what the camera is really looking at. Such is the power of shadows in helping us to 'read' the picture!

- *Structural shadows* falling across surfaces reveal their contour, texture and form. Shadows can reveal the outline or structure of objects that are hidden from our viewpoint (Figure 4.17).
- *Locational shadows* help us to assess time, weather, and location; e.g. we can interpret from the leafless shadow of a tree on a wall that it is a sunny day in winter.
- *Symbolic shadows* such as shadows of a crucifix or a swastika have immediate associations.

Dappled backgrounds

Irregularly-shaped shadows or light patterns can transform a plain background.

1. If these dappled background patterns are sharply focused, the effect can appear crude, incongruous, unconvincing.
2. As dappling is defocused, it becomes more effective.
3. Dappling has a greater impact when it is used selectively, and is not over-obtrusive. Where background patterns are sharper, they draw attention to themselves, and the effect becomes more dramatic.
4. The indistinct light break-up of soft dappling can produce very attractive, even beautiful effects.

1

2

3

4

1

2

3

Dappled key lighting

1. An unobstructed frontal key light provides a clear-cut picture of the subject.
2. But when light from the key is broken up with a cookie to produce slight dappling, it can produce an attractive 'atmospheric' impression. Fill light reduces the strength of the dappling. Strong dappling can be used for dramatic effect (especially in low-key mysterious situations), or to suggest unseen tree branches.
3. Heavy dappling can be used to isolate and emphasize parts of the subject, or direct attention to specific areas; e.g. using a cut-out sheet of metal foil in front of a key light.

1 2 3

Dappled key and background lighting
1. If the background alone is dappled, it forms an interesting environment for the foreground action, with the subject standing out from its surroundings.
2. Where the subject is lit with dappled light against a dappled background, it will tend to become *integrated* with its surroundings. (Under dappled lighting, a person matted into a background photograph of a forest will appear to be *within* the scene.)
3. Very heavy dappling can suggest thick foliage, or very confined surroundings.

- *Atmospheric shadows* may be used in a number of ways: *dramatic* (e.g. the floor shadow of the waiting mugger), or *comically grotesque* (e.g. the shadow of a round window cast on the floor . . . appears as a spider's web, in which a cat sits alert), or *interpretative* (e.g. shadows on a window-blind show action in silhouette).
- *Abstract shadows* are used for dramatic, decorative and environmental effects. They range from the shadows traced by grilles, screens, latticework to the broken-up dappling cast by cookies.

Atmospheric treatment

How often have you looked at a picture and thought, 'What a great atmosphere this scene has!' The shot has an immediate emotional impact. It has a happy uplifting quality . . . or a sense of foreboding; it feels inviting . . . or there is a chill mystery about the place.

Part of the total effect is due, of course, to the way the setting has been designed, its furnishings and decorations. The environment will probably have associative overtones. The way the shot is composed will influence our response. *But* take that same shot and flood the scene with soft light . . . and any vestige of 'atmosphere' disappears. On the screen we see walls . . . windows . . . furnishings . . . coldly and unemotionally displayed. The magic has gone.

So how do you go about building up such emotional illusions? 'Atmosphere' begins in the imagination of its creator and culminates in the imagination of the audience.

We have been looking at ways in which arrangements of light and shade can influence. If you closely analyze the photographs in this book you will see how they bear out these ideas. Look at the subtlety with which tonal contrast, gradation, shadow formations, contribute to the overall effect. See how details are emphasized or hidden by light direction and by light restriction. Guided by such principles, one can develop a range of different atmospheric impressions.

The way we interpret any picture must depend, to some degree, on our own understanding and experience. While one person may find a particular effect unusual another will accept it unquestioningly, for experience provides them with an immediate explanation. Even completely natural effects can appear strange if we have not met them before:

● A window shadow stretching across the ceiling from a street-lamp below.
● A room lit by the strong reflected light from sunlight on fallen snow.
● The appearance of a familiar room by candlelight.

Where visual clues appear to us to be 'normal' we accept what we see as natural. Where they are slightly unusual, unexpected, unexplained we tend to be uneasy. There is something wrong. We are apprehensive. We look at shadows and wonder what they are hiding . . .

We have said that 'atmosphere' begins with the imagination, so, as an exercise, let's use the mind's eye to build up atmospheric impressions of a scene.

■ *Restful relaxation*   The shot shows us a sitting room at night. In its comfortable welcoming atmosphere a person listens to quiet music. Our attention moves to where a large table lamp forms a pool of light within the darkened room. Its illumination is reflected in the waxed surface of the antique table on which it stands. On the floor beneath lies a richly patterned oriental rug. A streak of light reveals a shelf with a carved figure, silhouetted against faded tapestry. As you read, you are forming a mental image of this situation.

But if this imagery is to become real we need to be practical. What equipment are we going to use, and where do we position it? Beginning with the table lamp, we could fit a photoflood bulb to provide a pool of illumination. But would its light spill uncontrolled over nearby surfaces? Would it result in a prominent hotspot on the table? It may be preferable to use a regular low-wattage lamp in it (to establish that the fitting is lit) and *simulate* the pool of light with separate lighting fixtures that we can restrict to illuminate exactly the areas we select.

Heavily diffused spotlights softened to avoid hard shadows can pick out items and materials that suggest richness and quality: the embroidered silk of the chair covering, the gleaming sheen of polished wood, the rich carpeting. Light half-reveals the attractively shaped forms of the lamp base, deep carving on the table, a silver candlestick, the folds in the window drapes. Flowers in a ceramic pot are unobtrusively side-lit, detail being dimly visible without giving them over-prominence. A glimmer of backlight from a projector spotlight reveals the beauty of a sparkling crystal vase. There is just sufficient fill light for us to perceive forms within the shaded areas; just enough and no more of the rich surroundings.

■ *Romantic*   We might change the entire balance and direction of the lighting. With the surroundings lit to a lower level, features in the room are less accented. The lovers are illuminated by soft firelight glow. The table lamp no longer dominates, but appears to be the source of the subtle back light that separates them from the shadow surroundings.

■ *Tragic death*   Now shadows are dense; perhaps concealing the assailant. Little can be seen of details in the room. Light from the table lamp falls across the victim's body, but the head remains in shadow. We cannot identify him. The more we tell, the less we leave for the audience's imagination to work on. Too few visual clues will have them baffled rather than intrigued. Too many, and the sense of mystery evaporates.

■ *Neglect*   The room has been long neglected. The light is meager, and what there is, appears stark. Any soft light is from a single central space-light. Lighting is hard, crude and angular, emphasizing imperfections. Oblique hard light on the walls reveals all wrinkles and irregularities. Long, ugly, downward shadows are cast

Street scene
1. *Daylight.* A strong key light is blended with a soft base light.

2. *Night.* At night this sleazy street is
   dimly lit in localized patches of
   light from unseen gas lamps.

3. *Mysterious night.* Now the street
   is treacherous and sinister, with
   deep shadows.

Cave scene
Basically just a few molded fiber-
glass 'rocks', and a straw-strewn
floor.
1. Under overall soft light, the result
   is flat and artificial-looking.

2. Lit with a single 10 kW lamp, the
   setting appears to show an
   exterior scene.

3. Patchy lighting transforms the setting into a convincing cave *interior*, apparently lit by oil lanterns.

4. Restrict the lighting further, and we see the 'unlit' cave.

Hospital ward

1. When flooded with soft light, there is no feeling of an *interior* scene. The way in which the scenic artist has 'blown down' the walls (dark spraying), to create texture and shading, is clearly visible.

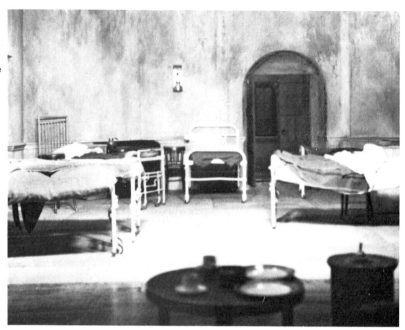

2. Localized hard lighting, relieved by low-intensity soft light suggests a day *interior*.

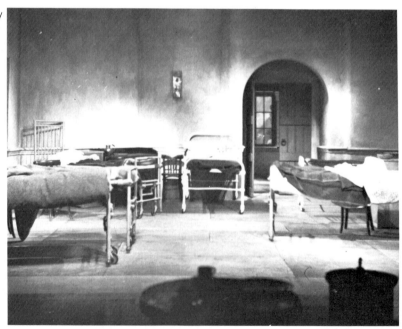

3. When hard light alone is used, we see a 'closed-in' atmosphere.

by wall fittings, pictures and ornaments. Shading is high on the walls. There is an absence of back light. Window drapes are flatly lit to conceal their modeling, or side-lit so that the folds appear crude. There are no patterns of sunlight on walls. Corners of the room are dingy.

As you realize, these are just random examples. There are many ways of tackling such situations. But they should help to start you thinking 'atmospherically'; to see how by selective enhancement and coarsening, by controlled shadows, you can develop a range of persuasive treatments.

Darkness

A *blank screen* tells us nothing. It does not even reproduce successfully. We are usually aware of a distracting background of dirt or damage on the film print, processing variations, video noise, etc. Very dim pictures are no more satisfactory, for their detailless muddy tones are barely distinguishable, and have no pictorial appeal. Only rarely will a Director want to use a blank screen, for although the audience may envision what is going on, guided by the soundtrack, darkness alone cannot sustain their interest.

When a story-line calls for 'total darkness' the problem is to discreetly introduce light of some sort that will be accepted by the audience without a moment's thought. Where the dialogue emphasizes how dark the surroundings are, any over-generous lighting will simply make the characters look stupid! Fortunately, given half a chance, the audience suspends disbelief. Only if the light is over-emphatic or strongly directional will they be puzzled about where it is supposed to be coming from.

If a scene that takes place in the dark lasts for several minutes it is advisable to find an excuse for introducing environmental illumination of some kind. This may be light from a street lamp, a flashing sign, or the moon coming into the darkened room through the window or an open door. Perhaps someone has a reason to use a flashlight or light a candle. There might be firelight. Anything to relieve unbroken darkness.

However, as you will see, carefully controlled 'darkness' can stir the imagination.

'Total' darkness

Where 'complete darkness' is unavoidable the best solution is to pick out enough visual clues with light to allow the audience to maintain a sense of environment. Let's look at an actual example from a play.

The entire action involved a man who was digging a well, which collapsed, trapping him underground. Occasionally, he struck a match to look at his watch! If he had been lit only by fragmented light against a black background the audience would have realized that he was in 'darkness' but nothing more. But by allowing odd slivers of light to catch the debris around him they could feel that he was trapped within a confined space. The dramatic difference was considerable.

As far as possible, use localized, broken-up rim light in these circumstances. Side light may reveal too much of the subject. Any frontal lighting should be very dim, with indistinct irregular dappling. The darker the surroundings are to seem, the more limited the light fragments should be. Unless there are any giveaway shadows its direction will generally be undetectable.

Use fragmentary light to give the audience little clues about distance, size, and scale. If a feature is unlit, it is not there as far as the camera is concerned. Even a small streak of light helps them to make spatial judgments. Odd random blobs of light are distracting and tell us nothing.

The value of darkness

'Darkness' provides you with both dramatic and practical opportunities, and you can use it for several purposes:

● *To obscure.* To prevent the audience from seeing the surroundings clearly. Only a bush or two are visible within the darkness, and the audience assumes the rest of the forest. In fact it does not exist in the studio.

- *To hide information.* In a shadowy doorway, we see the feet of an unknown watcher . . . but the rest is hidden in darkness.
- *To intrigue.* The camera pans over the darkened Victorian nursery. The hoof of a wooden rocking-hose looks like a groping hand in the half-light; its tail appears to be the long hair of someone crouching in the dark.
- *To mystify.* At night an escaping thief runs into a deserted warehouse . . . As a camera moves into its shadowy depths we catch brief glimpses of the surroundings and wonder what sort of place this is.
- *To enhance.* Low back light can impart a magical quality to some subjects in dark surroundings. Trees and plants are strangely transformed as they are rimmed by hidden sources. Light glistens and reflects within the foliage.
- *To threaten.* Distorted shadows, jagged streaks of light, harsh modeling, areas of deep shadow, create surroundings in which danger may lurk.

Moonlight

Moonlight is one thing to the eye, another to the camera. In reality, moonlight is simply reflected sunlight; a low-intensity hard light source of similar general color quality.*

As you saw earlier, our eyes' color response changes at lower light levels (Purkinje shift). In daylight it is most sensitive in the yellow-green region but in dim light the eye undergoes a spectral shift towards blue. Hence the apparent blueness of moonlight and the illusion that there is no color in the moonlit scene.

Blue light is widely accepted as a convention to simulate moonlight; for it imitates this subjective blueness and tends to 'kill' scenic color. (Remember the effect of colored light falling on a colored surface.) Many Lighting Directors prefer light-blue and/or pale-green color medium for this purpose. If you use an excess of blue illumination, deep-blue filters, or a blue filter over the camera lens, the results are very phoney, especially on faces.

The optimum light direction for moonlit scenes is from behind subjects, rimming and kicking their edges. Low-intensity soft filler from side or front positions relieves shadows on faces; some Lighting Directors preferring to use steep fill light and to exaggerate any modeling. Tonal contrasts are high. Large areas of the background scene are left in shadow and selectively outlined. Skies are left dark or illuminated (in dark blue) to a low level.

If exterior action is taking place near buildings at night any warm-colored light (e.g. orange-yellow, light amber) streaming from

---

*Theoretically, a very long time exposure or ultra-sensitive film would produce pictures with a color quality resembling daylight. But reciprocity failure in photographic materials upsets the color balance with very long exposure times.

Dappled interior lighting

1. When walls are evenly lit, the setting is clearly visible, but unatmospheric.

2. Slight dappling can reduce the stark, artificial appearance of newly-constructed flattage, giving it a more convincing appearance.

Heavily-dappled interior

1. Heavy dappling produces a highly dramatic effect that must be used with care, to avoid its looking crude and mannered.

2. Very heavy localized dappling can be used to build up a low-key effect.

Low-key scenes

In a low-key scene, light is restricted, and shadow areas are dense.

1. The unbalanced tones in this shot suggest that the person is sitting beside a light source (e.g. outside a window at night, near a camp-fire). Regularly used for vehicle interiors, to suggest passing lights.
2. A light-toned floor tends to destroy low-key effects; e.g. night shots, 'unlit' surroundings. (Overlit floors are virtually unavoidable when using hung lighting for low-key scenes.)
3. Darker floor tones will improve the effect, but most floor materials reflect considerable amounts of light.
4. Using lamps at a much lower vertical angle and shading off the lower part of the shot, the overall effect improves considerably, giving a feeling of enclosure.

1

2

3

4

1

2

3

Very low-key scenes

In very low-key scenes, restricted patches of light give us strategic visual clues within the darkness. (Do not rely on *very low* light intensities. It simply produces under-exposed gloom.) Too little light, and it is hard to see. Too much, and the illusion is lost.

1. Fragmentary lighting barely reveals the surroundings, creating a sense of mystery and suspense.
2. As the dappling becomes more extensive, the effect becomes increasingly expectant, and more decorative.
3. If the lighting for both subject and background is considerably fragmented, it can be very difficult to see what is going on.

windows or doors will help to emphasize the coldness of the moonlight.

White light can be used to imitate moonlight provided it is only rimming subjects and there is negligible filler. But there is the danger that from some camera angles, subjects may appear too fully lit, with poor lighting continuity in cross-shots (see Figure 6.17).

Dramatically, there are several variants on moonlight, from the *romantic* with a deep blue sky, visible moon, and widespread back light picking out tall grasses and leaves sparkling on their branches . . . to the *threatening* qualities of a black sky, predominantly unlit surroundings, and negligible filler suitable for smuggling and similar dirty deeds.

Dawn and sunset

'Dawn' and 'sunset' are easily confused pictorially unless there are other clues about the time of day, such as cock-crow or nocturnal owls.

On location

In the early morning or the late afternoon lighting conditions change quickly. The light level, the sun's angle, the color temperature of the light, the sky's appearance, the amount of fill from sky light are continually altering. So when on location all wider viewpoints (wide-angle shots/long shots) need to be shot quickly and continuously to avoid picture-matching complications during subsequent editing. Closer shots are easily shot later, filtered and filled to suit the wider establishing shots.

You can simulate 'sunset' or 'dawn' sequences during broad daylight. Typical subterfuges include sky-filters, graduated lens filters (colored and neutral density), gold or blue-silver reflector boards to tint reflected light, and, where necessary, large scrim diffusers ('butterflies') to reduce the lighting contrast on the subjects.

In the studio

When a dawn or dusk effect is required in a studio drama it is usually seen incidentally, beyond the window of a room, rather than as a full-screen panoramic effect. One approach to *dawn* could be an unlit skycloth or light-blue cyclorama, with a pale yellow groundrow. As a yellow glow grows on the horizon . . . pale blue light on the main sky increases . . . gradually overwhelming it . . . until subsequently the yellow fades, and it is day.

In equally broad terms one could suggest *sunset* by a mid-blue sky fading to dark blue . . . while a red-orange glow in its lower half gradually fades . . . to leave just a border of light outlining the horizon. If the horizon is hidden from view then, for some reason, the effect appears more convincing when the lower lighting is partly to one side.

Although such approaches are, admittedly, pretty stereotyped, you will find that any attempt to strain after the subtleties of greater realism will normally be lost on camera.

Sunlight

Is reality enough?

There is a paradox we have to face whenever we imitate natural lighting and that is a continual trade-off between accuracy and attractiveness. Steep-angled tropical sunlight, for example, with its hard-cut unrelieved masses of shadow, creates coarse, harsh facial modeling. But if you add a lot of fill light (or alter the key angle) to improve portraiture the result looks less like tropical sunlight!

Single shadows

In the real world the single sun casts a single shadow. Where we find the resulting illumination unattractive, we redirect light with reflectors or add compensatory lamps. These improve the effect but cast shadows which are strictly unnatural. Exteriors scenes (real or studio) in which a person stands near a background accompanied by half a dozen shadows lose credibility, and even an unobservant audience is likely to notice that the effect is abnormal.

Any setting in the studio that is nominally lit by 'daylight' should, strictly speaking, have only a single key light. Complete realism requires a single shadow. But even where this is physically possible, one modeling light direction over an entire setting can restrict pictorial opportunities considerably.

The trick is, wherever possible, to have one primary shadow in shot at a time. The fact that the next shot has a main shadow in another direction will probably go unnoticed, especially if it falls onto a dark or broken-up surface.

Visual continuity

In everyday life our own viewpoint changes relatively slowly, and we subconsciously adjust as we move from one position to another. We find nothing unusual about a subject being strongly lit from one direction and silhouetted when seen from a reverse angle. But if a Director *cuts* between these viewpoints we find the instantaneous changes in the subject's appearance distracting, disorienting, annoying. It takes a moment to realize what has happened.

Even when lighting is strongly directional we can still adapt to shot changes if

● Shots are intercut over a very limited angle, or

- The camera itself moves (tracks/dollies) between diverse viewpoints.

Otherwise we have to avoid such intercutting, or make the lighting
more compatible.

Multi-directional lighting

Unless you have a highly dramatic situation, most Directors are
unlikely to be happy about a close shot of a face that is in shadow or
half-lit. So where people turn away from the light (in this case the
sun) you really have little option but to add a secondary key light or
filler to make this new head position acceptable. This supplementary
or 'cheated' lighting may not be justified environmentally, but
artistically it is essential; although it should not, of course, be
overdone.

Interior scenes are perhaps less critical. If, for instance, a person
stands at a window lit by the sun, then moves to another part of the
set and is strongly lit in their new position, an audience will accept
the incongruity. Presumably, they assume the existence of another
window. Only rarely do they lose a sense of orientation due to
changing sun directions.

Changing light in interiors

Where an interior is lit by natural light its appearance changes
gradually and imperceptibly throughout the day. Although very few
drama productions contain scenes representing continuous action
over a span of many hours (e.g. as in the film *Rope*), many include
settings that are seen at different points in time. Beyond the broad
extremes of 'day' and 'night' are many subtle variations; and these
distinctions can make a major contribution to atmospheric effect.

Let's imagine ourselves within a room, watching how approaching
day modifies the appearance of our surroundings.

It is night. Outside the window, trees and distant skyline are
vaguely silhouetted against a velvet sky. At first, as the sun rises,
little light penetrates beyond the window. The walls are darkest
there. The outlines of objects within the room are barely
distinguishable. General masses merge into soft-edged shadows.

Dawn begins to brighten the horizon. Outside, trees become firm
silhouettes. Gradually, light seeps into the room; the ceiling lightens.
We begin to detect the form of lighter-toned objects in the room.

As the sun rises, walls opposite the window brighten. Darker
objects in the room now become clearer. We can discern detail in
shadows.

Daylight becomes established. By now, the sun has a steeper
elevation. Light scatters and reflects within the room, and it becomes
brighter overall. The window-wall is now well lit. Objects within the

Atmospheric changes (1)

1. Under overall soft light, the scenery has a very fundamental appearance. We can see the structure and its set dressings clearly, but there is no associated 'atmosphere'.

2. *'Early day'.* The main light enters from imagined windows on the right, and we see tantalizing glimpses of its furnishings, which encurage our imagination.

Atmospheric changes (2)

1. *'Early afternoon'*. Here the room is
   fully lit, with less sense of
   restriction. Details are clearly
   visible. The skylight implies the
   sky beyond.

2. *'Late afternoon'*. Sunlight picks
   out details by the window, but
   shadows give the room a
   closed-in, claustrophobic feel.

Atmospheric changes (3)

1. *'Evening'*. The low evening light spreads across the room. (The potential window shadow is broken up by the shape of the room.)

2. *'Night'*. Moonlight. Pools of light, apparently from the hung practical lamps, reveal the character of the room. Deep shadows hide information from us, and add to our interest. Furniture merges into uncertain shapes. It is a cozy comfortable home.

Atmospheric changes (4)

1. *'Harsh, sordid night'*. The room lighting is more extensive, but its coarse high contrast creates a discordant, crude effect.

2. *'Deserted unlit room'*. The room appears to be 'unlit', deserted, mysterious.

room cast defined shadows. Light patches form on walls. Sunlight reflects, and we see specular reflections from polished furniture and glass.

As the day develops, the sun gradually moves round, and its vertical angle grows steeper, so that it no longer penetrates the room. Now the walls furthest from the window have become quite shaded. Light levels fall as we move away from the window.

Effects vary with the weather, the orientation of the room, the number of windows, predominant tones, and so on. You would find it an interesting exercise to make detailed notes of the ways in which light changes in your home throughout the day.

Lighting changes

Here we come to one of the most fascinating aspects of production lighting – the ease with which a scene's appearance can be transformed by lighting alone. You can introduce lighting changes at various stages during a show:

- Off camera – altering the pattern of lamps or the lighting balance while cameras are shooting another setting.
- On camera – making visible changes during the scene; on cue, or as an effect, or as a visual development.

There are right and wrong moments to make a change. In a musical number, transformations should, wherever possible, relate to the music, and to the shots. If you alter the cyc lighting from green to red during a cutaway close-up of a singer your audience is likely to be taken aback when they discover the change in the next long shot!

On the other hand, there are times during continuous action when close shots will give you the opportunity to unobtrusively readjust the lighting balance in another part of the set. So, for instance, while taking a close-up by the door you can change the lighting over by the table (which is out of shot) to suit the next action there.

Atmospheric changes

Let us look in some detail at ways light can be used dynamically to enhance imaginative production. The examples here show just a few of their potentials and practical aspects we need to consider.

Rapid change: total

This is the sudden lighting change that transforms the entire scene. It can be used as an act of *revelation*.

- It may reveal nothing more than the fact that it is morning, as someone throws open the bedroom window drapes. Sunlight streams through window, walls lighten, a window pattern stretches across the bed.

● More dramatically, as clouds part and moonlight shows the moorland traveler his surroundings, it reveals notices informing him that he is entering a minefield!

■ *Concealment*   You can use a rapid total change for concealment; to prevent further knowledge of the action. E.g. The gunfighter shoots out the lights, plunging the place into darkness.

■ *Mood change*   Here the mood of a scene alters in a moment. E.g. The soft dream-like atmosphere of a candle-lit scene is shattered as harsh cruel room top-lighting is switched on.

You can even achieve a lighting transformation without making a lighting change at all! For instance, a person who has been gently lit by diffused light from a table lamp . . . steps forward and is illuminated with stark menacing underlighting as he leans over it.

■ *Room lights*   The most common rapid change occurs when someone enters a darkened room and switches on the lights. Fundamentally, this is a straightforward switching process between two conditions. You can have the change remotely actuated by the actor operating a normal wall-switch beside the room door. (The switch operates circuits connected to the lighting board.) This method works perfectly well. However, it has proved unreliable for *live* TV transmissions. On one embarrassing occasion an actor forgot to operate the switch, entered the room and sat down. The scene ground on in 'darkness' while lamps were surreptitiously faded up to

Shadows create the environment

Here shadows convey dramatically, and with powerful economy, the activities in a blacksmith's forge. Yet it would need little more than a wooden wall and a single lamp to produce this effect!

Implied lighting
The lantern held by the boy
illuminates his face. But it lights little
of the surroundings. Instead, extra
skillfully controlled lamps are used to
build up the total effect.

Decorative shadows
This is a clever blend of naturalistic
and decorative effect. The shadows
give the picture a compelling appeal.
Try analyzing how 'real' such lighting
is. Join the shadows to the objects
causing them, in order to trace how
the scene is lit. The steep lighting on
the doorway, the banister shadows
from a lamp at floor level, and the
well-lit painting may not be strictly
rational, but the effect is superb.

Passing time

1. In this sensitively lit scene, there is a great deal of attention to detail. Controlled depth of field, careful tonal balance, effectively convey the mood and the time of day. Check out the window patterns: are they projected or cast? See how successfully the darkened walls emphasize them.

2. The same room 'hours later', in which the lighting captures not only the passing of time, but the worried and depressed mood of the characters. We can see no light sources in the shot. How is the room lit? The upward shadows indicate that floor lamps have been used.

correct the error! On-air, it is safer to have the actor put a hand over a dummy wall switch until the lights are switched on at the control panel. This avoids the incongruity of having the switch click or the hand move away before the light change has been made.

Methods of transformation

You can set up an atmospheric change of this kind in two different ways.

- *Added groups*. Here one set of lamps provides the low-key scene and, on cue, another group is *added*. This has the advantages that it switches a smaller power load and maintains good lighting continuity.
- *Switched groups*. With this method two *different* sets of lamps are used, and you switch between them on cue. At times this can produce rather odd effects; for example, when one sees a patch of moonlight suddenly appear on the wall as the bedside light is turned out. In addition, the transition during inter-switching can look somewhat ragged, for whereas low-power lamps reach maximum output instantly, high-power lamps take a moment to warm up.

■ *Flame cues*   Where a light cue has to coincide with someone lighting or extinguishing an oil lamp or candle there are useful tricks to bear in mind. As an oil lamp is lit the light tends to increase in several stages: the match is lit . . . it is applied to the wick . . . the wick begins to burn . . . the glass funnel is fitted. You can treat the operation as a gradual light increase, but it appears that much more convincing if done in steps. You can treat candle lighting in two stages: lighting the match . . . the wick begins to burn.

It is exceedingly difficult to synchronize a lighting change to someone who blows out a candle! So, wherever possible, have them extinguish it with a wet-finger pinch, or by snuffing.

■ *Car interior*   A pattern of lamps of moving shadow projectors positioned around a static car body can be rapidly switched or interfaded to suggest that a car is moving through lit streets or shady countryside. The old idea of moving a small branch past a lamp works well *if* it is done convincingly.

Rapid change: partial

In this situation localized lighting is added or removed from a lit scene – perhaps as a character moves around the room, switching on table lamps, or putting out candles. To be successful you have not only to synchronize with the actions but also to ensure that the effects of the lighting changes are reasonably proportional at each step.

■ *Realism*   Remember, whenever we see someone switch on any light source, you need to create the impression that –

Fig. 8.4 Lighting transformations

1. Lighting transformations enable you to redirect attention.

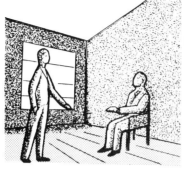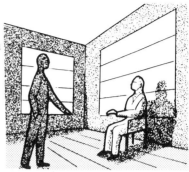

2. Transformations also allow you to modify tonal balance and change the mood.

3. Localized light patches can be used in various combinations and achieve a series of transformations. Similarly, you can alter relative prominence, influence spatial relationships, perspective, solidity, modeling, texture, etc.

- Light of convincing intensity and direction is now falling onto the person from the new source.
- An appropriate pattern or pool of light is now falling onto the surroundings.

■ *Moving light sources*   This is the situation we meet when someone walks around with a light source. While they remain more or less constantly illuminated, parts of their surroundings grow brighter and darker as they pass (see *Traveling key*).

Gradual change: total

In the theater, where performance is continuous and the audience can watch the overall visual development, very slow lighting changes during the action can be particularly effective. However, they are normally only successful in film and television where there are sufficient long shots to show the effect; e.g. during an extended dance routine or a song.

When scenes are brief, shot out of order, and mainly in close-up, continuous gradual changes are really impractical. Instead, it is preferable to alter lighting in distinct stages between shots.

■ *Failing light*   When you want to simulate the changing light of dusk or an approaching storm it is unwise to do this simply by general dimming. Reduced light levels will produce inferior pictures (under-exposure). Stopping the lens down would achieve the same result without lowering the color temperature.

If dialogue in a drama draws attention to the failing light, instead of lowering key light levels try reducing the back light and filler, and light patterns on walls. If there is already some source of light in the setting, such as a hanging lantern or a fire, you can improve the overall illusion further by gradually making it stronger as you fade the daylight.

■ *Mood development*   Slow lighting changes can be used to alter the entire mood of a sequence. For example:

At the start of a scene the happy reunion of the lovers is reflected in the high-key lighting treatment (*low-angle keys; generous low-angle filler*), the sunshine (*wall patterns*), the glamorous *back light*.
But as they talk there are accusations, and rising tension. The lighting coarsens (*steep, reduced filler*). The room loses its charm (*sunlight fades*). Faces become more crudely drawn and less attractive (*steeply angled keys; no eyelights now from the camera light*).

■ *Center of interest*   By gradually increasing the light in one area of the setting, and perhaps slightly reducing it in another, it is possible to shift attention selectively. For example;

The young girl at her first dance sits in a row of chairs with companions awaiting a partner. One by one the others go, leaving her alone. The wall is no longer evenly lit . . . and she sits isolated in a pool of light.

■ *Increasing effect* Here an effect gradually becomes more pronounced as the scene continues. For example:

It is night; beyond the window, light from a small fire grows stronger and stronger.

The firelight here could be anything from joyous campfire festivities to the conflagration of 'The barn's ablaze!' so you need to keep a sense of proportion or the whole effect becomes exaggerated and theatrical.

Gradual change: partial

This is another localized effect that needs careful handling, and is best demonstrated by an example. Imagine that the scene is a darkened room. The door opens slowly, and light streams through onto a seated figure.

You can treat this situation in several ways. The most obvious is to simply let light shine through the doorway onto the chair. In a continuous scene this may not be possible; e.g. the lamp through the door would be seen by the camera within the room.

Instead, you might cheat the effect with a second spotlight; one through the door, one on the chair. The latter is faded up slowly as the door opens. Will its rays illuminate the room, or will it form a confined rectangular pattern of light? The impact is different; the choice is yours.

Decorative lighting

Decorative lighting has become such a regular part of pop shows, band shows, dance, singers, etc. that the dilemma is to produce anything fresh or original. Most permutations have been over-used through the years. In television, video effects are increasingly introduced to create visual variety – even to the point of subjugating the actual performance!

Decorative lighting may be applied to the background, the floor, and to scenic pieces. Occasionally, you can project patterns over the performers. The very brief list that follows is just a reminder of the range and diversity of treatments in regular use:

● *Scenic lights:*
  Rear-lit translucent panels (plain, shaded or patterns).
  Front-illuminated panels.
  Attached border lighting, strip lights, miniature spots, etc.

Fig. 8.5 Decorative light effects
In this selection of decorative lighting we have:
1. Cyclorama lighting (barndoored spotlights on the floor shooting upwards).
2. Hanging festoons of decorative lamps (shaped bulbs).
3. A tubular metal skeletal frame with lamp sockets.
4. Lamps fitted to the edges of scenery (normally low-wattage bulbs in batten-fitted holders).
5. Reflective panels (embossed or stencil-painted mirror-plastic, sequins, etc.).
6. Internally illuminated treads and platform (e.g. plate glass panels with plastic facing).
7. Hidden strip lighting that edge-illuminates panels.
8. Slash curtains (reflective PVC strips) or tinsel.

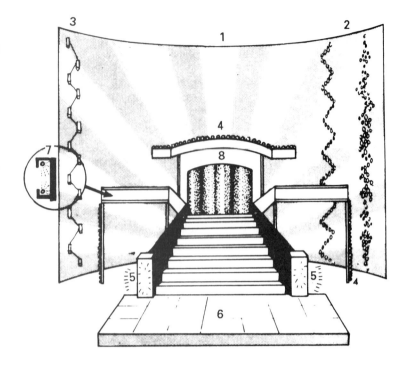

Fig. 8.6 Dappled backgrounds
Random dapples on a background create interesting abstract effects suitable for a wide range of productions. Produced by cookies (dapple-sheets), cut-out color-medium, or projected patterns. Most effective when very defocused.

- *Cycloramas:*
  General washes of color (single or intermixed).
  Shaded color (groundrow).
  Multi-tones, colored dappling, blobs.
  Light-columns, arcs, streaks, reflections.
- *Projected patterns:*
  Realistic silhouettes (tree branches, city skylines).
  Decorative designs such as fleurs-de-lis, sun-rays, flowers, checker-boards, toys, etc.
  Abstract motifs.
  Animated patterns (inter-switched gobos).
- *Cast shadows:*
  From cut-out stencils; skeletal screens; suspended rope, nets, trellis, etc.
- *Light patterns:*
  White or multi-colored flashing lamps: in strings, festoons, clusters or strung out along scenery.
- *Light movement:*
  Chaser lights.
  Moving patterns – rotating, swirling, jittering, glittering, twinkling, fluttering, exploding . . .

Although most decorative lighting effects are controlled manually, others are actuated automatically by switching/flashing circuits on the control board, or by sound-controlled relays.

1

2

3

Background light patterns
1. A background light pattern may be used to concentrate the attention on a particular subject, or part of a subject. (The key light and the pattern are from different lamps.)
2. Where patterns are projected onto decorated backgrounds, they are much less effective.
3. Plain backgrounds display shadow effects most successfully.

1

2

3

Pattern sharpness
1. If you project a pattern onto a surface from an angle, it becomes distorted, and parts become defocused. Notice how your attention goes to the sharper areas.
2. Sharp contrasty patterns create a dynamic, vigorous impact.
3. When the same pattern is defocused, it produces a vague, nebulous, or delicate effect.

1

2

3

Strong background patterns
1. Here the projected pattern dominates the scene. Its shapes are arresting, strong, powerful.
2. This pattern cast on the floor and wall provide a decorate background for action.
3. Here the same pattern becomes a menacing, forceful effect, when coupled with the shadow of the figure.

Animated lighting

Moving light can give a scene life and a sense of realism; e.g. water ripple reflections falling onto someone leaning over a ship's rail. It can also be used dramatically, to increase pace and tension, as changing light flashes, fluctuates, sweeps across the action area. Animated lighting can be introduced in many ways:

- As a *decorative effect* – the fleeting spots of light from a mirror ball; the rotating colors of a flicker disc.
- As an *environmental effect* – flickering flames from the hearth; rain running down a skylight throwing streaking shadows into the room below; a rhythmically flashing street sign; the passing window lights from an unseen train.
- Through *action* – the ceiling-light set swinging during a fight.

9 Light sources

You can use light creatively yet know little about the nature of the light sources themselves. However, you will be able to use them much more effectively if you appreciate their potentials and limitations.

A variety of luminants

A number of different *light sources* or *luminants* are used in film and television lighting:

- Regular tungsten lamps;
- Overrun tungsten lamps;
- Internal reflector lamps;
- Tungsten-halogen lamps;
- Gas discharge/metal-halide lamps;
- Fluorescent lamps.

Each has its particular merits, but it is important to fit a lamp designed for the particular fixture you are using. Not only are there a couple of dozen or more different forms of lamp base and connections but the arrangement and support of lamp filaments varies with the purpose of the lamp and the optical design of the fixture. Some lamps may be burned at any angle, while others are limited to e.g. 30–45° maximum, cap-up, or cap-down positions.

The American National Standards Institute (ANSI) has formulated a comprehensive three-letter code that specifies the form of each type of lamp, including its base arrangements, bulb size, and coverage. In addition, various manufacturers have their own identification systems.

Regular tungsten lamps

Various terms are used for this familiar form of light source, including 'filament lamp', 'incandescent' lamp (which *all* TV luminants are!) or 'tungsten' lamp (tungsten is used in quartz lights too!). Traditional tungsten lamps are still used in all forms of lighting but are increasingly being superseded by tungsten-halogen/quartz lights. The main advantages of tungsten lamps are that they are relatively cheap, have a reasonably long life, exist in a wide range of power ratings (from a few watts to 20 kW), are generally reliable, and can be mounted in many types of fittings.

The tungsten filament is generally constructed in a coiled-coil form and supported within a sealed gas-filled glass bulb (argon or nitrogen gas). When a voltage is applied the tungsten filament heats and becomes incandescent. As the current through the filament is increased, its light output and color temperature increase. If allowed to exceed the working rating for which it is designed, the filament melts or 'blows' (around 3400°C) and the circuit is broken. This is a hazard to look out for when using freshly charged batteries as a power source, or a poorly regulated supply. Exceeding the lamp's rated voltage for just a short time may noticeably reduce the lamp life.*

The lamp's glass envelope may be blown or molded to various shapes, and is usually clear; although opal and frosted internal coatings or surface treatment are used in some lower power tungsten lamps.

Disadvantages

Regular tungsten lamps have several disadvantages:

- A considerable amount of the electrical energy is wasted as heat.
- The color temperature of many tungsten lamps is relatively low, and falls in use. The light is deficient at the blue end of the spectrum.
- The tungsten filament is proportionally large, so does not allow the optical efficiency that is possible with point light sources. (Greater light loss, and less clear-cut shadow formations.)
- During use, the filament metal slowly evaporates, and despite the argon or nitrogen gas filling, condenses on cooler parts of the bulb as a progressively obscuring black deposit. Eventually, the gradually thinning filament becomes increasingly brittle and fails without warning.

When shooting with color film balanced at 3200 K, fixtures using regular tungsten lamps are best fitted with CT (color temperature) types. Their life is about 100 hours, and like all regular tungsten lamps, color quality deteriorates in use. When it has fallen to around 3050–3000 K, use them to light less critical areas (e.g. backgrounds).

Care with tungsten lamps

To overcome the problem of progressive bulb blackening the larger standard 5 kW (five kilowatt) and 10 kW lamps actually contain loose abrasive tungsten granules. Periodically, you should remove the lamp, invert it, and gently swirl it around to scour blackening from the bulb wall. (Tilt the lamp carefully, to prevent granules falling on the filament.)

---

*Lighting equipment is often fitted with a *fuse* which melts under excess current (e.g. due to a short-circuited lamp), to protect the system from overload.

If a lamp is set at too steep an angle for its design it is liable to premature failure. Apart from overheating, and filament sag causing optical misalignment, the most common problem is premature blow-out. Here the bulb becomes distorted, forming a hotspot blister which bursts, ruining the lamp. Excessive heat may also cause corrosion and welding of the lamp connections.

Lamp life is markedly reduced by vibration or shock; especially if the lamp is alight or still hot from use. So it is advisable to avoid rough operation, dropping, or shaking. If possible, when turning on lamps from cold it is preferable to fade them up slowly rather than apply immediate full voltage. Particularly where filaments have sagged during use, the mechanical shock at switch-on can destroy or weaken them.

If for any reason the bulb's airtight seal is fractured (e.g. due to careless handling or water splashing) air will enter the lamp, which will discolor and fail; perhaps with an explosion. The failure can arise during use, or at switch-on.

Lamp operation

As you can see in Figure 9.1, when a tungsten filament lamp's voltage supply is gradually increased from 20% below normal to 20% above, several things happen:

- Its light output increases.
- It takes more power (higher current).

Fig. 9.1   Tungsten filament lamp performance

1. As the supply voltage is varied from the designed value, various factors change:
   Its light output (*L*).
   Its life in hours (*H*).
   The lamp's efficiency.
   Its consumption in watts (*W*).
   The current flow (*I*).
   The color temperature (see Chapter 11).

2. A comparison between the light output and life of a typical 5 kW tungsten lamp and a 5 kW tungsten-halogen lamp.

3. The color temperature of a tungsten lamp rated at 3200 K (240 V) varies as the supply voltage alters.

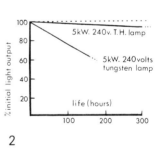

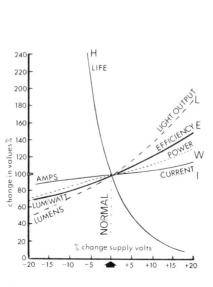

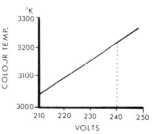

1

2

3

● Its efficiency improves.
● Its color temperature rises (the light becomes bluer).
● Its life expectation falls dramatically.

From this graph you can see that by slightly reducing the voltage supplied to a lamp you can increase its life expectancy. In a large TV studio, where equipment is switched on for long periods, *underrunning* lamps by dimming them slightly can have real advantages. The reduced light levels and lower color temperature (e.g. 2950 K) can be compensated by adjusting the camera channels.

Conversely, when the voltage supplied to a lamp is higher than normal (e.g. 10–15%) its light output and color temperature increase, but its life is correspondingly reduced.

Overrun lamps

The filament of an overrun lamp is designed to pass a higher current than normal in order to achieve a much greater light output at a high color temperature; although at the price of a shorter lamp life. A type of overrun tungsten-filament lamp that has been used for many years in photography is the *Photoflood*. In its smaller form it is rated at about 275 W, yet provides the light output of a normally run 500–750 W tungsten lamp – for about 2–3 hours of intermittent use.

Overrun tungsten and quartz (tungsten halogen) lamps are used in situations where the benefits of high light output and a high color temperature (3400–3500 K) outweigh the disadvantage of a shortened life. Blue glass versions rated at 4800 K are available for use with daylight.

Overrun lamps are particularly useful on location when you need a high light level from a compact lightweight fitting and the available power is limited. A *Colortran* unit, for example, which uses only 3 kW of power, can provide the light output of a regular 10 kW tungsten rig. Overrun lamps in hand-held *sun guns* offer intense illumination although run from belt-batteries. However, always remember to have spare lamps standing by! Color quality and the light output will fall off during use.

Although the life of an overrun lamp is somewhat uncertain it can be prolonged by *underrunning* it at a lower voltage while making preliminary adjustments and setting them up; and only switching to full power for the actual take. Using a dimmer, multi-tap transformer, or series-parallel switching also overcomes the shock of the high-current surge that occurs when switching on.

You can fit overrun photographic lamps (*Photoflood, Nitraphot*) into simple reflectors, hidden behind furniture or scenery to provide broad illumination for the background. But their regular application in film and TV lighting is in *practical* light-fixtures such as wall brackets, table lamps, pole lamps (standard lamps), where the normal light bulbs would appear too dim under bright studio lighting.

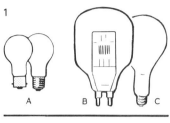

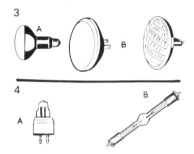

Fig. 9.2 Typical light sources
1. *Tungsten lamps*
   Relatively cheap, but the light
   output and color quality
   deteriorate badly during use.
   (A) Low-power regular tungsten
        bulbs and overrun tungsten
        lamps.
   Typical studio tungsten lamps
   (B) Fresnel spotlight bulbs.
   (C) Scoop.
2. *Tungsten-halogen lamps/quartz
   lamps.*
   Compact adaptable light sources
   of constant output and color
   temperature.
   (A) Fresnel spotlight.
   (B) Lensless spotlight.
   (C) Cyc light.
   (D) Scoop.
   (E) Effects projectors and
        ellipsoidal spots.
3. *Internal reflector lamps.*
   The bulb's integral silvered
   reflector directs and focuses the
   light.
   (A) 'R' type;
   (B) 'PAR' type.
4. *Metal-halide lamps.*
   High-efficiency, high-output
   sources requiring ignitor/ballast
   units.
   (A) CSI type.
   (B) HMI type.

Internal reflector lamps

Most lamps require a fixture of some kind to focus and direct the light. However, in the *R* (*reflector*) and *PAR* (*parabolic aluminized reflector*) types of lamp the bulb itself incorporates an internal reflector, formed from a silver coating on the inner surface of its parabolic-shaped envelope. This reflects and focuses the emergent light, so the lamp needs no fixture or housing, except to provide protection or support any add-on fixtures. The lamps are available in regular and overrun forms.

The internal reflector lamp is a remarkably cheap, adaptable light source; especially useful in confined places:

● It may be hand-held.
● When fitted into a clamp or a clip-light it can be attached to furniture, a door, or to room fittings.
● Held in a wall-plate, it can be attached to any convenient surface with gaffer tape, or hung from a pipe.
● It can be mounted on a lightweight telescopic stand.
● It can be fixed to the end of a long pole to light inaccessible places, or provide high-angle lighting.
● Internal reflector lamps can be grouped to form a multi-lamp floodlight bank.

Incandescent 'R' types are available in hard-focused and diffused forms, and use clip-on accessories (spill-rings, barndoors, flags) to control or restrict the light. *PAR lamps* have ribbed molded-lens front glass (spot, medium, wide beam-spread) or clear-front glass (flood) and in daylight versions incorporate blue dichroic filters (internal or external) to convert the color temperature. Clip-on adaptor lenses (beam-spreaders) or filters too can be used.

Wide-angle versions provide a very broad coverage, enabling much of the scene to be illuminated from a single position. The intense concentrated beam of the narrow-angle forms is particularly useful when the lamp has to be located some distance from the subject.

Tungsten-halogen (TH) lamps/quartz lights/quartz-iodine lamps

A regular tungsten lamp has the disadvantages that its light output and color temperature deteriorate considerably with use, as its filament evaporates and blackens its bulb. The tungsten-halogen lamp largely overcomes this problem.

The halogen family of elements includes iodine, bromine, chlorine and fluorine, and has the property of combining with tungsten in a reversible temperature-controlled reaction. By adding a halogen vapour, such as iodine or bromine, a regenerative cycling process is

Table 9.1   Typical light sources

Regular tungsten lamps	For decorative fittings (wall brackets, table lamps), strip-lights, localized illumination. Typically 60–150 W. 2600–2900 K. Life: 1000 hours.
Overrun tungsten lamps (Photoflood, photobulb)	Used to boost existing decorative fittings' output. Fitted in lightweight portable floodlights. Typically 275/500/1000 W (equivalent of 800/1600/3000 W with life of 2/6/10 hours. 3400 K.
Studio tungsten lamps	(a) For fresnel spotlights. 100–10 000 W. 3200 K. 25–100 hours. (b) For scoops. 1000–2000 W. 3200 K. 500–1000 hours.
Tungsten-halogen studio lamps (quartz lights)	(a) For fresnel spotlights. 1/2/5/10 kW. 3200 K. 150–750 hours. With 'hard glass' of 'quartz' envelopes. (Some dual filaments.) (b) For reflector (lensless) spotlights. Typical 250 W/30 V. 650 W/120 or 220 V. 3200/3400 K. 15–75 hours. (c) For cyc lights, broads, soft-lights (scoops). Linear filaments 300–1500 W. Clear and frosted. 2950–3350 K. 100–2000 hours. (d) For scoops. 200 W. 3200 K. 500 hours. (e) for effects projectors (profile spots), ellipsoidal spots, follow spots. 1000–3000 W. 3200 K. 200–500 hours typical.
Internal reflector lamps (R lamps)	For decorative fittings (e.g. 60 W), lightweight clip-lights, multi-lamp floodlight banks. 100/300/500 W. 3200 K, 100 hours *and* overrun versions 275/375/500 W. 3400 K. 3/4/6 hours. Available in focused (hard) and diffused forms. Not for external use without protective hood.
PAR lamps (Parabolic aluminized reflector)	Used for individual lightweight clip fittings, and multi-lamp floodlight banks. 75/150/200/300/500/650/1000 W versions with molded heat-resistant optical or clear fronts (focused or flood). Versions with elliptical beam shape (e.g. 6 × 12/12 + 28/24 × 48/30 × 40 degrees). Daylight 5200 K (int. or ext. dichroic filter); studio 3200 K. Life 30/100/200/400 hours. Suitable for external use.
Metal halides	(a) CSI (compact source iodide) CID (compact iodide daylight). High-efficiency source for fresnel spotlights, effects projectors, follow spots etc. 1/2½ kW. 5500 K. 500 hours. Available in bare bulb and sealed-beam forms. (b) HMI (high-pressure mercury metal iodide). Primarily used in fresnel spotlights, follow spots. *All* metal-halide sources require ballast units. 200/575/1200/2500/4000 W. 5500 K. 300–750 hours. 6000/8000/12 000 W. 3200 K.
Carbon arcs	Mainly used as high-intensity sources for large-area or exterior lighting. Available in 225 A (Brute) down to smaller fresnel and follow-spot versions (65 A) 6000/3350 K. ½–1 hour burning time.

introduced, so that the evaporated tungsten now becomes redeposited onto the filament, avoiding both blackening and undue thinning of the filament. As the redepositing process is uneven (being greater on the cooler parts), the brittle filament does eventually break – usually due to mechanical shock while operating, or heat-up expansion at switch-on.

Evaporating tungsten leaving the filament reacts with the halogen vapour to form a tungsten-halide, at between 250° and 800°C. This is streamed back by convection currents to the hot filament, where at around 1250°C it is broken down again into tungsten and halogen, the tungsten being deposited on the filament, while the halogen is released to recycle. Although the tungsten-halogen lamp's output is identical with that *initially* from a regular tungsten bulb of the same wattage, it maintains its performance for very much longer. Always avoid handling *quartz* bulbs (use gloves, cloth, paper) for they are blackened or weakened by skin acids. Clean them with alcohol if accidentally touched. 'Hard-glass' envelope versions are cheaper, but larger and shorter lived.

Burning at a much higher color temperature than regular tungsten lamps, quartz lights provide a more compact high-efficiency source, enabling lamp housings and their optics to use the light more effectively.

Twin-filament lamps

Tungsten-halogen lamps for fresnel spotlights are available with independent dual filaments (2.5 + 2.5 kW or 2.5 + 1.25 kW). These can be switched on separately or together. This effectively provides a high power/low power light source in the same lamp housing (luminaire). So if the light intensity with two filaments is greater than you need (this would normally necessitate dimming, thereby lowering the color temperature), you can switch to either single filament instead, to obtain a lower intensity at 'high Kelvins'. Alternatively, you can light with single filaments, and use them paired where you need a particular boost in output.

Where a large studio is lit with a 'saturation rig' of identical lamps this facility helps to maintain a more even high color temperature than might otherwise be practicable.

Carbon arcs

Carbon arcs are low-voltage, high-current sources developed in motion-picture studios to provide high-intensity light of excellent color quality. The arc being a very concentrated point source, produces very sharp, crisp illumination with enhanced modeling, textural and shadow formation. With the advent of other powerful, more compact, and less demanding light sources, the use of carbon arcs has lapsed in both film and television studios.

Although the shortcomings of arcs make them less convenient for many studio applications, they remain invaluable where extremely high light levels (as from 24-in lens 225 A 'Brutes' for example) are needed to cope with large acting areas. They are also valuable to balance natural sunlight intensities in the open air, to simulate sunshine or for night shooting, where other sources are inadequate.

The carbon arc is necessarily heavy and bulky, particularly with its associated resistance unit (*ballast unit; 'grid'*) and large cables. As the unit needs the continuous attention of an electrician to check accurate burning and renew the carbon rods periodically it cannot be hung in isolated positions.

The unit typically requires a 115 V DC supply; so you may need a separate motor generator unit where this supply is not available. The resistance unit in the supply line reduces this to 73 V DC across the carbons. The two carbon rods ('trims') which are angled at 45° are touched briefly, and part to form a flaming arc.

The color temperature of this arc will depend on the material (cerium) packed within the central core of the positive carbon electrode. White flame (6200 K) and yellow flame (4000 K) carbons are available.

In use, the rods are continuously burned away; an automatic motor-driven feed mechanism rotating the quicker-burning positive pole to sustain a symmetrical concave crater. The distance between the positive rod and the shorter bullet-nosed negative rod is controlled to maintain a constant gap.

The unit requires skilled individual adjustment and frequent checks to preserve the flame shape and intensity. The carbon arc burns for a period that depends on the current and the length of the electrodes. Two positive electrodes and one negative electrode typically last for about an hour. Because the unit gets extremely hot, one must turn off power and allow cool-down time before re-trimming with new carbons.

To reduce the high ultra-violet content of the light, and correct its color to match the rest of the lighting, color filters are fitted to the arc (see Chapter 12).

Metal halide (MH) lighting/gas discharge lamps

Metal halide lamps today are compact, very efficient units with high light output and relatively low heat dissipation. The highly efficient HMI light source is particularly convenient for use on location to fill shadows in exteriors and to light within large daylit interiors, for its color temperature blends well with daylight.

As the HMI lamp is a compact *point source*, its light produces sharp, well-defined shadows. To achieve shadowless illumination, you would need to rear-light panels of frosted plastic sheeting, or bounce light from a white reflector, ceiling, walls, hung cloth, etc.

An HMI lamp provides some three to five times as much light as tungsten-halogen/quartz light of equivalent power. The heat produced may be reduced by 80%. So electric power and air conditioning needs can be around 20% of those for similar wattage tungsten-halogen lamps.

Typical power ratings range from 200 W (with a light output similar to a 1 kW quartz light) through to 1200 W (equivalent to 6 kW quartz light), and can be run from normal 120 V 60-cycle household supplies (mains). Larger units of e.g. 2.5–4 kW are valuable for single-source large-area lighting, especially at night.

In the gas discharge lamp an arc is struck between two tungsten electrodes in an atmosphere of mercury vapor within argon gas. Unlike tungsten lamps, which cover a *continuous spectrum*, earlier gas discharge sources only radiated restricted parts of the UV-to-blue region of the spectrum, and this *multi-line spectrum* was inadequate for good color rendition. If a light-source covers very restricted parts of the spectrum only surface colors that reflect those particular hues will look bright; others will appear dark!

Modern metal halide lamps have now largely overcome earlier color deficiencies. By adding various rare earths (metallic iodides) within the quartz (fused silica) envelope the lamp's output has been increased over the entire visible spectrum to provide a luminant with either a near-daylight color quality (5600–6000 K) or 3200 K light to blend with incandescent lamps. Additional halogens (bromine) prevent any evaporated tungsten from the electrodes, blackening the lamp envelope.

A metal halide lamp's output includes a high proportion of ultra-violet light, particularly for the first 250 hours or so of its life. (Many MH lighting fixtures include burn-time meters.) So take care to avoid direct radiation (conjunctivitis) and use a UV filter over the lamp. Otherwise where the film or TV system is sensitive to UV this could cause gross overexposure.

Several types are in current use, including:

- The *HMI* lamp (hydrargyrum medium arc-length iodide; halogen metal iodide) – Correlated color temperature 5600 K or 3200 K.
- The *CSI* lamp (compact source iodide) – 4000 K.
- The *CID* lamp (compact iodide daylight) – 5500 K.
- The *xenon* lamp – 6000 K. Mainly used as projector light sources.

Metal halide sources are available in bare-bulb, linear (tubular), and sealed-beam (PAR light) forms. They are used for a wide variety of applications, including area lighting, effects (pattern) projectors, ellipsoidal spotlights, follow-spots, and daylight booster lamps.

The actual metal halides used (there are some 40 available) modify the light's color quality and constancy. The uneven spectrum of the luminant may cause some false color rendering, but this is seldom obvious. The *HMI type* of lamp is available in ratings of up to 18 kW. Its color temperature of around 5600 K blends well with daylight. A single 2.5 kW version can give as much light as *two* color-corrected 10 kW tungsten lamps; while a 6 kW HMI source has a similar output to a 225-A carbon arc but uses only a quarter of the power. Lightweight portable battery-operated 200 W versions in a reflector housing are increasingly used for mobile lighting set-ups. Particularly where power supplies are very limited, such high-efficiency low-consumption units are extremely welcome.

The *CSI type* uses a single-ended bulb, with a sodium, thallium, gallium filling. Its color temperature equates to about 4000 K – a value that makes it readily adaptable, with slight color-corrective filtering, to both daylight and quartz lighting. Sealed-beam versions provide a compact fitting that is waterproof for external use and can be grouped for large-area illumination. As a 1 kW version has an output about five times that of an overrun, short-life tungsten lamp of the same power rating and has a 1000-hour life, its benefits are considerable.

The *CID type* is available in single-ended bare-bulb and sealed-beam PAR verions of 1 kW, 2½ kW ratings, with a correlated color temperature of 5500 K (±400) in standard or hot-restrike forms, 500–1000 hours life.

However, gas discharge lamps have various disadvantages too. Their high color temperature is not suited to mix with typical tungsten-halogen lighting, and corrective filtering necessarily reduces their efficiency.

The metal halide lamp has an associated high-voltage *ignitor* (*starter*), which produces 20–50 kV across the electrodes and strikes an arc. A current-limiting choke/reactor *ballast unit* prevents excess current flow as the arc strikes and regulates it while burning, controlling small supply variations. Because the HMI arc requires an initial voltage of 230 V, countries using lower AC power supplies (120 V) include a step-up transformer within the ballast unit. Power losses due to ballast units can range from 10% to 15% of the rated lamp wattages, so this modifies the efficiency of the lamps (85–108 lumens/watt absolute).

Warm-up time from cold to full intensity can take 1½–3 minutes or more for an HMI lamp, ½ minute for a CSI, and 1½ minutes for a CID lamp. During warm-up, the metallic iodides enter the arc stream, the arc shape changes, and some variations in light output are normal.

A discharge lamp requires an AC supply and auxiliary equipment in the form of an ignitor and a portable ballast unit. The latter limits excess current flow as the arc strikes, regulates the current supply during the lamp's life, controlling small supply variations. However, a standard choke ballast produces rhythmical voltage fluctuations (hence a brightness ripple) at the supply frequency, and unless a film camera is synchronized to these changes the resultant pictures will reveal stroboscopic flicker, beat, dark-frame. These problems are much less obvious with TV systems.

Under controlled conditions, such defects can be alleviated in the film camera –

- By adjusting the exposure period (shutter opening angle, camera speed);
- By feeding lamps from different supply phases;
- By using squared-waveform supplies (which permit dimming);
- By using a high-frequency alternator supply.

If the supply voltage to a metal halide lamp is lowered, or conditions become very cold, the lamp runs colder, fewer metal halides are ionized, and its correlated color temperature *rises*. The 'fill-in effect' of the rare earths is lessened, and the spectrum becomes more spikey, revealing very pronounced peaks. Below a certain voltage the arc fails.

Should the supply voltage rise above normal or conditions become very hot, the correlated color temperature falls, the electrodes erode unduly, striking becomes more difficult, and the lamp's life is shortened.

A metal halide arc can only be dimmed electrically by a limited amount (e.g. 30–40% max.) using square-wave control or a variable transformer (Variac) unit. A regular SCR dimmer cannot be used to control it. Instead, a lamp's output can be reduced with a multi-blade shutter or iris.

Overall changes of color temperature from 4500 K to 9500 K during use are now unknown, and suitable compensatory color filters may be needed to bring it to 6000 K ($\pm$500 K).

Some types of HMI lamp allow instant hot restart if the arc is extinguished deliberately or accidentally, taking up to 45 seconds before reaching full output. (This can require an ignitor pulse of e.g. 30 000 V or more.) Conventional HMI lamps, however, require a 10–15-minute cool-down period before restarting.

Clip-on lenses can be attached to HMI sources housed in sealed-beam (PAR) envelopes to give different beam-widths (spot, medium, wide, very wide) and adjust beam shape.

As with quartz-lights, one is advised to clean HMI lamps with alcohol before use, and to use clean, lint-free gloves when handling them.

Fluorescent lamps

The familiar fluorescent lamp is a gas-discharge light source, giving roughly three times as much illumination as a tungsten lamp of equivalent power. Basically, a fluorescent lamp consists of a sealed glass tube filled with argon and a small quantity of mercury. When lit, ultra-violet radiation from the mercury vapor within the tube strikes the phosphor coating on its inside surface, causing it to glow (fluoresce). The type of phosphor mix used determines the lamp's color.

Circuit designs vary, but typically at switch-on, tungsten filament electrodes at the ends of the tube are heated and electrons from them flow through the tube. As a small glow-tube starter switch in the circuit warms, its contacts part, causing a current-limiting ballast choke to generate the high-voltage pulse, which ignites (ionizes) the mercury vapor. The initial light output drops after some 150–200 hours and then declines only gradually for the rest of its life.

Because the energizing AC voltage reverses 120 times (60 Hz supply) or 100 times (50 Hz supply) a second, the light inherently pulsates. Activated by each pulse, the coating fades slowly between pulses (a slow decay or die-away time). The resulting flicker is not obvious to the eye (except by our peripheral vision). However, flicker problems can arise when filming, due to these pulsations, unless the camera speed and shutter angle are suitably adjusted.

The color quality of fluorescent lighting is not really satisfactory for film or TV cameras if you are aiming at accurate color reproduction. Its spectral spread consists of a series of strong 'spikes' arising from the mercury vapor, plus a weaker continuous spectrum from the glowing phosphor coating. The correlated color temperature of a phosphor is a guide to the corrective color filters needed to avoid the greenish cast of fluorescent lighting. Internationally, color quality classifications can vary – e.g. for 'warm white (3000 K); cool white (4800 K) daylight (6500 K)' tubes. Opinions also differ regarding optimum filtering, for correction is very subjective.

10 Lighting equipment

Over the years, many types of light fixture have evolved to suit a wide range of situations: from the lightweight portable equipment needed for on-the-hoof news gathering to robust heavy-duty gear able to withstand the rigors of daily studio and location work.

Assessing equipment

When assessing any lighting equipment we are interested in several features:

- *Light quality* – whether the source provides hard, softened, or diffuse light.
- *Intensity* – the amount of light provided by the fixture.
- *Efficiency* – the light output relative to the power consumed.
- *Spread* – the maximum coverage of the light source.
- *Control* – how readily the light output can be restricted and controlled.
- *Size and weight* – how heavy and cumbersome the fixture is in use. Its balance and stability. Easy of storage.
- *Type of mounting* – Whether it is rested on the floor, slung, supported on a stand, etc.
- *Adaptability* – whether it is a one-purpose unit or can be used for several applications.
- *Auxiliary devices* – are diffusers, color medium, etc. easily fitted?
- *Reliability* – consistent performance, lamp life, deterioration.
- *Robustness* – easily damaged, bent, misaligned, etc.? Does it withstand rough handling, weather?
- *Ancillary units* – does it require additional equipment to operate (e.g. ballast units)?
- *The type of light source* – the kind of luminant used in a light fixture can directly influence its performance.

Thumb you way through any comprehensive catalogue of lighting equipment and you will be overwhelmed with the plethora of fixtures and accessories available; some have generic or slang names, others have differing trade names. The forms of equipment we shall look at here are typical, although designs vary, of course, between manufacturers. Most Lighting Cameramen and Lighting Directors develop preferences for particular types of units to suit different projects.

The kinds of units you use will mainly depend on shooting conditions. The smaller fixtures ideally suited for confined spaces would be of little help for large-scale action. The powerful units needed for area lighting would be an embarrassment in confined spaces.

Basic reflector design

A bare *lamp* (*bulb, globe, 'bubble'*) radiates light in all directions. Very occasionally, you may need that kind of illumination but it is wasteful and undirected. Usually, you want to be specific, and light a chosen area. The simplest way of directing light is to place a *reflector* behind the lamp. This will redirect rays from the back of the lamp forwards and increase its effective output. Unless a small metal shield is positioned in front of the lamp to mask off direct illumination, the result will be a combination of direct light from the lamp itself *and* the reflected light.

As you realize, the type of surface used will affect the nature of the reflected light (Figure 10.1):

- *Specular reflection*. If a highly polished metal mirror is used it will reflect a bright image of the lamp. Specular reflectors are used to concentrate the light in spotlights.
- *Spread reflection*. Here the reflector's surface can range from a dull etched finish to a white stove-enameled surface, depending on the amount of light dispersion needed. The softened illumination provided by this form of reflector is found in various broad sources.
- *Diffuse reflection*. When the reflector has a coarse rough surface, or is ridged, dimpled, or corrugated the emergent light is widely scattered.

A point to remember here is that the performance of all reflectors will vary in use. A surface that reflects up to 80% of the lamp's light when new can deteriorate with dust, grime, and discoloration to as little as half that value. So the overall characteristics of the light fixture can change. Then not only will its output fall, but where illumination was originally shadowless, it may become noticeably 'harder' or produce multi-shadows.

Reflector shape

As you can see in Figure 10.1, by adjusting the shape of a reflector the light rays can be concentrated or dispersed, according to the purpose of the fixture. The commonest forms are

- *Spherical* (*spheroidal*)   (4)
- *Parabolic* (*paraboloid*)   (5)
- *Ellipsoidal* (*elliptical*)   (6)

The performance of these curved reflectors changes with the position of the lamp. In many fixtures the light source is fixed at the

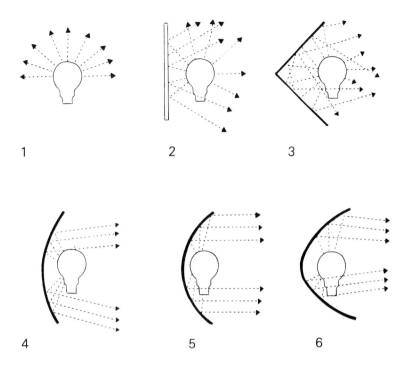

Fig. 10.1 Reflector design
A *reflector* can collect, concentrate and direct the light emitted from a source and increase its overall efficiency. Its size and shape determine how effectively it does this. Most reflectors are carefully shaped to concentrate light at a specific *focal point*. The illumination directed towards the subject is a combination of direct and reflected light.

Reflectors may be matted, rifled (grooved) or dimpled to soften the light beam.

1. A *bare bulb (open lamp)* radiates light in all directions.
2. A *flat mirror* simply provides a virtual image of the source. Its light combines with direct illumination from the source, and is radiated forwards in all directions.
3. A *conical* reflector offers basic light restriction, and multi-images (360 ÷ angle) interreflect between the surfaces to reinforce forward illumination.
4. The *spherical* reflector is mainly used in spotlights (in combination with a fresnel lens system), and in *scoops*. When the source is further from the reflector than its focal point (focal length), emergent light is broadly spread. When the source is closer to the reflector than this position, emergent rays form a parallel beam of light. In a *focusable scoop* the lamp position can be adjusted to change the light spread.
5. *Parabolic* reflectors produce a light beam with parallel rays; for localized illumination, concentrated distant illumination, and pre-lens light control. Light rays are parallel when the source is at the reflector's focal point. (Beam spread can be controlled – as 'spherical'.) This type of reflector is used in *internal reflector* bulbs – 'R' and 'PAR' lamps. Certain soft light fixtures use a combined spherical and parabolic reflector shape to collect light and then control the emergent beam.
6. *Ellipsoidal* or *elliptical* reflectors have two focal points. Light from the source at the primary focal point is reflected to concentrate at the secondary focal point (rays then diverge). In *lensless spotlights* (open-face fixtures) the lamp's position is adjusted relative to the primary focal point to 'spot' and 'flood' the unit. In projector systems a converging beam concentrates light onto a small area (e.g. gobo or slide) which is focused by a lens system and projected. Combined with a plano-convex lens, this is the bais of the *ellipsoidal spotlight*.

reflector's focal point. In others, the reflector or the lamp can be moved to alter the fixture's performance.

A *focusable scoop*, for instance, can be 'flooded' or 'spotted' to adjust its coverage. The problem with this simple method of adjusting the spread is that the emergent light will not be even overall but will show hotspots or shadows.

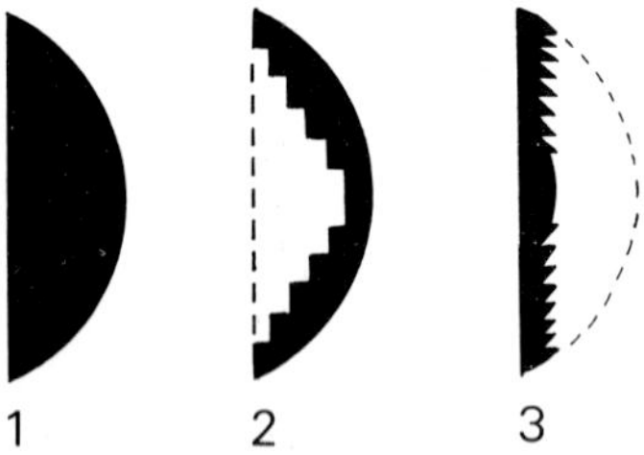

Fig. 10.2 Lens design
1. *Plano-convex lens.* A simple lens comprising a flat face and a convex surface. The shorter its *focal* length, the thicker and heavier the lens will be and the wider the emergent light beam. The light beam tends to be uneven (hot center), and suffer from color fringing. Used in basic spotlights, follow spots, effects projectors.
2. *Step lens.* Basically a plano-convex lens, in which a series of progressively deeper concentric rings are formed in the plano side of the element. Lighter, more efficient, and used in certain spotlights.
3. *Fresnel lens.* Here the convex surface becomes a series of concentric rings of triangular cross-section (prismatic), which are progressively smaller towards the edges. The central area is virtually a small plano-convex lens. The result is an even light beam, with a soft edge.

Basic lens design

Most lenses used in lighting fixtures are plano-convex in shape, i.e. they have a flat surface on one face and a spherical surface on the other. Basically, as you can see in Figure 10.1, if a beam of parallel light rays strikes its curved surface each ray is slightly deflected inwards according to its distance from the center of the lens. As a result, all the rays come to a common *focal point* on the far side of the lens. If you reverse the process, and place a lamp at this focal point, its light will be directed by the lens into a beam of parallel light rays.

Because the *plano-convex lens* is formed from a solid piece of molded glass it is generally thick and heavy, and there is an appreciable light loss. Its light may have a greenish tinge, with distinct color fringes. Under intense heat from a high-power lamp it is liable to crack. The light beam has a sharp, well-defined edge.

To overcome these limitations, the *fresnel lens* was developed. This widely used lens is much lighter and thinner, and has lower light loss. It is very durable, even when used with powerful light sources. Formed from a series of concentric prismatic rings, its central area is dimpled to avoid the central filament image found with the simple plano-convex lens and to produce a more even light beam. The risers of the prismatic rings may be darkened or frosted to reduce light-scatter.

A *step lens* resembles a plano-convex lens, with a series of concentric steps recessed into its flat face. The focused light is rather harder than that from the fresnel lens, with a less even intensity overall.

Light fixtures/luminaires

The simplest form of lighting fixture consists of an open lamp (bare bulb) in a base or holder. But with light scattering in all directions, this arrangement is too inefficient and vulnerable for normal use.

Sometimes the lamp (bulb) itself incorporates a reflector as part of its design (as in the PAR lamp). Or a clip-on accessory may be used to restrict the emergent light. But in most cases, the lamp is designed to fit in a specifically designed housing, which will concentrate and control the illumination.

Fixtures generally take two forms:

- *Open/open-face types*, where the lamp is fixed within a trough, or a box open on one face; or
- *Enclosed types*, where the lamp is fitted within a box or tube behind a lens system.

Reflectors of various types are usually fitted and can be combined with a lens system. The light-intensity from the fixture may be adjustable (e.g. by lamp switching) and its light-spread may be variable.

Many fixtures have attachments that restrict the light in various ways: spill rings, louvers, gratings, shutters, and hinged flaps. Effective with hard light sources, they are less successful when used with diffused light sources – unless they are particularly large or protrude some distance from the fixture. They only shade off or localize the illumination to a limited extent.

Brackets (lugs) at the front of fixtures support diffusers or color media. Unless a lamp housing (lamp head) is specifically designed to rest on the floor, it is generally supported within a U- or Y-shaped *yoke* (*stirrup*). This may be tubular or strip steel, or cast aluminum. At the ends of the yoke are rotating joints that allow the fixture to tilt up and down. In the center of the yoke is a short post. This *spigot* (*spud*) is used to support the lamp, and allows it to be turned from side to side.

The spud fits into a socket in a *bracket* (*receptable*) or a *C-clamp*. This fixture enables you to suspend the lighting fixture from an overhead bar or rail. Alternatively, you can insert the spigot into the socket at the top of a *lighting stand*.

Adjusting fixtures

There are several methods of adjusting lighting fixtures:

- *Manually*. You release lock-knobs at the end of the yoke and on the retaining bracket; turn/tilt the lamp and tighten the locks. With this hands-on method you must have access to a lamp. When a lamp is suspended, this necessitates lowering it, or using a step ladder or wheeled platform – difficult when lamps are high or awkwardly placed.
- *With a lighting pole*. Standing beneath a lamp, you fit the hook at the end of a telescopic aluminum pole onto the appropriate control and turn it to adjust the lamp. This method is extremely flexible, even when adjusting high lamps.
- *By remote control*. Motorized remote controls (wired or radio activated) may be used to operate lamp adjustments. Particularly handy for isolated inaccessible fixtures, this method is too costly and overelaborate for general use.

Soft-light sources

In nature, scattered diffused shadowless illumination comes from cloudy overcast skies or is reflected from rough surfaces. Lighting fixtures that provide soft light artificially rely on:

- Internal reflection within large light fixtures;
- Large-area sources, such as multi-lamp banks;
- Heavy diffusion in front of light sources;
- Bouncing hard light from large white surfaces, such as matte reflector boards or a white ceiling or wall.

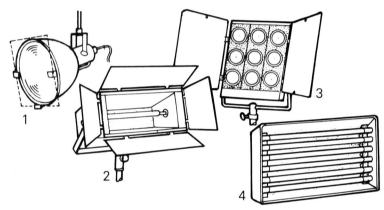

Fig. 10.3 part 1   Soft-light sources – open-lamp types
1. *Scoop* (1–1½ kW) has spun aluminum bowl of diameter 0.3–0.46 m (1–1½ ft). Emergent light is not particularly soft (or controllable) where an open lamp is used (tungsten or quartz). Spun-glass scrim fits into supports, aiding diffusion. The lamp is obtainable in fixed-focus and adjustable (ring-focus) forms; the latter has adjustable light-spread. (Versions with lamp-shields are available to prevent direct light emergence.)
2. *Small broad* (½–1 kW) is a tubular quartz light. Flaps or barndoors restrict the light spread. The emergent light from an open bulb is fairly hard. That from versions using a light shield is softer. Broads of 1 kW, 2 kW, 5 kW are widely used.
3. *Floodlight bank (nine-light cluster)* consists of groups of internal reflector lamps (650–1000 W each). They create soft light through multi-source light overlap. Its sections may be switchable and adjustable. Widely used for location lighting.
4. *Fluorescent bank* is a group of fluorescent tubes providing diffuse illumination.

Fig. 10.3 part 2   Soft-light sources – internal reflection types
1. Internal reflection from shielded tubular quartz-lights produces diffused high-intensity illumination.
2. *Large broad* 'Softlight' (1–6 kW typical) has reflected internal lamps (tubular quartz or fluorescent). A diffuser may also be attached.

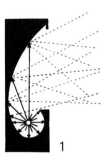

Truly shadowless illumination only comes from a non-directional diffused light source that is considerably larger than the subject. The larger the area of the 'soft light' source, the more diffused its light can be; but the more it spreads uncontrollably over the whole scene.

It is obvious, therefore, that any single lighting fixture of a convenient size and shape must be something of a compromise. The illumination from smaller fixtures is localized and directional, and hence fairly controllable. But it inevitably produces discernible shadows – however diffuse they may be. The light from a small source must, at best, be 'softened-off' rather than truly diffuse.

In practice, any fixture emitting light from an area smaller than about a meter square is going to cast a perceptible shadow of a person onto a nearby background. The solution is to use a series of soft-light sources, so that they illuminate and dilute each other's shadows.

Open-fronted soft-light fixtures

The scoop

Once the universal soft-light source in television studios, the *scoop* is still widely used, for it is cheap, simple, robust and lightweight. As a light source, it is very basic and barely controllable. The scoop comprises a spun matte aluminum bowl of 0.3–0.46 m (14–18 in) diameter, with a central lamp of 1, 1½ or 2 kW. Originally scoops used large frosted tungsten bulbs that enabled them, when suspended, to produce acceptably diffused light, especially when used in pairs or groups (although with tarnished or dirty reflectors, the emergent light 'hardens').

In the *lensless scoop* the matte inside of the spherical-shaped scoop serves as a reflector. With the lamp fixed slightly forward of its focal point (the center of the sphere), the emergent light scatters.

In the *focusing scoop* the lamp position is variable. With the lamp set close to the reflector, the emerging light rays converge ('spot'). Positioned at the reflector's focal point, they form a soft forward 'beam'. Further away from the reflector, the light diverges ('flood').

With the advent of quartz lighting, small compact light sources (point source or linear filaments) are now fitted to standard scoop units which, unless shielded, result in appreciable shadows. Because the emergent light cannot really be restricted (even by top-half gobos) scoops tend to flood the scene, overlighting backgrounds and causing distinct boom shadows. The adjustable, focusing, or 'ring-focus' form of scoop has variable coverage (e.g. 27–50°), although this feature has arguable practical use, for the unit often provides a peripheral spill of light (10% of maximum intensity) over an angle some 35° more than the main beam, which is itself uneven. *Diffuse* light cannot be focused. Spun-glass sheet diffusers or scrims can be clipped in a frame in front of a scoop to soften the illumination – but at the price of a 40–60% light loss – and for safety, should its bulb explode.

The broad

The *small broad* (broadside) has a short trough reflector with a linear-filament (tubular) quartz light of ½–1 kW. Both open-lamp and shielded-light versions are used. The housing often has a two-leaf or four-leaf barndoor shutter to restrict light spread, and can produce a fixed or variable-width beam. This type of fitting is lightweight and can be used suspended or on floor stands, so it has proved very adaptable on location and in the studio as a 'soft light' source (its fairly hard light is improved by clip-on diffusers), to illuminate confined areas, backgrounds, etc. A smaller version (*nook light*) can be used for very small areas, while a *large broad* using 2–5 kW lamps or a *double broad* (two separately switched 1 kW lamps) are used for larger applications.

Internal reflector soft-light source ('Softlites/softlights')

This popular type of soft light unit is powered by hidden linear-filament (tubular) tungsten-halide lamps. Their light is reflected from a curved aluminum sheet, with a matt or white enamel surface. Power ranges from 750 to 4000 W up to super 8 kW versions. The separate 1 kW quartz lamps may be separately switched to adjust the light output.

The near-shadowless light scatters over a wide angle unless an 'egg-crate' grid is fitted. Rather bulky, these softlite units can be suspended or stand mounted. Because they rely on reflected light only, efficiency is comparatively low. Smaller softlite units are more compact, but their light is more directional and less diffuse.

The same principle of using reflected light from a hidden light source is used with folding *umbrella reflectors*. Here a central lamp (or two) reflects from the aluminized inner surface of the umbrella unit. A typical 600 W unit produces 180 fc at 8 ft (1940 lux at 2.5 m) covering 80 ft^2 (7.4 m^2); equivalent to a standard softlite. An invaluable highly portable facility for localized set-ups.

The floodlight bank/cluster/multi-lights

In this arrangement a number of identical PAR lamp units are grouped to form a multi-source fixture. The 650 W *PAR36* lamps fitted (3200–3400 K) have inbuilt parabolic aluminized reflectors and are available with 'medium', 'spot' or 'wide' coverage; 'medium' being the most popular. Stippled or diffused versions can also be used. For use in daylight, the lamps may be dichroic coated (FAY – 5000 K); or suitable filters can be clipped on when necessary.

Floodlights are available in three-, four-, six-, eight- and twelve-lamp fixtures. Larger fixtures include 24 × 1000 W *PAR64* lamps, grouped in fours.

Lamps can usually be switched (either individually or in vertical rows) to alter the overall light intensity without affecting the color temperature. Each vertical row may be pivoted to adjust the overall horizontal light angle; to provide a higher intensity in a concentrated area or greater coverage with less illumination. When a floodlight bank is used with a 220 or 240 V supplies, 120 V lamps are often fitted, connected in pairs with their filaments arranged in series.

Each lamp in a multi-light unit casts a shadow which is illuminated by its neighbors. The more lamps a fixture has, the greater its output and the softer the overall illumination. However, if you switch off a number of lamps in a fixture to reduce the light level multiple shadows become more apparent.

Floodlight banks are often identified by the number of lamps they contain, so we have a 'nine-light' or a 'twelve-light'. Some have colloquial names such as the 'mini-brute' or 'maxi-brute' (24-light). Floodlight banks are used to provide –

● Overall soft light for large-area situations;
● Broad, intense fill light to illuminate shadows in strong sunlight;
● Booster light for exteriors on overcast days;
● Simulated daylight through windows or windshields (windscreens);
● For general soft light in confined areas.

Floodlights are usually supported on lighting stands, from low turtles to large telescopic motorized units (*Molevator*). They may also be slung or fixed to an hydraulic platform/hoist.

The most common accessories for floodlight banks are large metal-sheet side-flaps (light shields used to restrict the sideways light spread) and diffuser frames fitted with spun-glass sheets.

Overall soft light

Although overall soft light is normally an anathema there are situations where one needs diffused overall illumination in the studio to give built studio settings a convincing authentic look:

● 'Street' scenes;
● 'Open-air' scenes;
● Simulating environments which would have a continuous ceiling of fluorescent lamps;
● Simulating sky light through glass roofs, skylights, etc.;
● High-key area lighting for dance (e.g. ballet).

Typical approaches to this situation include –

● Very diffused soft light bounced from large cloths or reflectors;
● Rows of floodlights with strong diffusers;
● Spotlights rear-illuminating large sheets of cloth, scrim or other diffusers;
● Direct lighting from small overhead groups of bare lamps, surrounded by a tubular framework or netting to which diffuser material or white netting is attached ('space lights', 'chicken coops');

● Internally reflected light from overhead troughs with downward pointing front-silvered lamps (1000 W silver bowl globes).

Cyclorama lighting

There are many occasions in the studio when one needs a broad light source resting on the floor, lighting upwards to illuminate a backdrop, cyclorama, groundrows, or profiled planes. The light shades off towards the top and provides an attractive decorative effect. The lighting fixtures used for this purpose are known variously as *strip lights, battens, cyc lights, troughs, floodlights, groundrows.*

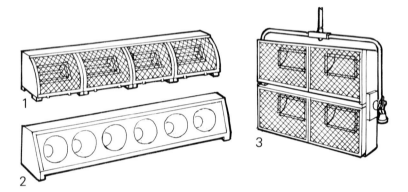

Fig. 10.4 Cyclorama lighting
Special fittings used to provide an even spread of light over a vertical surface (cyc, backdrop). Fitted with color medium where colored light is required.
1. *Groundrow (trough)* containing tubular quartz units (625 W–1500 W).
2. *Strip-light (border light)* consists of a row of internal reflector lamps.
3. *Suspended cyc* light is a single or multi-unit fitting (for color mixing) with strip quartz-lights.

The simplest form of fixture is the *batten* or *strip light*. The most basic arrangement is just a horizontal board carrying a series of vertically mounted open bulbs behind a light-shield. (This obscures the lamps from the camera.) Alternatively, it may use a row of internal-reflector lamps. The illumination is uncontrolled; hottest nearer the bottom and falling off rapidly with height. Taking the unit further from the background improves the light spread but at a lower intensity.

Undoubtedly the most satisfactory units consist of individual open-fronted troughs, each with a linear-filament (tubular) tungsten-halide lamp, which is reflected from a curved aluminum sheet. In the best fixtures the reflector is designed so that the illumination is spread evenly in an elongated lobe for several meters over a vertical surface, with lower areas not appreciably brighter than higher ones. The reflector can be tilted to control the coverage.

These units are made in one-, two-, three-, or four-light combinations (larger six-light groups are also available, but can prove cumbersome). A series of 500 W, 650 W, 1000 W or 1250 W units are usually arranged side by side, cabled to be switched/dimmed together.

You can fit color filters to the units to provide single-color or color-mixture arrangements. If, for example, red, green, and blue filters were fitted to each three-group it should be possible to achieve a wide range of hues and shades simply by adjusting dimmers to alter their proportions. In practice, there is always the possibility that, low down on the background, the differing colors of individual units may be discernible. In addition, one often finds that wherever individual units are switched off the coverage looks uneven.

Lighting from the floor is fine when you want a top-shaded background. However, where you want it to appear evenly lit overall, the low-angle lighting needs to be supplemented by overhead units.

For overhead background lighting, a suspended form of the floor unit is used – singly, or in two, three, or four groups. The quad units are in-line or square 2×2 form. The last are grouped for improved color-mixing; red, green, blue, white.

Some forms of overhead cyc-units have specially shaped reflectors that enable you to light a background evenly with hung units alone. Critically positioned, they produce flat overall illumination when placed 3 m (10 ft) from a background at 2.5 m (8 ft) intervals.

Fluorescent lighting

Fluorescent tubes have a higher efficiency than tungsten lamps, are economical, have a long life, and can be grouped to provide banks of soft light. They are, therefore, still often used in small TV studios. They have major disadvantages, however, including a relatively low light output, bulk, and a light spread that is difficult to localize. More important still, the light quality of fluorescent sources is not really suitable for optimum color reproduction.

Although 'equivalent' or 'correlated' color temperatures of 3700–3800 K are often quoted (according to their internal coating), the output is very uneven throughout the spectrum (Figure 2.18), so causing some colors to reproduce lighter or darker than normal and considerably distorting color mixtures.

In fact, the color rendering of subjects under fluorescent lighting is so unpredictable that although 'compensating filters' are available, many lighting cameramen prefer to switch them off where possible and use their own portable equipment, or at least overpower the existing illumination. Small portable fluorescent units on floor stands are sometimes used to provide eye-level fill light to supplement 'toppy' fluorescent sources on location, an appropriate lens filter being used to compensate as far as possible for the color distortions.

Hard-light sources

As you know, *hard light* is produced by compact light sources. The more concentrated the luminant, the 'harder' the light, and the more sharply defined the shadows. So while the modeling from arcs is crisp and clear-cut, that from large tungsten lamps with a much bigger luminant area is less definite.

In most hard-light fixtures a reflector behind the lamp collects and reflects its rays forward. This makes maximum use of the radiated light and increases the overall efficiency. In addition, many hard-light sources use a fresnel lens or a plano-convex lens system to concentrate or focus the light beam.

Some hard-light units are general-purpose (*lensless spotlights, fresnel spotlights*) and can be used as key lights, back lights, set lights, effects lights, to light backgrounds, etc. Other fixtures have more specific applications (*ellipsoidal spots, follow spots, hand-held lamps*).

Internal reflector fixtures (sealed-beam units)

The *internal reflector lamp* provides us with a cheap, versatile, self-contained hard-light fixture. Incorporating an inbuilt reflector, and perhaps a molded face lens, it is highly adaptable, as we saw earlier when looking at *floodlight banks*. While some PAR lamps have clear or dimpled glass fronts, others are molded to provide prefocused 'spot' or 'flood' coverage over a circular field. Some have an *elliptical* coverage, spreading the light over a field of e.g. 6° × 12° to 30° × 40°.

Fig. 10.5 Internal reflector lamp (sealed beam)
Available in regular tungsten and quartz (tungsten halogen) forms. Used individually in clamps or grouped.
1. The glass front may be clear, ribbed or frosted, to give a wide, medium or spot beam, *hard* or 'softened' quality. A color-compensating filter (integrated, or attached) matches the luminant to daylight.
2. Widely used *PAR lights* incorporate a parabolic aluminized reflector (650–1000 W, 3200 K).
3. *Profiled wall plate:* Tapes to any firm surface; hangs on pipes.

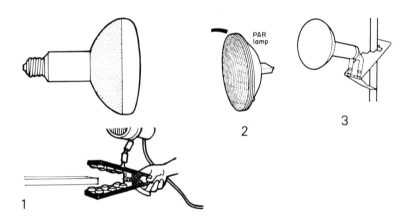

Lamps of this kind, such as the PAR lamp and the internal reflector lamp (R), come in standard and overrun versions. An overrun lamp has a brief bright life of e.g. 2, 6, or 10 hours, and uses, respectively, 275, 500 or 1000 W. The boosted light output is the equivalent of a regular tungsten lamp of 800, 1600 or 3000 W.

Screwed into a simple holder, an internal reflector lamp can be fastened to any suitable support – using a wall plate/pipe plate, clip-light, or similar lightweight attachment. Although lamp units cannot be adjusted (you have to choose the right type for the job), clip-on light shields can help to localize the illumination.

Lamps in simple supports are very vulnerable and easily broken, with little protection if the bulb explodes. Where possible, therefore, they should be used in a tubular metal or plastic housing, which helps to protect the lamp against accidental damage. Because PAR lamps are weatherproof, they may be used externally in appropriate watertight lamp sockets.

Lensless spotlight

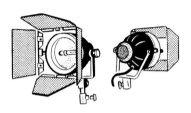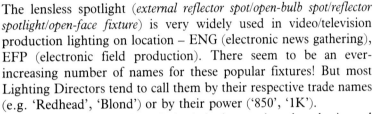

Fig. 10.6 Lensless spotlight/ external reflector spot
A highly efficient, very portable lamp with a beam angle of e.g. 44–82°.

Typically:
250 W, 30 V (life 12 hours).
650 W, 120 V (life 100 hours).
800 W, 220 V (life 75 hours).

The lensless spotlight (*external reflector spot/open-bulb spot/reflector spotlight/open-face fixture*) is very widely used in video/television production lighting on location – ENG (electronic news gathering), EFP (electronic field production). There seem to be an ever-increasing number of names for these popular fixtures! But most Lighting Directors tend to call them by their respective trade names (e.g. 'Redhead', 'Blond') or by their power ('850', '1K').

Its open-fronted reinforced plastic body contains a bare horizontal linear quartz-light (tungsten-halogen lamp), e.g. 80 mm (3.18 in) located within a compact adjustable reflector. The absence of a lens results in a small lightweight fixture (e.g. 1.5–4 kg (3.3–8.8 lb)) with high overall efficiency (no lens light-losses). The beam spread and intensity are adjustable by moving the lamp position, with a screw knob or a sweep lever (slider) at the rear of the fixture. Typical lamp sizes include 250, 650, 800 W ('Redhead') and 1 kW, 2 kW ('Blond'). A number of accessories are available, including dichroic daylight filters, stainless steel mesh 'scrims', and barndoors.

Because the fixture is so portable, adaptable and economical it is a valuable lighting tool; especially where power and ventilation are limited. A typical output for an 800 W lensless spotlight (spotted) is only half a stop down relative to a 2 kW fresnel tungsten spot covering the same area.

Another major advantage when working in confined spaces lies in the fixture's considerable light spread – over 80° compared with the maximum 60° of many fresnel spotlights. When the lamp-to-subject distance (*throw*) is limited it can be a major embarrassment to find that your single key light will only cover part of the action area and needs to be supplemented.

On the other hand, performance is something of a compromise, for the light beam is uneven, varying as the fixture is focused. There are noticeable performance differences between designs. In some there is a tendency for the light beam to be of lower intensity in the center than at its edge, or to develop hotspots as the lamp is spotted (moved away from the fixed reflector). The beam of one design of fixture may have a more gradual fall-off either side of its central axis than another.

For many situations the barndoors on lensless spotlights are largely ineffective. Although they will gradually shade off the light they will not provide the sharp cutoff that is so often needed to avoid shadows or overlit areas. To produce a well-defined beam edge you will need to place a gobo or flag a meter or two in front of the fixture.

The lensless spotlight can be fitted onto a heavy grip (*gaffer grip/gator clip*), mounted in a wall-plate, supported in a lightweight floor stand, or suspended.

Fresnel spotlight

In film and TV studios, where most lamps have to be positioned a fair distance from the subjects (e.g. over 15 ft/6 m), the heavy-duty fresnel spotlight is universal, suspended from ceiling bars or battens, on spot rails and floor stands. Although fresnel spotlights are heavier and more bulky than lensless fixtures these disadvantages are less important in the studio than on location, where there are greater transport and rigging problems.

Fig. 10.7 Fresnel spotlight
Designs cover a wide range from e.g. 100 W to 10 kW, with lens diameters of 8–38 cm (3–15 in).
1. Sliding the lamp/mirror assembly to/from the fresnel lens alters the light-spread (spot-to-flood). Adjustment is by a hand crank or sweep lever, lighting pole (loop, ring, or T-fitting), or remote electronics. *Controls:* Focus, tilt, pan, filament-switch (where dual-filament lamps are used). Barndoors may be permanently fitted.
2. A hook-on frame supports color medium or diffuser/scrim in front of the fixture, yet allows barndoors to move freely. In many spotlights the medium is held in a flat support frame inserted between the barndoor fitting and the lens. (More likely to overheat.)
3. The light output and the beam spread change as the fixture is adjusted from *full flood* to *full spot*, as this polar diagram shows.

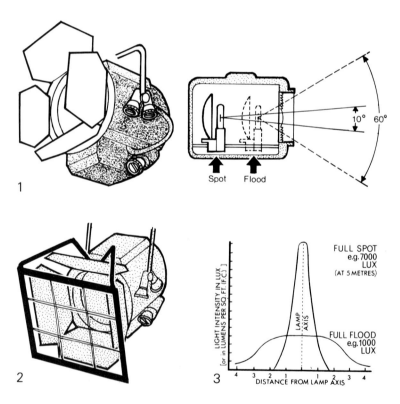

The fresnel spotlight is an enclosed lighting fixture. Its optical system is designed for maximum light control; an essential for creative lighting, where you want to precisely adjust light coverage. The light source in the studio is usually tungsten-halogen/quartz light, or a tungsten lamp. On location, metal halide (HMI, CSI, CID) or carbon arcs are also used.

The lamp is fixed at the focus of a spherical metal mirror, the resulting parallel light rays passing through a plano-convex, fresnel, or stepped-lens system. The reflector/lamp assembly is moved along support bars by turning a screw knob, hand-crank or sweep lever. As the lamp travels towards the lens, the spot's light beam spreads ('floods') up to an arc of perhaps 60°. Moving it back from the lens, the beam narrows, and concentrates to a minimum coverage of e.g. 10° ('full spot').

The intensity of the light beam alters as the fixture is focused, and may be anything from five to thirty times as great when fully spotted as when fully flooded (depending on the design and power of the fixture). So one method of adjusting the intensity of a fresnel spotlight is to floor or spot it a little, rather than use a diffuser or a dimmer. (And, of course, its color temperature remains constant.) The disadvantage of this technique is that the lamp's coverage changes at the same time as the light level!

Most fresnel spotlights are designed so that the edge of the light beam is diffused, and falls off in brightness. This 'soft edge' enables you to slightly overlap the beams from adjacent lamps so that they merge and blend to maintain even illumination over a large area.

Mounted in a stirrup or yoke, the *lamp head* can be turned, and tilted up and down over a wide angle. However, do not be tempted to tilt the housing too steeply ('on its nose'), for this can result in the lamp overheating and failing through filament sag or bulb blisters.

Fresnel spotlights using tungsten and quartz lamps cover a power range from about 100 W to 10 kW; the most widely used being 1 kW, 2 kW, and 5 kW ('*1K, 2K, 5K*').

The HMI light fixtures most widely used in the field are 200 W, 575 W, and 1200 W. Where more powerful units are necessary, 2.5 kW, 4 kW, 6 kW, 12 kW fixtures are available with either fresnel lenses or glass (clear or stippled).

As with all light fixtures, for maximum output it is important to keep the lens and reflector clean, and to check the lamp condition and its alignment periodically.

Dual-purpose fixtures

Dual-source light fixtures have been developed to save cost, space, and time in a busy studio. By making simple adjustments, they can be converted from a hard-light source to a soft-light source. How effectively is sometimes arguable.

The simplest example of a dual-purpose fixture is the *dish reflector*. This consists of an open lamp (e.g. 5000 W) fixed in the center of a shallow aluminum pan reflector. When used in this form ('sky pan') it can light backings and cycs with a spread of hard light. When a hemispherical light shield is clipped in front of the lamp, direct illumination is cut off, and only softened reflected light emerges. An overall diffuser may be added. Because the light spills around uncontrolled, this type of fixture has limited use.

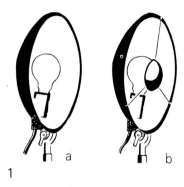

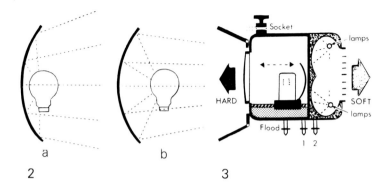

Fig. 10.8 Dual-light sources

1. *Dish reflector – 'Pan', 'Sky pan' 'Sky light'.* Used to broadly light backdrops and backgrounds.
(a) The direct hard light from the central bulb is augmented by softened light from the reflector.
(b) When a central light shield is clipped on, only reflected illumination emerges, to produce a softened light source.

2. *Long-range scoop.* The focusing long-range scoop adjusts the bulb-to-reflector distance, to change the emergent light from (a) wide-beam softened light to (b) a narrow-beam soft-edged spot.

3. *Dual-source luminaire.* A combined fixture providing a fresnel spotlight on one face, and an internally-reflected soft light source from the other. While some designs ingeniously use a

single bulb with repositioned reflectors the usual form is a pair of back-to-back units. The fixture is pole-operated, and its controls include: Focus. Mode switching (hard or soft). Filament switching (2½ or 5 kW) in *hard mode;* lamp switching in *soft mode.* A four-leaf barndoor is permanently fitted. The fixture has a flag-holder socket and a color/ diffuser frame can be attached.

A *long-range* or *focusing scoop* is again simple in design. In this case the lamp's position is adjustable. At the focus point the rays of the emergent beam are parallel, and it serves as a hard-light source. Moved closer to the reflector, the rays diverge, and it becomes a wide-beam softened source.

Dual-source fixtures have developed for use in TV studios which provide a fresnel spotlight at one end of the unit and a soft-light source at the other.

● The fresnel spotlight has a permanently attached barndoor, with lugs to hold a diffuser/color frame, and flag-support socket.
● The soft-light end of the fixture has concealed linear halogen lamps, their light being internally reflected from curved matte aluminized surfaces. An 'egg-crate' lattice restricts the light spread.

Although there have been several ingenious forms of dual-source fixture using a single lamp and moveable internal sections, the design now widely used is virtually two separate back-to-back units with switchable controls. The fixture has several pole-operated controls:

● *Mode switching.* Determines whether the hard or soft end of the fixture is live.
● *Power selection switch.* In the *spotlight* position this switches on one or both filaments of a dual-filament quartz-light (e.g. 2½ + 2½ kW). That allows light levels to be increased or decreased without altering the color temperature of the light. In the *soft-light*

position the power-selection control switches one or two linear quartz lamps to provide half and full-power soft light.
● *Spot focus.* The normal spot/flood adjustment, moving the lamp/reflector assembly relative to the fresnel lens.

In a studio rigged with dual-source fixtures it takes little time to alter lamp power and to change the type of illumination. Where space is limited, or for economy, separate single-purpose units (spotlights, or soft lights) are also available.

Special-purpose spotlights

The various light fixtures we have considered so far cover most situations. But there are occasions when only special-purpose units will achieve the effect you want.

Projection spotlight

This type of fixture projects an intense, even beam of hard light with a sharply defined edge; useful for spotlighting subjects in a circle of

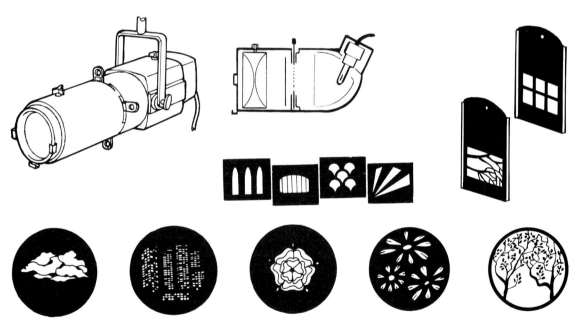

Fig. 10.9 Ellipsoidal spotlight (profile spot)
Provides precisely shaped hard-edged beams which are controlled by shutters, iris diaphragm, or metal plates (profiling blades). Shadow patterns can be projected, using an etched stencil plate of stainless steel (gobo, mask). Typical ratings 500 W, 750 W, 2 kW.

light. (When used for this purpose, the edge of the beam may be defocused and softened slightly by adjusting the projection lens.) The unit usually incorporates shutters for shaping the light beam, and may have facilities for projecting light patterns.

Projection spotlights come in several forms and are known variously as pattern projectors, profile spots, ellipsoidal spotlights. The simplest design is a two-element projection lens system which can be fitted to the front of a regular fresnel spotlight (200–2000 W). The most sophisticated is the ellipsoidal spotlight, which comes in a range of power ratings: 750 W, 1000 W, 1500 W, 2000 W. This fixture uses an ellipsoidal mirror system which, you will remember, has two points of focus (Figure 10.9). The lamp is set at the inner focus point, its position generally being preset to give an even, circular light beam. At the other focus point a set of four internal metal blades (*'framing shutters'*) is located. A two-element plano-convex projection lens system at the front of the fixture focuses on this second focus point, and varies the sharpness of the projected image. By angling the *framing shutters* you can alter the shape of the light beam, producing slits, quadrilaterals, triangles of light. When projecting a square or rectangular patch of light onto a surface from an oblique angle the internal blades enable you to compensate to some extent, for keystoning or shape distortion.

To change the coverage of the spotlight you may need to replace the projection lens with another of a different angle – narrow, medium or wide. Some fixtures have varying lens positions, or interchangeable zoom lens tubes, providing a range of beam angles from around e.g. 10–50°.

Many ellipsoidal spotlights have a slot in the top of the fixture, just in front of the framing shutter, at the second focus point. Here you can insert a drop-in variable iris to adjust the beam size, or a metal plate with a hole cut into it (*'matte'*) to change the area or shape of the light beam. This allows you to light very selectively; isolating a subject from its surroundings, or even picking out a small part of it – e.g. a slit of light across a person's eyes!

A regular use for this type of ellipsoidal spot is to project detailed light patterns onto a background. Simple geometrical shapes can be cut direct into thin sheet steel or a metal plate. More elaborate designs such as windows, decorative motifs, leafy branches, tracery are etched into thin stainless steel foil. This *gobo* (*cookie*) slips into the guides on a support plate ('dip stick'), and, with care, can be removed and re-used many times. Where the design is particularly intricate a fine mesh may be overlayed to give it stability.

The size of the projected pattern will depend on the distance of the lamp from the background, the angle of the plano-convex lens (e.g. 5–40°), and the size of the gobo. Although the pattern is less likely to become distorted when the projector has a narrow-angle lens and is some distance away, it is not always possible to ensure an unobscured line of sight ('clear throw'). Hanging lamps or scenery may get in the way. You can use fixtures with wide-angle lenses proportionally closer for the same pattern size, but there is always the possibility that the image will be geometrically distorted (barrel, pincushion), and its sharpness uneven.

Very small or dense gobo patterns can be a problem. For instance, a gobo with just a few pinpricks to simulate a starry sky lets little light through, so the projected image is correspondingly dimmer. Fine patterns may show color edge-fringing and aberrations.

When you need a large pattern, or a decorative feature spreading over a wide area, it is often better to use a series of projectors to build up the effect in sections rather than attempt it with a single gobo. Some gobo patterns are actually designed in sets. The component patterns are superimposed, to provide multi-color or animation effects (stained glass windows, flashing neon signs).

Unlike fresnel spots, the hard edges of ellipsoidal spotlights cannot be blended unobtrusively. So instead, you have either to use internal shutters to square-off the light beams and butt-join them or accept overlaps or blanks in their coverage.

Although you can use these special-purpose spotlights as key lights – e.g. to pick out an individual in a group – the hard illumination and sharply defined edges may not be as acceptable as a fresnel spot with its soft-edge fall-off.

Scenic projector

Some forms of scenic projector use glass slides or large revolving glass discs with photographed or painted subjects. Unlike metal gobos, their projected image can include a full range of tones. At best, the result can be extremely effective, but under typical film and TV studio lighting, spill light badly dilutes the image unless you can shield off that part of the background.

Animation discs are increasingly used. These are rotating stainless steel gobos, used individually, or in conjunction with a stationary gobo.

As well as geometrical patterns, flicker and similar decorative effects, the system can simulate flames, snow, rain, clouds, water ripple with varying degrees of realism. The speed of the gobo rotator can be controlled to suit the situation.

Follow spot

The familiar *follow spot* is mainly used whenever you want to follow moving performers (singers, skaters, dancers) around in a confined pool of light. The large, carefully balanced fixture is mounted on a portable floor stand or scaffolding, free to pan around smoothly and accurately as the operator follows action.

Because the follow spot needs to be at a high vantage point in order to have an uninterrupted view of the action it usually has to be located a considerable distance from the subject. For such a long 'throw' it therefore needs a powerful light source and a large lens system for maximum efficiency. All forms of luminants are used in these units – tungsten, tungsten-halogen, metal halide, carbon arcs. However, HMI lamps (1–4 kW) have the particular advantages of greatly increased light output, with lower relative power consumption (one-fifth of a tungsten-halogen lamp and half the heat output).

The follow spot may be fitted with several operational controls:

- A focused variable internal iris or diaphragm, which can be 'irised in/out', to alter the size of the spotlight beam.
- A second iris which adjusts the intensity of the light beam (to avoid over- and underexposure).
- A hinged flap ('douser') which can be swung over the spotlight to black out the beam. (So the light-beam can be presented at the correct intensity and diameter on cue.)
- Two adjustable framing shutters to restrict the top and/or bottom of the beam.
- Color filter frames that allow the color temperature of the follow spot to be matched to other lighting. Color filters for effects lighting are seldom used in follow spots.
- The edge of the spotlight is normally hard, but can be softened by readjusting the front lens.
- Some designs can be fitted with a large gobo to adjust the beam shape or give it a decorative edge (scalloped, castellated, serrated).
- There are several forms of sighting device, which allow the operator to accurately prejudge the spotlight's position before opening the beam on cue.

Hand-held lamp ('sun gun')

Colloquially often called 'sun guns' after an earlier popular model, several forms of lightweight hand-held lamps are widely used, ranging from compact *lensless-spotlights* (external reflector), to *internal reflector* lamps (PAR and R fittings). Miniature fresnel spotlights tend to have a more restricted light-spread.

Small, portable quartz-lights are regularly used on location, either hand-held or attached to the top of a video camera. A focus knob at the rear adjusts the beam width. Some designs are fitted with a cooling fan; those without tend to overheat if used for long periods. Many units include a hinge-over dichroic or color-compensating filter, and barndoors.

The size and power of units that can be hand-held is necessarily limited. Typical power ratings include

- 100 W, 250 W, 300 W, 650 W tungsten units, and
- 125 W, 270 W, 575 W, 1200 W HMI units.

Thirty-volt lamps can be powered from a *battery belt*, or a shoulder *battery pack*. While the latter may provide power for one and a half to four times as long, it is much heavier and bulkier.

Portable supplies allow the camera greater mobility, but one needs to be alert to battery life. A 250 W/30 V lamp may run for about 30 minutes from a battery supply, while a 350 W fixture would only last for around 20 minutes.

When the hand lamp is being used in the vicinity of a utility/mains supply it can be powered directly, using suitable 120 V or 220/240 V bulbs, or an AC adapter which converts the supply to 30 V.

As you will remember, we discussed the techniques and problems when using hand-held lamps in Chapter 7.

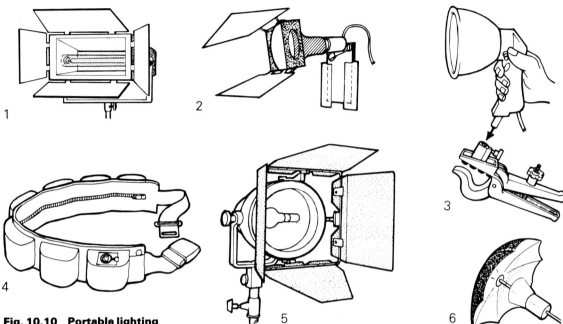

Fig. 10.10 Portable lighting
Typical portable lighting units include:
1. Small broad (600–1000 W).
2. Internal reflector lamp on plate (gaffer taped to wall, hung, hooked on bar) or fitted into clip-light.
3. Sun-gun (250 W) hand-held or clamped (gaffer-grip, or alligator clip). Powered from utility supply or 30 V battery belt (4).
5. External reflector spotlight, lensless spot.
6. Umbrella head. Central lamp shooting into collapsible white umbrella reflector.

Reflector units

Reflector units/reflector boards offer power-free illumination on location. By redirecting sunlight you can –

● Provide keylights, filler, and rear kickers;
● Illuminate dark shadowy areas;
● Direct light under canopies, through windows or doors;
● Reduce overall contrast.

Reflective materials

Strictly speaking, any highly reflective material can serve as a reflector, but the color and character of the reflected light will depend on the surface you select. Typical materials reflecting 'hard light' include: plastic mirror sheets, silvered plastic sheeting (e.g. 'space blankets'), metal sheet, foil-faced boards, silvered roller blinds, aluminized screens. Typical materials reflecting 'soft light' include: white cloth surfaces, white styrofoam (polystyrene) panels, white paper sheets, matte-white painted panels.

Double-sided boards give an instant choice of reflected light quality; 'hard' on one side, 'soft' on the other. As you would expect, the reflected hard light projects over some distance, is easily directed, and provides distinct modeling. The reflected soft light falls off rapidly, and is less controllable. By adjusting the surface shape of a flexible metallic reflector you can alter the coverage of the reflected light to some extent: bowing inwards to concentrate and outwards to spread the light.

The area covered by the reflected light will depend on its distance from the subject, and the size of the reflector board. Larger reflectors (e.g. over $1.2 \times 1\,m/4 \times 3\,ft$) are increasingly difficult to support; but for very close shots even a small hand-held 'bounce-card' can be extremely useful.

Limitations

Reflectors have their limitations, of course. You are completely reliant on the sun's angle, direction and intensity being suitable. However, there are some useful tricks that may help:

● When the sun is not in the optimum position you may be able to use a 'relay-mirror' to reflect sunlight onto the main reflector, which is set at the required angle.
● When the intensity of the reflected light is too great, single- or double-density woven 'scrims/nets' can be stretched over the reflector surface.
● Most reflectors are used with sunlight, so the reflected light can appear 'warm' compared with the illumination from *sky light*, which has a high color temperature. To reduce this disparity use a 'cool blue' type of reflector to raise the effective Kelvins of the reflected light – particularly when shooting at early morning or sunset.

Supporting the reflector

Reflector boards can prove awkward and cumbersome, posing handling and transportation problems. Very lightweight flimsy reflectors are easy to carry but are readily damaged, and flap in the slightest breeze. Nothing draws the attention to a reflected light patch more than movement or unsteadiness.

The simplest way to support a reflector board is to rest its lower edge on the ground; assuming that the reflected light angle is appropriate. A reflector board can be fitted into an adjustable yoke, and supported in a portable telescopic stand, raised several feet above the ground (winds permitting!). It will need to be tilted and turned onto target, and the adjustments tightened off, so an apple box/step-box is a useful facility. It is essential to ensure that the fixture is well bottom-weighted with sandbags (typically weighing 7/11/23 kg (15/25/50 lb) to improve its stability, and to have it manned at all times.

Reflector board surfaces

Several forms of reflective sheet plastic have been developed (e.g. by *Rosco, Berkey*) that can be attached to reflector boards and other reflective surfaces such as walls and ceilings. These provide an excellent variety of light qualities.

- *Mirror* – a mirror finish providing strong reflections and capable of reflecting light over a long distance. Can effectively reflect bright sunless daylight over short distances. May be used as a relay mirror.
- *Hard* – reflecting a hard strongly directional light. The character of the light source is evident, but the reflected light is rather less intense than with a mirror finish.
- *Soft* – providing slightly diffused, rather less directional reflection, without the harshness of 'harder' materials.
- *Supersoft* – reflecting very soft scattered light. Excellent for wide-angle fill light.
- *Gold tint* – a warm reflected light quality, used to create 'atmospheric' early morning or late afternoon effects.
- *Blue tint* – converts reflected tungsten light (3200 K) to daylight color quality. (Less transmission loss than when using blue conversion filters on tungsten lamps.)
- *Cool blue* – raises the color temperature of the reflected sunlight, to mix more effectively with blue sky light. It avoids reflected light used to illuminate shadow areas, appearing unnaturally warm.
- *Flex/featherflex* – thin, soft, tearproof sheeting, attached to surfaces on location to provide diffuse bounce light in inaccessible places.

The camera-light

Since the early days of motion pictures some sort of *camera-light* has proved invaluable. Known variously as an 'eye light', 'basher', 'headlamp', 'camera fill light', 'Obie light', it is a familiar appendage to both film and television cameras.

Forms of camera-light

Camera-lights take several forms, including small fresnel spotlights, direct and internally reflected strip lights, mini-scoops with photoflood lamps, small broads, multi-bulb troughs (3 × 650 W).

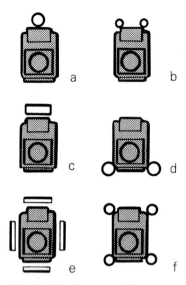

Fig. 10.11 Camera-lights
Camera-lights take several forms:
(*a*) Single spotlight, (*b*) Spot bar,
(*c*) Broad source, (*d*) Twin headlamp,
(*e*) Strip lights, (*f*) Spot frame
(garland) circle of photofloods.

A small spotlight (e.g. 75–200 W) is regularly clamped to the top of the television camera *on location*; especially the work-alone single camera unit, using a 'combo' or 'camcorder'. Apart from any attached scrim/diffuser, its intensity is not controlled.

In *television studios* the intensity of the camera-lamp is usually adjusted remotely by a dimmer on the lighting control board. So where necessary, the Lighting Director can continually vary its output to suit each shot variation while watching preview monitors.

The intensity of lights on *film cameras* is usually modified by attaching scrims, stainless steel wire mesh, fiber-glass sheet, etc. over the lamp. Ingenious designs using adjustable wire-mesh shields or internal reflection from vari-tone rods allow on-the-spot adjustment. The advantage of these arrangements is that the color temperature of the light remains constant.

Flexibility

The camera-lamp can be used for a number of purposes:

● As an *eye-light* – even a low-intensity lamp on the camera will reflect in the eyes of a performer, giving them a lively gleaming appearance. Experience suggests that eye-lights (catch-lights) can create a vivacious, interested, alert, glamorous impression; while without such reflections eyes can look dull, uninterested, tired, secretive.
● As a low-angle fill light for deep-socketed eyes, and shadowed necks.
● To reduce facial modeling; to light-out wrinkles, bags under eyes.
● To compensate for steep lighting, or a downward-turned head.
● As a key light when other lighting is not available, or cannot reach the subject.
● As a key light for a brief item (e.g. title card), to save rigging a special extra lamp for that purpose.
● As a mobile key light, to continually cover a moving subject.
● To light out and disguise momentarily unattractive portrait lighting.
● As brief fill light for a subject not generally needing it.
● As fill light on closer shots (where longer shots are to be unfilled).

Limitations

Although, as you can see, the camera-light is an extremely useful facility, it has its weaknesses too:

● Its light is clearly reflected in any shiny surfaces facing the camera, including spectacles, where a prominent specular can be very distracting. A profusion of reflections from multiple camera-lights is usually most unattractive.

- The coverage of the camera-lamp is usually limited. This localization may be obvious in a longer shot, or a wide-angle shot. Subjects may be seen entering and leaving its beam. By diffusing the lamp, one can increase its spread, but this will also reduce its output.
- The lamp may be of little use when a camera uses a narrower lens angle for close shots of distant subjects (telephoto, long focal length lens).
- As a camera pans or changes its viewpoint the direction of its lamp changes too, and this may be visible on the background.
- As a camera dollies closer or further from the subject, the effective intensity of the camera-light alters. Where two camera are so close to the same subject during continuous production that orthodox lighting would result in camera shadows, a light on each camera may be the solution. *But* if one camera moves away while the other remains on shot, the lighting effect alters!
- One camera's light can strongly illuminate another camera's subject; usually as a side light or a kicker.
- When the camera is close, even slight changes in distance can result in marked differences in the effective brightness of the camera-light. So it can be difficult to maintain a constant light level during a dolly shot. When a camera is working on a wider lens-angle this effect is more troublesome, for the light is proportionally closer to the subject.
- Although a camera-lamp may be intended for a specific subject, its light will also spill onto others nearby. Anything closer to the camera may be lit more strongly, especially if it is lighter-toned.
- Where a camera-light is dimmer controlled its color temperature will, of course, vary with its intensity. (Particularly noticeable when using overrun lamps.) It will be high at full intensity when 'reaching out' for distant subjects and much lower when dimmed for close subjects.

Lamp supports

In order to light with precision you must be able to place a lamp exactly where it is needed, and to support it there firmly and safely in any chosen position. Various systems have been developed that enable us to do this. Each has its advantages and limitations.

The ways lighting fixtures are supported will not only affect your lighting techniques but will directly influence the amount of time and labor required to rig a show. The principal methods include:

- Floor lamps (ground lamps)
- Lighting stands
- Boom light
- Auxiliary supports
- Support pole
- Suspension systems

Floor lamps (ground lamps)

These are simply lamps resting on the ground or fixed to a very low support (e.g. a bracket, plate or turtle). They may be used on the open floor or hidden behind scenery or furniture.

There are a number of situations where we need very low lighting:

- *Groundrow lighting units* (*cyc lights*) to bottom-light backdrops, cycloramas, cutouts (scenic groundrows).
- Hidden spotlights or broads lighting under ceilings, arches, and other overhanging planes.
- To provide localized illumination for furniture in the upstage area of a set.
- To illuminate subjects near floor level that are being shot on a low camera.
- To create an environmental effect – simulating firelight.
- As a dramatic keylight – e.g. horrific underlighting.

Floor lamps are not without their problems.

- The light direction is generally unnatural.
- It can cast upward shadows onto the background.
- We may see patches of light streaking across the floor.
- People may walk in front of the lamps and become brightly lit from the knees downwards!

Lighting stands

Most types of lighting fixture are fitted with a 'yoke' or 'stirrup'. This has a short central rod ('spigot, spud, pin') which fits into the tubular socket at the top of a support or mounting, such as a lighting stand. There are two regular sizes for spigot and socket – ⅝ in (16 mm) for lightweight fixtures and 1⅛ in (28 mm) for heavier types. Various adapters exist to fit the lamp housing (head) onto different mountings or accessories.

Basically, a lighting stand is a telescopic vertical tube with three horizontal bottom supports. The yoke spigot fits into the socket at its top and is held there firmly by a securing screw. (The screw is slackened temporarily to turn the lamp.) The central column consists of a series of independently adjustable concentric tubes (e.g. 'single-lift, 2-lift, 3-lift'). Columns are extended as required, starting with the inner tube, and held with their individual securing screws. The support legs may be attached, or detachable for easy transportation and storage. Very lightweight stands used for mobile or location work often fold. (Individual legs may be extendable, to suit uneven ground.)

Traditional lighting stands are heavy, particularly when supporting a lamp, and have casters to wheel them around. However, unless wheel-locks are fitted they can all too easily be knocked out of

Fig. 10.12 part 1   Lamp supports
Studio lighting is mainly suspended (from a pipe grid in this example).
1. Clamped directly to the grid (with a C-clamp fitting).
2. Lowered from an extendable hanger (sliding rod, drop-arm).
3. A movable trolley holding a vertically adjustable spring counter-balanced pantograph (extension 0.05–3.6 m/2–12 ft).
4. Telescopic hanger (skyhook, telescope, monopole).
5. In confined space, a spring-loaded support bar (barricuda, polecat) wedged between walls or floor/ceiling.
6. Telescopic floor stand (0.45–2.7 m/1½–9 ft).
7. Clip-lamp (spring-clamp. attached to scenic flat.
8. Scenic bracket screws to top of flat.
9. Camera light (headlamp, basher). Low-power lamp for eye-light or local illumination.
10. Grips, alligator or gaffer grips.
Safety bonds (wire or chain) are fitted to all hung fittings and accessories.

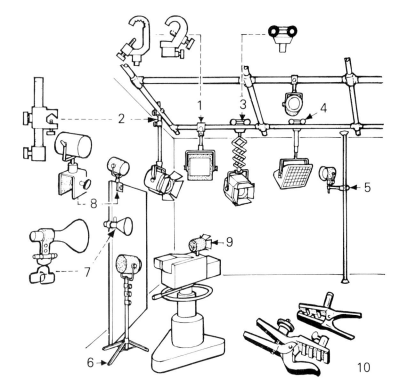

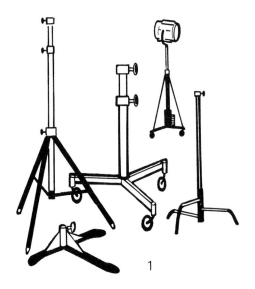

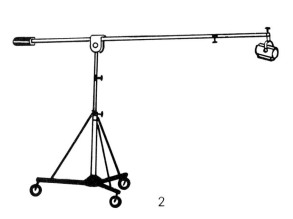

Fig. 10.12 part 2   Lamp supports
1. Lighting stands.
2. Boom light.

position. (Sandbags and wheel chocks help to immobilize the stand.) It is tempting to wheel lamps around on stands, but as they bump over cables or uneven ground, shocks are liable to shorten the life of the lamp filament.

Lightweight stands are invaluable on location, for they are easily transported, very compact when folded for storage, with maximum heights of up to e.g. 3.6 m (12 ft). It is advisable to take particular care with any stand lamp raised over about 2 m (6.5 ft) to prevent displacement or overbalancing.

The larger lighting stands used to support heavier fixtures (10-k's, arcs, softlites, floodlight banks) usually have a hand-cranked (wind-up) central column. The largest are motorized (e.g. 4.2 m/14 ft max.), or hydraulically pumped. Very low lighting stands support lamps from near floor level to a height of e.g. 0.6 m (2 ft) – *floor stand, turtle, spyder, T-bone, bracket, floor plate*.

Various ancillary fittings can be used with lighting stands, such as extension tubes, offset arms, multihead bars, flag holders. Adapters (reducers) may be needed to match the diameters of the spigot and the socket; particularly when using smaller fixtures.

Boom light

Here, a large wheeled tripod supports a long crossbar or boom arm on its central column. A small lighting fixture can be attached at one end of the boom, counterbalanced by an adjustable weight at the other. With a boom length of e.g. 2.4 m (8 ft) to 4.6 m (15 ft) this provides a great deal of flexibility in situations where there is no convenient method of suspending a lamp in mid-air. But the boom-light can be awkward to position and a menace in the wrong hands!

Auxiliary supports

There is a large miscellany of fittings that helps you to hang lamps in those awkward places in studio and on location. Extra care is necessary when using these methods of rigging lighting fixtures to ensure their stability and safety and avoid surface damage when fitting or removing them. Here are a few regular auxiliary supports:

- *Brackets* ('*scenery clamps*') that allow a lighting fixture (lamp) to be clamped onto the top or edge of a scenic flat.
- *Wall-hangers* such as '*trombones*', '*wall sleds*', '*hooks*' that fit over the top of a flat and hang down against it, supporting a lamp within the set.
- Devices incorporating a spigot or socket that allow a lamp to be fastened to whatever is at hand: rails, pipes, battens, poles, beams, framework, scaffolding, furniture, doors, windows, ceilings, etc.

These include *clamps* of several types – spring-jaw *gaffer clamps, 'gator (alligator) grips, space clamps, clip lamps, spring clamps, wrenches, furniture clamps.*

- *Fastenings* such as *screw-in spigots/spuds, face-plates* that can be screwed or nailed to surfaces.
- *Profile plates* that can be hooked onto vertical pipes.
- *Smooth-surface attachments* such as *suction cups, limpets, suckers.* These will temporarily attach lighting fixtures, bars, cameras, flags, etc. by suction to tiles, smooth metal, plate glass, car bodies and glasswork, plastic, polished marble, etc.
- *Gaffer tape (duct tape)* is an extremely useful multipurpose 50 mm (2 in) wide, strongly adhesive plasticized cloth tape. It will attach lightweight lamp fixtures, cables, and accessories to most surfaces.

Support pole

This is adjustable telescopic tubing, wedged securely between floor and ceiling (vertical pole) or wall-to-wall (horizontal pole), within corridors, arches, window openings, doorways, etc. It may be held in position by a strong internal spring or end-screws. Designs include *polecat, varipole, barricuda, jack tube, Acrow.*

Lamp fixtures can be fitted directly onto a vertical support bar or cross-beams (cross-tubes), using pipe clamps, C or G clamps, gaffer-grips, hook-on profiled plates, even steel wire loops (thongs). Where necessary, two or more tubes can be interlinked using 'couplers'.

Support poles have the advantage of being relatively quickly installed and removed, allowing quite elaborate lighting rigs to be improvised. Surface protector plates should be inserted at each end of a tube to avoid damaging the structure.

Suspension systems

A large proportion of lighting fixtures in studios are suspended, or clamped onto high structures. There are several advantages in this practice.

- The lamps will not get in the way of the action, cameras or sound boom(s).
- The lamps will not come into shot.
- More distant lamps can spread over a larger acting area.
- Fewer (but more powerful) light fixtures are needed.
- In many cases, fixtures can be adjusted from outside the setting area.

Single-point suspension

This method appears to offer maximum flexibility. Used for many years in film and some TV studios, in practice the system is pretty cumbersome and does not lend itself readily to rigging alterations. However, it is an approach that is still to be found, supplemented by other methods. Ceiling joists support wheeled trolleys or skids from which ropes are suspended on a tackle and pulley assembly. Each lamp hangs on a short length of barrel via a wire strop attached to the rope. These lamps can be suspended accurately over any point, positioned by thick cord brail lines; the lamp cables going to the nearest supply outlet.

Traditionally, film studios have a gallery or catwalk around the walls, with a tubular lighting rail supporting spotlights and arcs. Behind the scenery, custom-built gantries (parallels, gridwalks) with spot-rails, or light towers (scaffolding or timber) support lamps, while further lights clamp to the set walls. Various telescopic stands from a few inches high to hydraulic giants 4–5 m (15 ft) high provide the rest of the illumination. In television studios, however, where productions are generally simpler and continually changing, different approaches to lamp rigging are used.

If a large TV studio has a wall gallery/gantry at around 6 m (20 ft) this is only used occasionally to support lamps – e.g. for keylighting large areas, or for a following spotlight. Not only is the vertical angle too steep for normal use but all suspended fixtures in the vicinity have to be raised to avoid lamp shadows.

Fixed pipe grids

These are generally used in smaller studios for cheapness and simplicity and because relatively low ceilings (3.6–5.5 m/12–18 ft) preclude anything more elaborate. This system takes various forms, from a series of parallel tubes set across the studio to a regular pipework lattice, or ladder beams fitted about 0.6 m (2 ft) below the ceiling. Typical grid piping is 37–50 mm (1½–2 in) diameter and set at 1.2–1.8 m (4–6 ft) apart. Where there is sufficient room, there may be catwalks to give over-grid access to supplies, ventilation, and similar services.

Lamps may be attached directly to the grid using C-clamp fittings. Although simple, this method is laborious (requiring step ladders) and provides no height flexibility. Lamps may need to be hung some distance from the subject to avoid steep vertical lighting angles. Although some sort of height-adjustable lamp suspension is always desirable, where the ceiling is low there is usually too little headroom available, especially for larger light fittings such as effective soft-light sources. Smaller units such as broads have to be used instead despite their limitations. The maximum fresnel spot used is around 1–2 kW, with 2–3 kW soft light. Low ceilings bring further problems with ventilation, which, in turn, limits the lamp size and the maximum usable load.

Long battens

Suspended across the width of the studio, these are often fitted at regular intervals (e.g. 3 m/10 ft) over the staging area. Power outlets are available along each batten in the form of short individually numbered leads (pigtails) with in-line three-pin sockets, or twist-lock connectors. Each batten is suspended from wire ropes and hand-winched, motorized, or counterbalanced. With the last, removable cast-iron weights are added to match the lamps in use – a hazardous procedure that needs care in avoiding over- or underweighting.

This system nominally has the advantage that each battern can be lowered to the floor, so that lamps can be added, removed and serviced. However, this can only be done when the floor area is clear of scenery and equipment; otherwise a ladder or wheeled platform is necessary.

Barrel systems

These have proved versatile in large studios. They follow a similar principle, but use shorter lengths of tubular bar (1.2–1.8 m/4–6 ft long) in lines across the staging area throughout the studio. They can be raised/lowered by electric winches, individually or in groups from a wall switching panel. Each barrel has power outlets at its ends and may carry a 'standard' rig of two lamps per barrel (on sliding trolleys) or be rigged for a particular production's requirements.

All barrel systems have inherent problems when you want to rig lamps in positions between the barrel locations. The only solution is to attach clip-on crossbars, although this does result in a somewhat unwieldy arrangement in a large rig. Moreover, it adds to difficulties in adjusting/servicing lamps located over high scenery, ceilings etc., where the barrel networks cannot be lowered or reached. Pole operation obviates some of the adjustment hazards, and, on balance, barrel systems prove very practicable under difficult, high-pressure working conditions.

Parallel ceiling tracks

These provide another type of suspension system, in which supporting crossbars can be slid along to provide variations in rigging density. Whereas a fixed barrel pattern is spread over the studio, often leaving surplus lamps in one area yet insufficient in another, crossbars on ceiling tracks are readily adjustable.

Slotted grids

Slotted grids are used in the roof of a number of TV studios. Telescopic hangers (*telescopes, monopoles*) on wheeled trolleys, are suspended from slots at e.g. 50–60 cm (19.6–24 in) centers, giving considerable freedom when positioning lighting fixtures. Each telescope usually carries a single lighting fixture.

Trolleys and telescopes may be adjusted manually or by motorized control. In some installations the hoisting winches can be accessed from an overhead grid (false floor) below the ceiling of the studio. If additional lamps are needed, it may be necessary to remove an existing lamp in a channel to slide a new one in.

Basic supports

Each of these suspension systems has its particular advantages and limitations. Much depends on the type of productions being staged. The sort of equipment we need to light a large-area display is entirely different from that used for a complex drama production. A games show can pose quite different rigging problems from a musical program, or an interview. Studio lighting facilities need to be flexible to accommodate each type of show.

The basic method of attaching a lighting fixture to an overhead pipe, bar or barrel is by a *C-clamp*. Its spigot (spud) fits into a socket on the C-clamp (1⅛ or ⅝ in diameter) and is held firmly by a bolt, thumbscrew, or chained pin, then tightened (locked off) once the lamp's position has been set. In some designs the lamp is not locked in its suspension but can be rotated freely over a 360° arc (swivel, stirrup).

There are many occasions when you will want to vary the height of *individual* lamps. Where they are suspended separately (e.g. on monopoles) that is usually simple enough. But if you have several fixtures mounted on the same overhead pipe or bar you may need to vary their relative heights to avoid one obstructing another. A *hanger* enables you to hang a lamp some distance below its suspension point.

Fig. 10.12 part 3 Lamp supports
1. Telescopic hanger/slide pole.
2. Monopole/telescope.
3. Pantograph.
4. 'Trombone'.
5. C-clamp.
6. Wall-plate.
7. Suction cup.

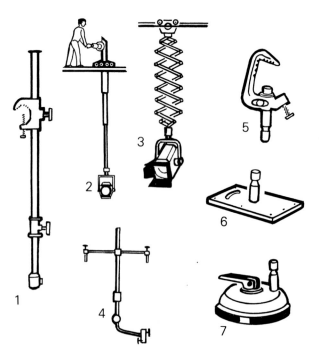

The simplest type of hanger (drop arm) has a *fixed length* (long or short). There are also two adjustable forms:

● In the first a vertical sliding rod passes through a special tubular C-clamp, and the lamp is held at the required height by a locking screw there. (An end-pin or bolt prevents the rod from falling out.) Hanger lengths can range from e.g. 1 m (3 ft) to 4.9 m (16 ft).

● The *telescopic hanger* is heavier, and is used where there is insufficient room above the C-clamp for the rod to extend fully upwards. This type has two parts; the vertical tube that is clamped onto the overhead suspension bar and a sliding rod within it, that holds the lamp fixture. A thumbscrew secures the rod.

Both forms of hangers are simple and cheap but can be dangerous unless securing pins are fitted through the rods, tubes, and C-clamps, and safety bonds (wire strops) or chains attached.

To alter the height of a lamp you need to support it with one hand (usually while standing on treads/steps), and with the other, release the securing screw, adjust the lamp height, and lock it off in its new position. So vertical adjustment can be tedious if the lamp is heavy or awkwardly placed.

Pantographs are widely used in TV studios to support lighting fixtures. These have a 'lazy tongs' construction, allowing the lighting fixture to be raised/lowered freely over e.g. 2 m (6.5 ft).

Pantographs come in two forms: the single-scissors version (which is lightweight, but rather flimsy and liable to sway), and the superior but bulky parallel-scissors construction. 'Counter-balance' strip springs are fitted within the pantograph. These are pulled down and clipped to crossbars to suit the weight of the light fixture being supported. When properly adjusted, the lamp can be raised/lowered easily with a lighting pole, and left at any height over its range.

Motorized telescopes (*monopoles*) are considered by some lighting directors to provide maximum rigging flexibility. Each 'telescope' consists of a set of interlocking tubes or parallel rods. The wire suspension cable, which hangs from a wheeled trolley on a ceiling track, passes through the tubes and is controlled by an overhead cable drum. Lamp height is altered by adjusting this support cable's length; manually by a hand power-tool, or a built-in electric motor. Telescopes can be rigged (where this is safe) from an overhead catwalk or grille (lighting grid) in the roof of the studio rather than from the studio floor. 'Transfer tracks' at intervals allow telescopes to be passed from one ceiling track to another. Power outlets are located above the tracks.

Lighting accessories

Having placed your lamp in the optimum position, the next step is to adjust its coverage. Within limits, you can vary the beam spread of a spotlight by flooding or spotting it; but, as we saw earlier, the effective output of the lamp changes too. If after spotting a lamp to restrict its coverage it proves to be too bright, you can dim it (but this lowers its color temperature), or use diffusers to reduce its intensity (but this softens and scatters the light).

Various ingenious devices have been created over the years that give the lighting specialist a much more precise control of lighting coverage.

Barndoor

This is undoubtedly the most widely used accessory for hard-light sources. It comprises a frame with two or four hinged black metal flaps. These can be individually adjusted to cut off parts of the light

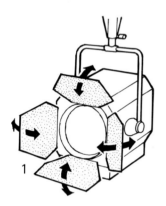

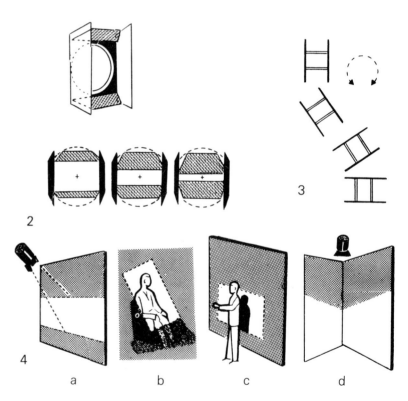

Fig. 10.13 Barndoors
1. The fitting consists of four (sometimes two) independently hinged adjustable metal flaps on a rotatable frame – two long, two shorter. The flaps may be used individually or in combination to selectively cut off the light beam in various ways.
2. The beam can be progressively 'boxed in' to provide a very localized area of light.
3. Rotating the doors will alter the angle of the cut-off.
4. Here are typical examples of the way the barndoors can be used. (a) To provide horizontal shading (long doors only). Light can leak from either end of the doors. (b) For selective illumination or subject isolation. (Tightly boxed in.) (c) For localized shading. (d) When lighting into an angle (e.g. the corner of a room) the shape of the light cut-off becomes distorted.

beam. The frame may slide into runners or lugs on the front of a spotlight or be permanently attached. On more flexible designs the frame can be rotated to allow the doors to be angled.

Barndoors are regularly used for a number of purposes:

- To restrict light to specific acting areas.
- To light a particular performer, piece of furniture or part of a setting without affecting other areas nearby.
- To light people, while leaving the background unlit or shaded.
- To light the background, while leaving nearby people unlit.

- To shade-off light from subjects that would cast unwanted or ugly shadows; e.g. people, scenery, furnishings, sound boom, practical lamps, etc.
- To provide isolated patches of light; e.g. around practical lamps.
- To produce atmospheric or decorative light streaks; e.g. shafts of sunlight on walls, light pillars on a cyc.
- To create shadowy areas within a setting.
- To avoid adjacent lamp beams overlapping and 'doubling', creating 'hotspots'.
- To prevent over-illuminating light-toned surfaces.
- To prevent light streaking along a surface that is parallel with the light beam.
- To shield the camera from back light that could cause *lens flares*.

Although barndoors are remarkably versatile, they have their limitations. You can only cut off the beam in a straight line. As one barndoor leaf is tucked in it can prevent another from being used effectively. Only rectangular shapes or slits of light are possible.

The sharpness of the beam cut-off can range from very soft to fairly hard, depending on the size and shape of the doors. Large doors give a sharper cut-off but they are cumbersome and take up a lot of space. If you have several closely grouped lamps the doors of one may get in the way of another, and it may not be possible to position them accurately. Moreover, it is not easy to restrict the light to small areas. Smaller barndoors, on the other hand, are less effective. The cut-off is very soft-edged, but you can achieve smaller isolated pools of light.

As the barndoors are closed, the effective output of the lamp is reduced, particularly as the light-slit narrows and edge-diffraction effects arise. The edge-shadow too becomes much softer. A lamp with the four flaps of the barndoors nearly closed wastes most of its light, and produces a small soft-edged blob, suitable perhaps to simulate the wall-illumination pattern of certain practical lamps.

Barndoors of the two-flap variety have excellent shading-off properties, but end-spill, or end-leak, may prove troublesome as uncontrolled light emerges from the corners of the doors. Similarly, light can leak between the flap-ends of four-blade barndoors if they have not been set together firmly, so creating distracting 'ears' of light on a background.

Light restriction

A number of devices can be attached to lighting fixtures to restrict the emergent light and prevent edge-spread.

■ *Light shield* This takes the form of large hinged plates, resembling vertical barndoor flaps, on either side of a *softlite* unit.

■ *Hood* A fixed four-sided box, similar to a large lens hood (ray shade), fitted onto a soft-light source.

Fig. 10.14 Light restriction

1. *Egg crate.* A latticework frame fitted over soft-light sources to restrict light spread.
2. *Spill rings.* A series of shallow concentric cylinders clipped onto a lamp to restrict the edge-spread of light.
3. *Snoot and funnel.* Metal tubes and cones of various diameters and depth are used to restrict a spotlight's beam to a localized soft-edged area. The beam size depends on the fitting's dimensions and the lamp's distance.
4. *Gobo.* A black wooden or cloth screen used to hide lamps or camera.
5. *Teaser* or *border.* An overhead gobo used to prevent fresh light from spilling onto the camera (causing lens flares) or back light spreading over the acting area.
6. *Flag.* A small gobo clipped in front of a lamp, or held in a *flag stand* to selectively restrict light, or cast a shadow.
7. *Targets, blades, dots.* Small rectangular, square, round or semicircular light shields (solid or translucent).
8. *Scrim.* A translucent flag used to reduce or soften local illumination. Very large versions (*butterflies*) are used for location exteriors.
9. *Cookies.* Irregularly shaped or translucent light shields, in both profiled and stencilled form. Used to create dappling, shadows, or general light break-up.

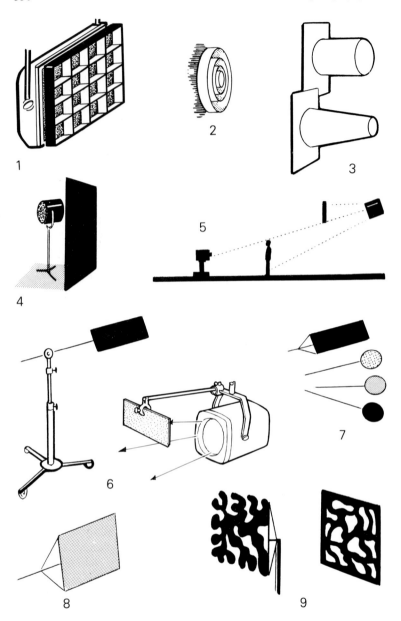

■ *Egg-crate* This is a lattice of criss-crossed vertical slats fastened over soft-light sources of all kinds.

■ *Spill-rings* A series of shallow concentric cylinders, usually clipped over an unfocused hard-light source, such as an internal reflector lamp.

■ *Snoots* A flat plate, with a large central hole, over which a cone or cylinder is fixed. *Malleable snoots* are seldom used but can be bent to provide a beam of any shape. The snoot is a useful device where you want to restrict the beam size of a spotlight, to isolate a subject from a group, or put a localized patch of light around a small subject, or a wall fitting. Because a snoot masks off a large area of the lamp there can be a considerable light loss. Flooding/spotting the lamp alters the intensity of the beam without affecting the light spread. Edge-softness varies with focusing, and the diameter and depth of the snoot. As the area of the spot of light depends on the snoot size and the lamp distance, a certain amount of trial and error is involved in getting a particular spot size. When the lamp is at an angle to the surface the spot will be correspondingly distorted.

1

2

3

Light restriction
When background lighting is localized, it can create a sense of confinement, and seem to restrict the subject. This impression is strongest when the area of light is sharp-edged.
1. The effect depends on the position of the lit area behind the subject. Here it is quite restrictive.
2. Where the subject appears to 'break through' the lit area, restriction is less, and the subject's 'strength' and prominence are increased.
3. If the area of light is soft-edged, the effect is less obvious.

■ *Gobos*   Originally, *gobo* ('go-between') was a general term for any large opaque black surface (cloth, board, plywood, metal plate) used to cut off light or hide lights/cameras. Nowadays, people regularly use the same term when referring to quite different things – for example a *camera matte*, i.e. a cut-out foreground plane, through which a camera shoots a scene. The stencil *mask* (*matte*) inserted into ellipsoidal spots to provide a light pattern is now universally called a *gobo*. And to confuse things further, it is often wrongly called a *cookie*!

Large gobos may be formed from flats, hung cloths, or drapes (drape frame) and used:

● *To screen a lamp*. You can place a gobo in front of a lamp and hide it from the camera's viewpoint. So the lamp can light part of the scene without being visible in the shot! A useful trick for long shots in the studio – particularly when shooting process work and chroma key treatment.

- *As a 'teaser'.* Hung in front of a backlight, to prevent light falling on the camera, and causing lens flares. Particularly helpful when shooting *towards* soft-light sources.
- *To cut off spill light (leak light, fresh light).* For example, to prevent unwanted light spilling onto backgrounds, from frontal soft light.
- *As a camera hide.* In picture the gobo appears as a black area but behind it another camera can be hidden, taking cross-shots of the action.

Several sizes and shapes of smaller gobo have evolved to suit various situations.* Made of thin black metal sheet, plywood, or wire-framed black cloth (Duvetyne), they are all used to mask off unwanted light, or to throw shadows. These devices are much more effective for those purposes than barndoors, and more versatile. But they are more trouble to fit and position, and require additional accessories. You can adjust the shadow's sharpness by varying the gobo's distance from the lamp.

- *Flag (French flag).* Typical sizes:
  45×30, 45×60, 60×90, 75×90, 120×120 cm
  18×12, 18×24, 24×36, 30×36, 48×48 in
- *Cutter.* Typical sizes:
  15×48, 30×100, 25×120, 60×152, 60×182 cm
  6×19, 12×40, 10×48, 24×60, 24×72 in
- *Blade.* A short narrow flag for very localized treatment.
- *Target.* Circular gobo, 25, 20, 15 cm (10, 8, 6 in).
- *Dot.* Small target, 7.5 cm (3 in).

■ *Cookies, cukes, cucoloris, ulcer* Skillfully used, cookies are a valuable tool in imaginative lighting. They are often 'home-made', and come in two forms: a sheet pierced with a series of randomly shaped holes ('dapple plate') and a sinuously shaped cutout profile. They have varying shapes and sizes, from a few centimeters across, up to e.g. 45×50 cm or 1.2×1.2 m (18×24 in or 4×4 ft).

Made of marine plywood or sheet metal (even perforated sheets of aluminum foil), the cookie is placed in front of a hard-light source to provide shadow patterns of various forms. Occasionally, a version painted on transparent plastic sheeting (acetate, celo, celloglass) is used. Whether these patterns provide slight dappling or hard clear-cut shapes will depend on the size and density of the cookie pattern, how close you place it to the lamp, and its distance from the background.

Cookies have a number of interesting applications:

- To suggest overhanging tree branches and foliage.
- To break up the light falling on plain surfaces, and give walls a more interesting mottled, uneven appearance. Flat overall

*The above dimensions are for *Mole-Richardson Co.* products.

illumination produces a blank featureless effect. Applied subtly, these variations look quite natural. Badly used, you land up with distractingly grotesque shadows.

- To provide decorative shadows on a plain background or cyclorama. Cookie sheets ('dapple sheets') can be slipped into the front of spotlights, in the supports used for color media and diffusers. Add a colored gel to produce colored patterns on a dark cyc, or color mixtures on a colored background.
- To reduce the harshness of a frontal key light, use a very diffused cookie pattern.
- To avoid the impression of strong or directional light sources; e.g. from a lamp beside the camera shooting down a narrow corridor.
- To provide fragmented light patches that suggest 'thick under-growth'. For such situations it is best to use a cookie with a framework having an open center.
- To avoid the impression of a flood of light when lighting action in dark surroundings (e.g. within a cave or mine) use a gobo plate with small openings and a solid center.
- To create small localized 'pools of light'. Using aluminum foil, you can cut holes empirically so that they are exactly placed for the action. One lamp can cast a series of separate pools of light onto objects in a display.
- To provide a specific type of shadow or light pattern, such as a window, venetian blinds. (Sometimes called a 'frame'.)

Accessory supports

Many light fixtures are designed with clips or other supports on their front edge that enable color frames, full diffusers, and metal cookie sheets ('dapple sheets') to be attached. But other smaller accessories such as flags or cookies usually need separate supports of some sort. The regular 'flag holder' involves *clamping discs* and support rods.

Clamping discs (*clamp heads*) consist of two thick discs with facing grooves and/or notches, held together with a thumbscrew. When the clamp is tightened it will securely grip rods, tubes, flags, cookies, scrims, etc. fitted into these recesses. So clamping discs are used as universal fasteners; to attach any accessory to a rod, a rod onto a lighting socket fitting. To hold accessories in front of a lamp on a floor lighting stand, you can either fit it with a supplementary offset arm or position a separate lightweight *century stand/gobo stand* nearby.

A *gooseneck* is a flexible steel tube, threaded at each end; to receive a spring clamp at one end and a spigot or socket at the other. Typically, the gooseneck is 45.6 cm (18 in) long, and 1.5 cm (19/32 in) in diameter. When fastened into the socket of a C-clamp, lightweight stand, or an offset arm the tube can be bent to hold a small gobo (or scrim) in place within the light beam.

Diffusers

Diffusers are broadly referred to as *scrims, wires, nets or jellies*. But in fact they are made of several different sorts of materials – sheets of spun glass, stainless steel wire mesh ('screens'), netted fabric (bobbinet) or frosted plastic medium. Earlier materials such as frosted/opal glass, oiled silk, gelatin wire mesh ('Windowlite') are seldom used nowadays, but each has its characteristic diffusion. Tracing paper or tracing linen can be pressed into use for brief periods, but these materials are inflammable, distort, and quickly become brittle.

All diffusers do two things: they reduce the intensity of the light and they scatter it, softening its quality. By selecting the appropriate material you can

● Diffuse with little light reduction (e.g. silks), or
● Reduce the light with little diffusion (e.g. wire screens).

Diffusion is always very relative. However heavily you diffuse a fresnel spotlight, its light will never become entirely shadowless; although shadow edges will become blurred and harsh modeling reduced. The further the diffuser is positioned from the light source, the greater its softening effect.

Unlike dimmers, diffusers allow you to reduce the effective output of a lamp without altering its color temperature. In fact you can use diffusers to balance the relative intensities of all the lamps on a set without the need for dimmers at all – a normal practice for location lighting.

Today, various grades of polymer, polyester, and acetate filters are available to provide diffusion in a range of densities. Lightweight versions are extremely effective, but because they are thinner and less durable than heavy-duty diffusers they should be fitted further away from a light fixture.

The term '*silk*' was originally used for the oiled silk material that provided a particularly pleasing diffusion. Today, it is applied to certain materials that give a similar visual effect: a woven cotton or nylon mesh termed white china silk or cornflower silk net ('lavender'), and certain translucent plastic sheeting.

Where a diffuser is to be left in position for long periods, or to be placed in front of intensely hot fixtures (e.g. open-face units), stainless-steel mesh is preferable, or a heavy-duty plastic laminate. Woven scrim or net is more vulnerable, for it distorts, shrinks and deteriorates in heat.

The light absorption and diffusion of a scrim depend on –

● The weave shape (square, diamond 'round');
● The hole size/weaving pitch;
● The tone and color of the material used. Black absorbs more light, with less diffusion. White is the reverse. Tinted (or burned) scrims modify the color quality of the light.

If you use multi-layer scrims, check the relative positions of their weave patterns for optimum results.

Diffuser design in plastic media is increasingly sophisticated. Some types (frosts) soften a light beam overall, while others ('silks') both soften and *spread* it. There is even an ingenious medium which has directional properties, and spreads the incident light beam in *one direction only (Rosco: 'Tough silk')*. Suitably placed, it will distribute light across a room without spreading upwards and downwards over the ceiling and floor. This property is a boon when space is restricted and the lamp is pretty close to the subject.

If you are using a diffuser in a lamp that already has a color-correction filter fitted to match a tungsten source to daylight their combined light loss could be as much as 75–90%! To reduce this loss, and avoid having to fit separate diffusers and color gels, *combination filters* are available. The most useful are the combined diffusers and blue color-correction 'boosters' that convert 3200 K tungsten sources to 'daylight'.

Using diffusers

There are several reasons for fitting diffusers to any lighting fixture:

- To *reduce* its light output (partially or overall);
- To *soften* a spotlight's beam, and reduce the edge-hardness of shadows;
- To *modify* a spotlight's beam: improve its evenness (eliminate a central hotspot, make edge fall-off more gradual); to make the edge of the beam less distinct; to spread and diffract the light;
- To *disperse or scatter* light from a broad source to make the illumination shadowless.

A typical single scrim may cut the light down by about one third, while at double thickness it is reduced to e.g. a half. (The actual proportions will depend on the density and nature of the material used.)

Instead of covering the entire lens of a spotlight with a diffuser you can reduce the intensity of just *part* of the light beam. A vertical or horizontal half-diffuser is regularly used. With a little care, you can even arrange graded diffusers – sections of increasing thickness – to provide progressive light fall-off. This technique allows you to light very selectively.

Normally, as a person moves towards a hung spotlight they become more strongly lit – even overlit. By inserting a bottom-half diffuser you can overcome this dilemma. At a distance, they are fully lit from the unfiltered top half of the light beam. Moving closer, they enter the diffused lower half, where the light level is controlled to a similar intensity. Conversely, you can use a top-half diffuser to restrain the amount of light falling on a person's face while allowing its full intensity to reveal modeling in their dark clothing.

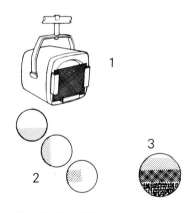

Fig. 10.15 Diffusers/scrims
Diffusing material can be put over a lamp to reduce the intensity of the light beam or to soften the light.
1. A full diffuser ('wire') gives overall light reduction or diffusion.
2. A partial diffuser allows you to alter the intensity (or softness) of part of the light beam; e.g. to prevent overlighting a specific area or intensity-doubling of adjacent light beams, etc.
3. A graded diffuser consists of a sandwich of single, double, triple thicknesses of diffuser. It provides progressive shading of a lit area; e.g. to shade a wall; to keep light intensity constant at various distances from a hung lamp.

You could, of course, use a *neutral density medium* for these purposes but it is less durable, and brightness steps may be more obvious.

Localized light reduction/diffusion

Many of the opaque ('solid') light control devices we have been looking at (flag, cutter, blade, target, dot, cookie) have an equivalent translucent form that provides localized adjustment of light *intensity*. These do not cast distinct shadows. A wire frame (usually open ended or open sided) supports an open-mesh gauze or net. This is classified according to its density; hence a 'single-scrim target' and a 'double scrim target'.

To reduce and soften harsh strong sunlight when shooting exteriors, or to avoid multi-specular reflections (e.g. when shooting glass or silverware in the studio), large overhead sheets of scrim or silk may be suspended as 'butterflies', e.g. 1.2×1.2 m to 3.6×6 m (4×4 ft to 12×20 ft).

Light control

So far, we have been looking at various ways in which you can modify the coverage, light quality and light output of individual light fixtures. If you are lighting a static subject such as a newscaster, then having set each lamp and diffused it where necessary for the optimum lighting balance, the job is complete. Apart, that is, from any afterthoughts or revisions.

However, as you have seen, there are many situations, particularly in television/video production, where you need to switch lamps or alter their intensities, and this requires some form of light control system.

Basic light control

The simplest method of controlling any lamp is to use a switch on the lighting fixture itself or in its power line. You should *never* 'switch' by inserting or removing a plug from a live receptacle (socket). This causes arcing and pitting on its contacts.

Where a number of lamps are to be switched on or off together a mobile *contactor unit/remote switchboard* can be used. This unit, which may be capable of handling several hundred amps, consists of large, heavy-duty switches which are actuated by electromagnetic relays. Cables are run to small remote-control switches that energize the electromagnets. Using this system, an actor can operate a wall switch in a studio setting 'to turn on the room lights' – and actually switch on many kilowatts of lighting fixtures.

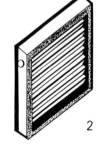

Fig. 10.16   Light control
1. *Iris diaphragm.* On a spotlight, an adjustable iris can be used to control beam spread (as in a following effects spotlight) or adjust the intensity of the lamp, according to its positioning in the optical system.
2. *Shutters.* Louvers or venetian shutters can be hand or motor controlled to adjust the light flow of any type of lamp, including arcs and discharge lamps.

There are mechanical methods of altering the effective outputs of some kinds of lighting fixtures:

- As we saw earlier, a follow spot may have an internal *iris*, which can be varied to alter its light output.
- Where a light source such as a carbon arc or a discharge lamp cannot be dimmed a variable *shutter* ('*venetian shutter*') can be fitted. This has a series of converging hinged blades, which can be adjusted by hand or remote motor control. Such shutters are now mainly used for effects such as lightning or explosions.

Why do we need light control?

Particularly in television production, one uses light control systems for several purposes:

- Preliminary lighting balance – estimated balance prior to camera rehearsal;
- Corrective lighting balance – revised balance made to suit shots monitored during camera rehearsal;
- Dynamic lighting balance – continual adjustments made while shooting to suit varying action and camera angles;
- Cues – in-shot lighting changes;
- Continuity changes;
- Routine switching;
- Load shedding.

Let's look at these in a little more detail. In the course of a production, or even a single scene, you may want, for example:

- To adjust the relative intensities of lamps lighting the action and the scene. Where time is short, it is quicker to adjust the overall lighting balance using dimmers than empirically fit individual screens/diffusers. If color temperature variations are unacceptable, they can be corrected.
- To compensate for different camera angles, and sizes of shot. The apparent brightness of surfaces can alter with viewpoint and distance.
- To provide decorative lighting changes – e.g. transforming the color-treatment or patterns on a cyclorama.
- To change to a different lighting treatment – e.g. switching a setting's treatment from 'daylight' to 'night'; or a different set-up for the next scene.
- For routine switching – e.g. an audience is unlit during an act, and illuminated ready for audience shots at its conclusion.
- To alter a scene's general atmosphere – e.g. from gaiety to a sinister mood.
- To provide action lighting changes – e.g. as an actor switches on room lights, or draws drapes.

- For lighting effects – e.g. fire, dawn.
- To adjust lighting loads. Switching off the lights on one set, when shooting another, to reduce total power consumption, heat, light leakage, etc.

Control requirements

Film making has traditionally made use of portable dimmers and contact breakers for lighting control. Early television often used a combination of film facilities and standard theatrical lighting dimmer boards. As TV production techniques developed, it became evident that more flexible facilities were needed. TV lighting control systems evolved, and today, highly sophisticated methods are available. Theatrical and, to some extent, film production now benefit from these developments.

Fundamentally, one needs switching and dimming facilities that will:

- Switch a single lamp on (or off);
- Switch a group of lamps on (or off);
- Switch some lamps, leaving others unaffected (i.e. on or off);
- Adjust the brightness of any lamp;
- Fade any lamp (or group of lamps) up or down at any speed;
- Fade some lamps up (or down) and others down (or up);
- Fade some lamps, while leaving others unaffected.

We want to be able to see at a glance the intensities of any lamp or group of lamps (by dimmer settings, or indicators); whether or not the lamps are actually lit at that moment.

Rather than have to continually reset the board by hand, readjusting it for every operation, we need some form of 'memory' or 'store' that will hold details of switching combinations and dimmer levels, that can be selected in an instant on cue, at the press of a button. Automatically timed fades are also a useful facility for some types of production; to control background lighting or color changes.

Dimmers

Dimmers control the amount of current flowing through a lamp and so adjust its light output. Over the years, several types of electrical dimmers have evolved.

Ideally, as you raise a dimmer from zero to maximum it should provide a gradual uniform increase in the light output. (And it should do this irrespective of the power of the lamps connected to it.) In other words, there should be a square-law relationship between the fader setting and the light output; for the eye's response follows a logarithmic law. In practice, this ideal is not easy to achieve, and several technical designs have been used over the years (see Table 10.1).

Table 10.1   Types of electrical dimmers

	Advantages	Disadvantages
Resistance dimmer. Lamp current varied by adjusting amount of resistance wire in series with lamp	Simple. Medium cost. Reliable Suitable for AC and DC	Power dissipated as heat in resistance wire. Fairly bulky. For even dimming, small load must not be fed from high load dimmer
Auto-transformer dimmer. Iron-cored coil across the power supply; variable voltage can be tapped off to feed lamp	Cool working; compact; reliable; smooth control with varying load. Excellent power economy	Suitable only for AC. Relatively expensive. Bulky
Saturable reactor (reactance dimmer) (choke control). An iron-cored coil in series with an AC-fed lamp will cut down its current; and hence lamp brightness By adjusting a DC control-current flowing through an overwound control-coil, the impedance can be varied	A small DC control supply can be used remotely to adjust large loads. Fairly low cost. Occupies small space	Suitable only for AC. Heavy. Dimming action can vary considerably with loading
Magnetic amplifier (low-current system). A derivation of the saturable reactor. An auxiliary DC amplifier adjusts control-coil current, using feedback developed from the load circuit	Smooth control for various loads	Suitable only for AC. Small dependence on load
SCR dimmer (silicon-controlled rectifier, thyristor). A semiconductor device which controls lamp current according to the timing of a stream of electrical 'gating' pulses applied to it. Suitable for AC supplies only	Lightweight; smooth control with varying load; compact; high efficiency. Medium to low cost	Can generate interference: (a) acoustical, as lamp filaments vibrate at audio frequencies; (b) electrical noise induced into nearby mic cables. Filter circuits and special mic cables reduce

Resistance dimmer

The basic *resistance dimmer* which is suitable for AC or DC supplies still has applications, particularly in location film making. Its construction is simple and robust, relatively cheap and reliable. On the other hand, it is fairly bulky and power is wastefully dissipated as heat.

If a resistance dimmer designed to handle e.g. a 2 kW load is used for lower power sources (150–500 W) the fader does not respond evenly over its range. Other systems such as the *auto transformer* do not have this particular shortcoming.

SCR dimmers (silicon-controlled rectifiers)

SCR dimmers are increasingly used for lighting control systems. This system relies on a semiconductor device which controls the lamp current according to the timing of a stream of electrical 'gating pulses' applied to it. The 'dimmers' or 'faders' adjusting this control voltage can be located remotely from the SCR rectifier bank, and this allows the lighting control board or console to be very compact. It can even be mobile: taken to the studio floor to switch and dim lamps as they are set, and then transferred to the production control area during rehearsals and recording.

Power supplies

To use electricity safely requires knowledge and experience. It is all too easy to overload circuits, to create hazardous situations, to unbalance supplies. So here we are going to look at just the broad principles and practices rather than installation details.

Away from the studio, power supplies for film and television/video production lighting can come from several sources:

- Batteries – battery belt, portable battery units, car battery, supplying low-power lamps (probably hand-held).
- Small portable generators – car-carried or towed generators (e.g. 4.5–6.5 kW).
- Household wall outlets – for loads below e.g. 3 kW AC.
- Main power-supply input points in a building – company supply bus-bars, sub-station.
- Large mobile generator (hired) – e.g. up to 1000 A AC or DC supply.

Cabling on location

When you are using only a few relatively low-consumption lamps you do not need comprehensive power-control systems. It is often sufficient to plug your lamps directly into regular wall outlets (power points) fed by general utility supplies. Each lighting fixture has a short length of attached cable fitted with a plug ('male' connector). If it is not long enough to reach the outlet, an extra length of supply cable (*extension or feeder*) will be needed, which has an attached plug and socket (male and female connectors). If an extension cable is fitted with a multi-socket plugging box (distribution box), several lamps can be fed from the same power outlet; provided, of course, that they do not exceed the permitted maximum current.

Most location lighting, however, is on a much larger scale and requires the services of specialist contractors. There are many pitfalls, especially where a three-phase system is involved, and circuits/equipment fed from the three separate phases must be kept well apart.

All AC power cables have three conductors: live, neutral and ground lines. (For DC supplies these lines become positive, negative and ground.) The live and neutral lines carry the power and the ground or 'earth' wire is attached to equipment casing to ensure that it remains electrically safe. Without this grounding there is always the danger that if the insulation becomes damaged and either conductor exposed, or moisture seeps into equipment, the fixture can become 'live', resulting in electric shocks when it is handled.

Cabling starts at the bus-bars of the incoming supply, where the *main feeder* is attached, using special lugs. (This may involve the permission and cooperation of the supply company.) At the other end of this feeder cable is a large *distribution box (spider, cable splicing block)*. From this central distribution point a series of *secondary feeder cables* are arranged. These provide the fused receptacles or sockets that supply the lamps. If a lamp is some distance from a secondary feeder's outlet an intermediate *extension cable* is needed, with a plug at one end and a receptacle for the lamp cable at the other.

If the lighting rig is widespread the cabling can stretch around over considerable distances and safety becomes of paramount importance. Wherever possible, cables are tied and suspended, laid along the bottom of walls, covered with sloped ramps to prevent damage from passing vehicles, etc.

Cabling in the studio

Some *studios* use a similar approach to cabling. Power is available at a few main wall outlets and distributed as needed by feeders with junction boxes and extension cables. In permanent TV studios facilities are usually organized to provide maximum flexibility with minimum labor. So built-in wiring provides power at a large number of distributed outlet points, allowing loose cabling to be kept to a minimum. As well as a series of wall outlets (power points) around the walls, catwalk, ceiling grid, etc., the hanging battens (bars, barrels) have permanently wired outlets from which suspended lamps are supplied.

While filament lamps can be powered from either AC or DC supplies, metal-halide fixtures (including HMI lamps) function only on AC. In some TV studios, where carbon arcs and/or camera cranes are used, a regulated DC supply is available.

Patching systems

For some reason, patching can mystify; but it sounds more complex than it really is. The principles are simple enough:

● Each lamp/lighting fixture is plugged into a nearby numbered outlet – e.g. '100'. This is usually a receptacle (socket) with a lock-on or screw fastening, attached to the studio wall.
● A permanently connected cable (tie-line) runs from this outlet to a central *patch panel (lighting patchboard, patch bay)* in a room near the studio. This is a simple unpowered interconnection cable, used to link the lamp to the power supplies. At the patch panel it appears as a numbered plug on the end of a *patch-cord* (in this

example, '100'). Overhead battens (bars, barrels) use either numbered or lettered outlets (e.g. 20,21,22,23 or Bar 5-A,B,C,D). These outlets may be in-built or in the form of three-pin or twist-lock connectors on short lengths of cable ('pigtails').

● The *patch panel* itself has a series of numbered receptacles (sockets) which supply *power* via the lighting control board (dimmer board, control desk, lighting console).

The point to bear in mind here is that the numbers on these receptacles have nothing to do with the tie-lines to the studio! The wiring numbering and the power supply numbers are quite independent. If you want to use power *channel* 23 on the board to control a lamp (or lamps) attached to studio outlet '100' you can do so. Nominally at least, any outlet in the studio can be plugged into any power supply channel on the dimmer board. Even when lamps are spread around the studio, and plugged into a variety of outlets (lighting a large cyclorama perhaps), free patching of this kind enables you to group them onto adjacent dimmers. When the patch panel has a *single patching* system you can only plug one patch-cord into each supply channel. If you are plugging 2 kW lamps into 2 kW supply sockets you are able to make full use of the available power.

But suppose you have a number of low-power decorative practical lamps (e.g. 100 W each) and plug each fixture into a *separate* 2 kW outlet. You could soon find that you have used all the available power points, and with power capacity in hand, are unable to patch the remainder of the lamps as there are no outlets left! The solution in this particular situation is to plug an extension cable with a multi-outlet distribution box into the studio outlet. A 'splitter-cable/splitter-box' used in this way not only makes full use of an outlet's available power but also allows you to control all the attached lights on one dimmer.

Multiple patching facilities reduce the need for all this extra cabling in the studio. Each supply channel on the patch panel has several parallel power outlets (two to four). You can plug a different patch-cord into each. Whenever you adjust that channel's dimmer or switch all these lamps change similarly.

Multiple-patching systems have their limitations. You can inadvertently overload a channel by plugging in lamps that together exceed its rated output. If you group together lamps of different power they will dim at different rates. So care is necessary.

Some patch panels include *switched-only* (*non-dim*) *power outlets*. These channels have no dimmers attached and so supply only full voltage. (They are switched at the lighting control board.) Apart from design economy and space saving, this facility is particularly useful if you want selected lamps to remain fully lit throughout a production, even when other lighting is being dimmed or switched. If the 'switched-only' lamps are too bright you will have to flood or diffuse them, alter their distance, or reduce their power.

Patchboards are simple to use and offer great flexibility. But the

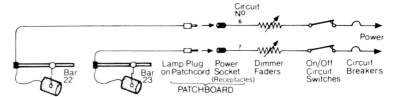

Fig. 10.17 part 1   Patching and control – basic patching

The *lamp* on the hanging bar/batten 'Bar 22' is plugged into a numbered *outlet* fixed to the end of the batten. All batten outlets are permanently routed to a central *patchboard*, where they appear as a series of numbered cables (*patch cords*), each marked with its bar/batten outlet number.

A patch cord is plugged (*patched*) into any one of a series of numbered power-circuit outlets (*channels*) – circuit No. 6 in this example. Each power supply incorporates its own *control dimmer* or *fader* and *channel switch*. Similarly, the lamp on bar 23 is being fed from power circuit outlet No. 7.

Each channel can be switched and dimmed independently. Avoid exceeding the power loading limit of a channel (e.g. 2 kW). Also switch off channels before patching lamps, to prevent burning contacts due to arcing.

Some installations use patch cords; other have permanent connections, switched patching, or miniature peg-board patching.

Fig. 10.17 part 2   Patch panel/patchboard

The cable from each studio outlet appears as a similarly numbered receptacle on the patch panel. Power from each channel on the lighting control board appears as a series of patchcords; numbered to correspond with their respective dimmer/switching circuits.

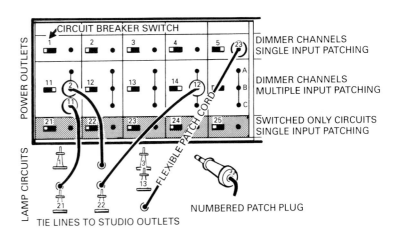

Fig. 10.17 part 3   Patching and control – group presets

A series of individually adjusted circuits can be grouped, to be dimmed/switched communally, e.g. fading cyclorama lighting up/down. A typical situation is shown here:

Preset A (fully faded up). Lamp 1 – Switched on, fully faded up (bright). Lamp 2 – On, faded down (dim). Lamp 3 – On, but faded out (out).

Preset B Faded down, reducing supply to all groups. Lamps 4 and 5 – On, individually faded up (but dim due to preset dimmer.) Lamp 6 – Channel switched off.

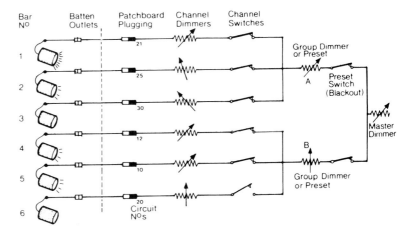

intertwining patch-cords can take time to disentangle whenever
changes are necessary. Remember, if the studio has 150 power
outlets, there will be 150 patch-cords – although they are rarely all in
use at the same time.

To avoid this 'knitting', various *automatic patch panels* have been
devised. Probably the neatest is the small *cross-point matrix panel,* in
which a punched board has the numbers of the available outlets
marked on a vertical scale and the power channels along a horizontal
scale. To patch a lamp, a small selector pin is inserted where the
required outlet and channel buses cross. One can see at a glance the
patching distribution, and repatching is straightforward. Other
versions using sliders or mini patch-cords are less convenient.

In all systems you should take care not to *hot-patch*, i.e. to patch a
lamp into a live circuit. The immediate current flow causes arcing
which burns contacts and can ruin them. Instead, use the channel's
circuit-breaker switch, or switch the lamp itself off while patching.
Take care, too, to push a patch-cord plug fully into its power outlet
socket.

Lighting control systems

Lighting control boards (dimmer boards, control desks, lighting consoles)
vary considerably in complexity from basic units with a handful of
faders to elaborate computerized systems with a wide range of
features. Even home computers can be used for lighting control!

The traditional *preset lighting board* we shall be looking at in detail
is widely used, and very suitable for many regular types of
productions, including interviews, newscasts, talks, demonstrations.
Where it will do all that you need, there is no real advantage in
having more elaborate facilities.

Simple control systems

The simplest type of lighting board (Figure 10.17, part 1) has a
single row of *dimmer controls (faders)*, each separately wired to a
corresponding power outlet on the patch panel. As you move a
dimmer over a 0–10 scale by its lever, slider, or thumb-wheel the
intensity of the lamp patched into that channel changes. You can
fade it up to maximum (10), fade it out (0), or leave it at any chosen
level. Using individual *channel switches* you can switch any lamp
on/off, or flash it momentarily. A *master dimmer* at the end of the row
of faders allows all lamps to be faded up/down together. A separate
blackout switch turns them all on/off.

In a TV studio it is a useful practice to work with faders around
'7', rather than maximum on the scale ('10'). This will not only
extend the life of lamps but it leaves power in hand to give an extra
boost wherever a lighting accent proves necessary. Although slight
dimming causes all lamps to work at a lower color temperature than

normal (around 2950 K rather than the maximum of 3200 K), you can easily compensate by white-balancing the cameras to this lower value.

As a rough rule of thumb you will find that the light output from a 10-step square-law dimmer is reasonably proportional to the fader setting *squared*: i.e. at '5', it is about $5 \times 5 = 25\%$ of the full output. (Remember, the eye sees the fader steps as proportional increases in brightness.) For most practical purposes there is no obvious change in color temperature if you fade up from 7 to 10 ($+150°$) or down from 7 to 5 ($-150°$).

Preset control systems

In a basic lighting board each channel has a separate switch and dimmer (fader). A *master dimmer* allows all the lamps to be faded up/down together. A *blackout* switches them all on/off. This

Fig. 10.18 Two-preset control system
Each lamp circuit has a choice of two alternative control systems or *PRESETS* (*A and B*). The presets appear on the board as two independent banks of faders and switches.

A PRESET SELECTOR SWITCH beside each channel decides which preset is in use for that particular circuit.

Each preset has a *GROUP DIMMER* (*SUB-MASTER*) that affects all lamps switched over to it. It also has a group on/off switch (*PRESET BLACK-OUT*).

Some boards have inter-preset faders that allow you to 'cross-fade' from the dimmer settings arranged on one preset to those set up on the other.

A *MASTER FADER* together with its *MASTER BLACK-OUT* control the board's entire output.

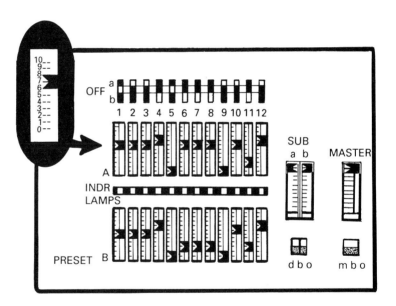

arrangement provides a straightforward method of switching and adjusting lamp intensities. But operationally, it is very limiting. Using several fingers, you may be able to control a few channel faders or switches at the same time. But consider the situation when you want to:

● Simultaneously fade or switch a number of lamps within a group;
● Vary the relative intensities of a number of lamps;
● Cross-fade between two or more different lighting conditions;
● Gradually fade some lamps up, others down, while the rest remain unaffected.

The *preset control system* offers extra flexibility. Fundamentally, you can think of a *three-preset system* as three identical basic lighting boards with arrangements for switching or fading between them. (There are designs with two to five presets.) These three units are referred to as *presets* or *banks,* and given letters (or colors) to identify them – e.g. presets A, B, C; or Red, Blue, White presets. Each of these presets has its own

- Group *dimmer* (now called a *sub-master, group fader, preset dimmer, or bank dimmer*), and a
- Group *power switch* (now called a '*preset black-out*' switch).

Looking at a typical three-preset board you will find that each numbered lighting channel has channel on/off switch and three separate faders – A, B, C; one for each preset. There is also a three-way *preset selector switch* to select which of the channel's preset faders is operative.

All the lamps that are switched to a particular preset can be controlled at the same time. So if some are switched over to *Preset 'A'* group, others to *Preset 'B'*, and the rest to *Preset 'C'* you have three independent communal controls. In a chat show you can put all key lights and filler on *Preset 'A'*, and at the end of the show, fade out *Preset 'A'* to leave people silhouetted against the lit cyc. When you want a particular lamp to remain at the same intensity at all times, whatever the preset switching, its channel must be at an identical fader setting on all three presets.

Some lighting boards have inter-preset faders to allow you to *cross-fade* between presets. If a channel's faders are at the same setting on each preset nothing will happen on cross-fading – a '*dipless cross-fade*'. Where '*timed cross-faders*' are fitted, these will automatically cross-fade at a chosen speed.

A *master fader (grand master)* with its *DBO switch (dead blackout)* has control of the complete lighting board, and is used to kill the board during a break or at the end of the show. It is good practice to use the *master fader* when bringing up the lights from cold. The instant shock of switch-on not only puts an immediate load on the power supply but may 'blow' lamp filaments due to rapid expansion.

The preset system has a number of *advantages*:

- You can get to the fader or switch of any channel and adjust it in an instant.
- The fader positions of all the channels are obvious at a glance.
- It is simple to adjust a series of faders in rapid succession (e.g. to change nearby background brightness as someone walks around a room with a candle).
- You can reset the standby presets while using the active preset.

There are *disadvantages*, too, including

- The proliferation of faders! If the board has 50 channels, with three faders for each, that is a total of 150 channel faders; plus

sub-master preset and master controls. This arrangement takes up a great deal of space.
- Only a limited number of faders can be operated at a time.
- It can be difficult to simultaneously operate nearby fader levers on the same preset (e.g. moving 1,3,5 up and 2,4,6 down).
- Repositioning a number of faders to different settings is time consuming and leads to inaccuracies.
- If you have a succession of lighting cues it may be necessary to alter fader settings and switching on a preset after use, in readiness for the next cue.
- Lamps may need to be switched from one preset to another during a show to avoid their being affected by later operations.
- You need to keep written lists of all patching, fader settings, and cues.

Preset variations

To reduce the number of faders on a preset board several ingenious design variations evolved. The *single fader preset system* has only one fader per channel. A *channel preset selector switch*, beneath each fader, determines which of the three preset circuits it is attached to and whether the dimmer is operational. This design is economical, and does allow inter-switching/fading between groups of lights. However, you cannot change a lamp's intensity by cross-fading between presets.

Some lighting control boards include a single *flash button (flash key)* for each channel, which allows you to flash it manually – either to identify lamps or for an effect.

Grouping buttons allow you to select a number of channels *within a preset* and control them with a separate *group fader*.

Chasers automatically switch or flash a series of selected channels in sequence at an adjustable speed to produce the effect of moving patterns of light. Timing variations give various effects, including forward, reverse, bounce, cycling.

Advanced lighting control systems

However convenient the preset system may be, it can prove limiting in bigger studios, where a couple of hundred lamps are to be controlled, with frequent lighting changes. To reduce the number of individual controls and provide a more compact layout, various computer-controlled systems have been developed. Unfortunately, their designs are too dissimilar to summarize here.

Some advanced lighting control systems have a central keyboard on which you punch up the required channel(s), using a multi-purpose fader lever to control the selected lamp(s). On other systems you use a VDU display unit (or two) with a touch screen or light pen to adjust fader settings and arrange lighting cues, recall data, display outlet usage, etc.

Computer memory allows you to store (*file, memorize*) all variations in switching and light intensity, using magnetic disc or tape, or solid state devices. At the press of a button you can cue gradual or instantaneous lighting changes, effects lighting, rebalancing, load-shedding, etc. In addition, there are usually facilities for automated light switching (flashing patterns, chasers), and 'timed fades' (fade-up/fade-down at a preset rate).

In some installations a display shows all studio power outlets, with lit indicators for live circuits. Flashing a channel on the lighting board immediately confirms the outlet to which a particular channel is connected.

Bigger and better . . .

How extensive your light-control facilities need to be clearly depends on the types of productions involved:

- The size of the staging area in the studio;
- The number of settings;
- The type of action;
- The spread of the action;
- Whether there are lighting changes;
- Whether these can be done while off shot or are seen in picture;
- How elaborate the lighting treatment is;
- Whether there are action lighting cues;
- Whether the same area or setting is used for several sequences with different lighting;
- Whether there are rapid turn-rounds; i.e. one show follows another on live transmission;
- Whether the show is planned in outline only or in detail;
- Whether action is rehearsed or off-the-cuff.

Theoretically at least, the greater the flexibility of the lighting control system, the more readily one can adapt to the unexpected.

The patching sheet

Although you can always leave the patching sheet to those rigging the show there are advantages to including it as part of the lighting plot. It is one thing to draw lamp symbols on a sheet and another to find that you do not have sufficient channels to power them. That is best discovered during the lighting plot! If you have drawn the plot well you do not need a list of notes detailing each lamp's purpose. It will be obvious.

As you saw earlier, a lighting fixture is plugged into the outlet socket of a nearby tie-line, which appears at the lighting patchboard as a numbered cord with a plug (e.g. 12B = socket B on barrel 12).

Table 10.2   Patching sheet

Studio:	A		Production: THE BIG SHOW
Settings:	5 March		Lighting Director: John Doe
Lighting rig:	5 March		Crew: 3
Rehearsal/taping:	6 March		

Channel[a]	Outlet[b]	Dimmer	Preset[c]
1	3a	7	A
2	3b	8	B
3	7	7	B
4	10	4	B
5	12	5	B
6	9	5	A

etc., etc.

[a] Dimmer control channel on lighting board.
[b] Power outlet point (tie-line) in studio.
[c] Dimmer/fader setting, and preset allocation, optional. (Useful on manual
     boards, for repeat productons.)

This is plugged into a power source which corresponds with a dimmer circuit on the lighting control panel. If you want a series of lamps that are spread around the studio lighting a cyclorama to appear on adjacent dimmers, so that you can fade them all in/out manually, then this can usually be arranged during patching. Although you can make the most use of power outlets by plugging several lower-power lamps into one socket (e.g. *all* low-wattage practical lamps into the same dimmer channel) remember this will prevent you from adjusting their individual intensities.

In many installations the outlet sockets of tie-lines are distributed around the studio. If you use all the sockets in a particular area you may need long extension leads to reach away to other sockets elsewhere. Where the number of lamps on a batten/bar exceed the available sockets this will mean that extension leads have to be run to wall sockets, or other battens. Matters of this kind are not simply academic. They affect the rig's complexity and the time/labor involved.

The cue sheet

A lighting *cue sheet* shows a list of the various lighting changes that are to take place during a production; lights that are to be switched on or off, or faded at certain moments. Typical cues include color changes, transformations, effects, day/night changes, room-light switching, etc. It can take the form of notes on a script or a running order or a list of shot numbers with the changes alongside.

How complicated the cue sheet is in practice will depend on the design of your lighting board. You may have a list of *presets*, or *memories (files)*, each with a different combination of lamps at various intensities. Then changing from one condition to another simply involves going to another preset or memory at the right time. If the lighting board is more limited you might need to rearrange presents or memories during the show, or adjust channels by hand.

In any event, there are occasions where you will want to operate a dimmer by hand during a shot – e.g. to reduce a lamp's intensity as someone gets closer to it, or to vary the brightness of a lamp that is simulating firelight.

11 Color temperature

If the color quality of the light and the color characteristics of the film or TV system do not match, the final picture will have a pronounced color bias. Here we shall discuss the theoretical and practical problems involved.

Identifying color

There are several ways of precisely identifying any color.

- You can make a list of the proportions of each spectral hue it contains. But this is a laborious process, and gives no sense of the subjective color effect.
- You can draw a graph showing their relative amounts as in Figure 12.1. With practice, this can be a very useful guide.
- You can identify its location on a *chromaticity diagram* (Figure 2.18), or quote a color notation system (e.g. *Munsell*).

However, during everyday picture making we need a rapid method of assessing the *color quality of light*, and none of these methods is really convenient. Instead we use a useful 'code' originally devised by the physicist Lord Kelvin; the concept of 'color temperature'.

Principles

Kelvin found that as you heat a carbon block the color quality of the light it emits changes with its temperature. As the block gets hotter, the light quality gradually alters from a dull red glow to a strong yellow-white, and contains increasing proportions of visible spectrum. If heated further, this light takes on a brilliant bluish-white color quality. You'll recognize these changes as similar to those you see on heating a metal rod in the fire.

The important discovery was that the color quality of the light was directly related to the block's (black body) temperature. So the idea developed of using these temperature readings as a 'code' to classify the color quality of any light source – its *spectral energy distribution*. Because the numbers involved here are large when measured on the conventional Celsius, Centigrade or Farenheit scales, a special Kelvin scale is used – an *absolute scale* in which zero °K is −273°C.

So if, for example, we have a lamp that produces illumination of reddish-yellow color quality we can theoretically check on the Kelvin scale and find a standard with a similar spectral spread (*spectral distribution*). If it corresponds to that of a body heated to 2000° on

Table 11.1   Typical color temperatures (K)

Standard candle	1 930
Household tungsten lamps[a] (25–250 W)	2 600– 2 900
Projector bulbs[a]	3 200
Studio tungsten lamps[a] 500–1000 W	3 000
Studio tungsten lamps[a] 2000	3 275
Studio tungsten lamps[a] 5 kW, 10 kW	3 380
Tungsten-halogen lamps[a] ('quartz lights')	3 200
Overrun tungsten lamps[a] (Photoflood)	3 400– 3 500
Fluorescent lamps	3 200– 7 500
High-intensity arcs	6 000
Sunrise, sunset	2 000– 3 000
Sunless daylight, early morning, late afternoon	4 500– 4 800
Midday sun	5 000– 5 400
Summer sunlight + blue sky light	5 500– 6 500
Overcast sky	6 800– 7 500
Xenon (arc or flash)	6 000
Hazy sky	8 000
Clear blue north sky	10 000–20 000

[a] Run at their correct, full voltage. Lamps using a supply voltage below 240 V operate at higher color temperatures (e.g. 50–100 K).

Kelvin's temperature scale, we refer to our lamp as having a *color temperature* of 2000 K. Remember, this 'code number' does not refer to the *real* temperature of our light source. We are simply saying that its *visual appearance* matches that standard. (Tungsten melts at around 3650 K, with a color temperature of 3600 K.) In Figure 11.1

Fig. 11.1   Color temperature
The color quality of a light source can be classified by comparing it with the standard of a black body radiator. The particular spectral distribution is measured on a scale of kelvin units which is the temperature, on the absolute scale, to which the black-bodied radiator is heated.

Although overall energy alters as well as spectral distribution (1) it is more convenient to center the set of curves (around an arbitrary 580 nm in this case) to compare their relative spectra (2).

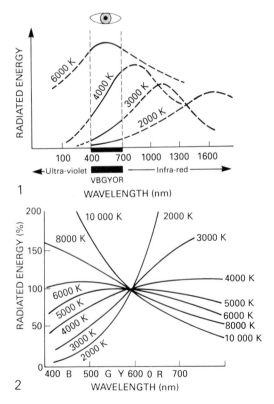

you can see how the proportions of different spectral colors change with the color temperature on the Kelvin scale.

Light sources

When you use this system to classify the color quality of the light from *tungsten lamps* you will find an extremely close visual match between the theoretical color temperature curve and the actual color quality of your lamps, for their illumination has a continuous unbroken spectrum. For many other types of light source, such as natural daylight, the actual proportions of spectral colors only very approximately resemble those of the Kelvin scale. So although we still refer to the *color temperature* of their light this is a less accurate, more subjective classification of its color quality. Close enough, though, when color matching pictures. As you can see in Figure 11.2, there are many variations.

However, when we analyse the illumination provided by *discharge lamps* (e.g. HMI and HID lamps, fluorescent lighting) we find that *their* spectral coverage is very irregular with pronounced 'spikes',

Fig. 11.2 Mixed color temperatures
For optimum color quality, the color balance of the system should match the color temperature of the prevailing light. So in daylight (A) the system balance should approximate to e.g. 5500 K, and in tungsten lighting (B) to e.g. 3200 K. When daylight and tungsten lighting are mixed (C) their incompatable color temperatures will cause major color errors, whatever the system balance.

The solution is to place compensatory color filters over the incompatible source, to convert its light to that of the system's balance (D) or (E).

Table 11.2   Standard luminants (CIE)

Physical sources: classifications used for calculation purposes

Luminant A[a]	2856 K	Whiteness of tungsten filament lamp
Luminant B[b]	4874 K	Whiteness of direct sunlight. (Not now generally used)
Luminant C[b]	6774 K	Whiteness of overcast sky (color TV picture tube) Representing average daylight

Spectral energy distributions: classifications used for viewing specification; suitable for fluorescing pigments and dyes

Luminant D_{5500}	5500 K	Sunlight plus sky light
Luminant D_{6500}[a]	6504 K	Standard, or typical average daylight
Luminant D_{7500}	7500 K	Typical north sky light
Luminant E		Hypothetical equal energy spectral distribution

[a] Standards used for most purposes. D_{65} replaces Standard Luminant C.
[b] Deficient below 400 nm (violet). Modified forms of luminant A.

and often discontinuous. So if, for example, the light output is strong in the blue region of the spectrum and weak in the red range we shall find that bluish surfaces appear lighter than normal, and red surfaces reproduce darker when illuminated with this source.

In this situation any reference to true 'color temperature' curves can only be a very rough guide. So instead we refer to their *correlated* color temperatures when classifying such sources.

System standards

Film

In *photographic/film cameras* the color balance of the film emulsion is designed to suit the color quality of two sorts of luminant:

- Daylight (average summer sunlight – midday): 5500 K or more.
- Tungsten lighting: 3200 K or 3400 K.

Providing the color temperature of the incident light is within about ±100 K of these figures the color quality should be optimum. Beyond this, corrective filters are needed. Because filtering results in light losses, such filters are sometimes given a *speed index* to indicate the effective lower film speed that results.

If you shoot in daylight with daylight-balanced film stock the color values in the picture are likely to be satisfactory. If there are discrepancies, e.g. as daylight varies, laboratories may compensate and correct if necessary during processing. Similarly, tungsten-balanced film stock is used when shooting in artificial lighting.

When filming outdoors with *film stock* balanced for tungsten lighting a number 85 or 85B filter can prevent the blue end of the spectrum from being dominant. Conversely, if you are filming in tungsten-lit interiors with stock balanced for daylight an 80A or 80B filter will compensate. (In both cases there is some loss of sensitivity through filter losses.)

Video

With *video cameras* the system is normally adjusted (set-up) to produce a neutral gray scale when used with a particular luminant – daylight or tungsten lighting. In the studio all lamps are usually dimmed a little to a working color temperature of e.g. 2950 K. (This improves lamp life, and gives an extra boost above normal output when needed.) These adjustments to video circuits can be made

- *Manually* – while shooting a special chart and checking video signals, or
- *Automatically* – pressing a *white balance* auto-circuit control while using a white reference surface.

This white balance adjustment can be made either while shooting a white surface or by putting a white translucent cap over the lens and pointing the camera towards the light. Where the camera is cabled to a separate remote CCU (camera control unit) or base station the relative balance of R, G. B channels may be adjusted there by e.g. ±500 K.

When moving between light of differing color temperatures with a *video* camera you have the option of

- Operating an inbuilt *filter wheel* to provide broad compensation (3200 K and 5600 K) which can be switched in as necessary, and avoid rebalancing the system, and/or
- Pressing a *white balance* button which automatically realigns the circuitry to suit the new color temperature.

Consistent color temperature

The color temperature of a tungsten lamp alters with the amount of current passing through it. When full voltage is applied it will reach its maximum Kelvin temperature, according to its power and design (e.g. 2600–3500 K). Lamps designed to operate on lower voltage supplies (120 V) produce light some 100 K above equal-power lamps designed for higher supply voltages (240 V).

As you dim an incandescent lamp (tungsten, quartz, tungsten-halogen) by reducing the supply voltage, its color temperature falls approximately 10°K per volt; changing from a cold bluish to a reddish-orange color quality. So a 120 V lamp which has a color temperature of 3200 K falls to 3000 K when its supply is reduced to 100 V.

As we saw earlier, TV studios color-balance cameras to a color temperature standard of 2950 K. This is achieved by dimming all lamps to about 50% at fader setting '7' (100% maximum output at around 3200 K, on fader '10'). In practice, the Kelvins at maximum output are usually nearer 3100 K than the theoretical 3200 K, due to lamp ageing, and voltage losses in dimmers and cables.

How noticeable color temperature changes are will depend on what you are lighting. Backgrounds and scenery may not be critical. The warming effect of a lower color temperature may even be quite acceptable on faces. However, unless you actually want a multi-colored effect avoid *inconsistent* color temperatures. For many subjects changes of up to ±100 K may not be noticeable, but when lighting large light-toned neutral surfaces even a deviation of ±50 K may be discernible.

If you light a large white cyclorama with a series of identical spotlights placed at very different distances away you will find that the close lamps have to be dimmed (low color temperature) and the distant lamps faded up to maximum (high color temperature) in order to achieve equal *intensity* over the whole surface. In monochrome, the cyc would look flatly lit, but in color, the warm and bluish patches of color temperature variations can be very noticeable. For everyday purposes it is widely accepted that a change of 150 K either side of 2950 K is not generally noticeable (3100–2800 K) – fader settings 10 to 5 (100% to 25% output).

Measuring color temperature

Why measure color temperature?

Television/video cameramen are fortunate, for they can instantly reset the system to suit changes in the prevailing light. Film cameramen may introduce 'standard' correction filters when working in daylight or tungsten lighting, and rely on the film laboratories to correct any remaining color inaccuracies.

But what if these typical filters do not happen to suit the lighting conditions? The color quality of daylight, for example, can vary considerably. How does one avoid color discrepancies or variations between scenes shot at different times? How do you know the exact color filter needed?

To introduce appropriate conversion filters you must be able to assess the color quality of the light accurately. And you can do this by measuring the *color temperature* of the luminant. Measurements enable you to

- Check whether the incident light is reasonably near the system's color balance (±100 K);
- See how much adjustment is needed to color-correct a luminant that you know is unsuitable (e.g. fluorescent lighting);
- Ensure that the color temperatures of various lamps lighting a scene are comparable. If the color quality of light varies throughout a setting some areas will appear warmer than normal (orange-red cast) and other cooler (bluish). Unless you want this particular effect it is best avoided by balancing all lighting to the same standard;
- Check whether the color quality of natural light has altered during shooting;

- Ensure continuity in color quality for shots recorded at different times;
- Decide whether to color-correct light sources or to use a corrective lens filter.

Color temperature meters

Several forms of meters are available to measure the color temperature of light sources. Two-color types rely on the fact that as the color temperature of a luminant changes, the relative red and blue content of the light alters. (Its spectral energy curve centers around 580 nm.) The simplest has a two-color filter disc. The red content of the light is measured, the reading adjusted to zero, and the blue filter substituted. The scale then indicates the color temperature. Another form of two-color meter uses two photocells, their balanced output indicating color temperatures directly on the scale.

Meters that sample just the red and blue content of light are fine when measuring tungsten/tungsten halogen sources (quartz light). But they do not give reliable readings when used to check fluorescent and metal halide (CSI, HMI, HID) light sources, which have a very uneven, discontinuous spectral spread. To measure their correlated color temperature, a three-color meter is used which measures the red/green and blue/green content of the light. From the readings you can determine suitable corrective filters.

Mired units

Supposing that you are shooting in daylight with a color temperature of 5000 K and using film rated at 3200 K. What is the compensatory filter needed? That is where *mired unit* (*micro reciprocal degrees*) can help you. This is a simple system for calculating the color shift required to match color temperatures.

Mired units are derived by dividing one million by the kelvin value. So, for example, 5000 K equals 1 million ÷ 5000 = 200 mireds. If you want to convert a daylight source of 5000 K (200 mireds) to 3200 K (312 mireds in Table 11.3) the required shift is 312 − 200 = 112 mireds. The term *decamireds* (1/10 mired) is also used. Regular color correction filters are given mired ratings (e.g. Wratten 85 = +112 mired units).

An important point to remember when using filters for color correction/color conversion is that the effect of a given filter depends on the *present* color temperature of a light source. The higher its 'kelvins', the greater the filter's effect.

A 50-mired filter in front of a 4000 K source will raise it to about 5000 K. (4000 in the table = 250 mireds. Subtract 50 mireds and in the table against 200 mireds we have 5000 K.) Place the same filter in front of a 3000 K source (333 m) and the resultant color temperature is 333 − 50 = 283, or approximately 3500 K. A 500 K shift. (A value within 10 mireds of the ideal will suffice.)

Table 11.3   Kelvin/mired conversion

Kelvin to mired shift

Kelvins	0	100	200	300	400	500	600	700	800	900
2000	500	476	455	435	417	400	385	370	357	345
3000	333	323	312	303	294	286	278	270	263	256
4000	250	244	238	233	227	222	217	213	208	204
5000	200	196	192	189	185	182	179	175	172	169
6000	167	164	161	159	156	154	152	149	147	145
7000	143	140	139	137	135	133	132	130	128	126
8000	125	123	122	120	119	118	116	115	114	112
9000	111	110	109	108	106	105	104	103	102	101

Example: the conversion number for 4500 K is 222 mireds.

Mired to kelvin shift

Mireds	0	10	20	30	40	50	60	70	80	90
100	10 000	9090	8333	7692	7143	6667	6250	5882	5556	5263
200	5000	4762	4546	4347	4167	4000	3846	3703	3571	3448
300	3333	3226	3125	3030	2941	2857	2778	2703	2631	2564

Example: a filter of e.g. 250 mireds converts to 4000 K.

$$\text{Mireds} = \frac{1\,000\,000}{\text{kelvins}} \qquad \text{Decamireds} \; \frac{100\,000}{\text{kelvins}} = \frac{\text{mireds}}{10}$$

Filters can be combined – e.g. a Wratten 82B at −33½ and a Wratten 82C at −44½, to provide the −76 mireds needed to raise a 2700 K light to an effective 3400 K. A yellowish filter will raise the mired value (positive shift); a bluish filter lowers the mired value (negative shift).

Thanks to color temperature conversion filters of this kind, it becomes possible for a Lighting Cameraman shooting under differing light conditions to compensate for variations in color temperature and obtain reasonably consistent color quality.

Mixed color temperatures

What happens if the color balance of the film or the video camera does *not* match the color quality of the prevailing light? Well, if you use a 'tungsten' balanced system in daylight, color pictures will have a strong blue cast. Use a 'daylight' balance in tungsten lighting, and it will produce a strong yellow-orange cast. When filming, you could change from 'tungsten' to 'daylight' stock as you relocated from an artificially lit interior to a daylight exterior. Shooting with a video camera, you would need to readjust the 'white balance'.

The alternative approach is to use an appropriate *color compensating filter* (*light balancing filter*) on the camera lens, or a *color correcting filter* (*light conversion filter*) over the light source to remedy this discrepancy.

What happens if you shoot in *mixed* lighting, i.e. tungsten *and* daylight?

● If the system is balanced for *daylight* then colors in areas illuminated by daylight will appear quite natural; but those lit by tungsten sources will have a pronounced yellow-orange cast. Whether these color inaccuracies matter will depend on the subjects and the relative proportions of the two kinds of luminant.
● If the camera system was balanced for *tungsten*, color values in areas lit by the tungsten will be satisfactory, while those lit by daylight will appear unnaturally blue.

When shooting in mixed lighting, some practitioners consider it preferable to balance the system for tungsten, and color correct the daylight sources (e.g. fit filter medium over windows). They argue that any filters over the tungsten lighting will reduce its intensity, and so waste available light.

All filter medium reduces the effective output from any source. Whether the light loss is appreciable depends on the color and density of the filter. If the main light source is filtered, the working lens aperture may have to be increased by a *filter factor* of 1/3 to 3 *f-stops* to compensate and avoid underexposure. Other practitioners maintain that it is easier and much quicker to put blue filter medium over the lamps than to correct incoming daylight. The losses when lamps are close to the subject are acceptable.

12 Filters

Filters have a regular place in creative lighting techniques. Here we shall discuss the various ways in which they allow us to compensate, control, and create.

Filter media

Color filters – or 'gels', as they are usually called – are available in several forms. Some are made to close color specifications while others are less precise, and intended for more general use.

Colored glass

Sheets of colored glass (surface coated or dyed) have a number of disadvantages. Not only is glass expensive and fragile but it offers only a limited color range and is not very adaptable. Where colored low-power light bulbs (below 40 W) are needed for decorative signs, etc. within a setting, they are usually coated in a special lamp-dip solution. This coating deteriorates and pales in use.

Gelatine

Gelatine is the cheapest form of color filter. In fact its low price and adaptability are really its main merits. It is quite suitable for short periods, and is expendable. It comes in thin sheets which are easily cut to any shape. You can fix the gelatine to a frame, attach it to wetted glass, tape or staple it to openings. The main disadvantages of a gelatine filter are that it soon becomes brittle in front of a hot lamp, it changes color in use, and may even melt or disintegrate (it is a potential fire hazard). Any water splashed onto the surface will blister and melt it.

Plastic sheeting

Plastic sheeting is available in a very wide range of hues and densities. Thin filter material (typically, acetate resin, acrylic, vinyl, mylar, polycarbonate laminate, polyester base) can be bought in rolls or cut sheets. There is a wide range of products including (alphabetically) *Chromoid, Cinemoid (Cinebex), Dura, Gelatran, Lee, Mole-Richardson, Roscolene, Supergel.*

Although plastic sheeting is relatively costly, it has many advantages, and is widely used in film and TV lighting. It is robust, adaptable, easily cut and re-usable. It only deteriorates, fades or changes color during prolonged use, and is unaffected by water (even submersion!). It is flame retardant, clear and undistorted.

Filter panels

Filter panels of acrylic, acetate, or polyester are typically (2.44 × 1.22 m) (8 × 4 ft) and 3.2–6.35 mm (⅛–¼ in) thick. Self-supporting, but rather cumbersome, they provide an absolutely flat stable surface over windows or other openings. Although subject to gradual fading in strong sunlight, they last well, and can be re-used many times.

Dichroic filters

A dichroic filter is formed by depositing a series of extremely thin semi-transparent coatings onto heat-resistant glass. Each coating has a different refractive index, and light waves are interreflected there, creating interference effects. The result is a color-selective filter, which allows only part of the spectrum to pass (e.g. blue-green) while another region (e.g. red-orange) is reflected. Typical dichroic filters are used

● To color-balance tungsten lamps to daylight;
● To suppress excess ultra-violet radiation – e.g. from arcs;
● To reduce infra-red radiation (to cut down heat).

Filter applications

Let us look now at regular applications for filters in their various forms.

Ultra-violet filter

When working at high altitudes, shooting open landscapes, seascapes, mountain scenery, or in misty conditions the extra UV in the prevailing light can degrade the clarity of the lens' image. Many cameramen fit an ultra-violet lens filter to overcome this problem, and achieve crisper, better contrasted pictures. Some leave UV filters fitted as a lens protector for all daylight shooting (e.g. a 'skylight filter' – Wratten 1A or Wratten 85B) for there is little overall light loss.

The light from carbon arcs can include a high proportion of ultra-violet radiation. The main hazard here is *conjunctivitis* ('*kleig eye*'), a particularly unpleasant inflammation of the eyelids' mucous membrane lining. To guard against this possibility, you can fit a UV filter ('white flame green'). If any any time you remove an arc's fresnel lens to provide sharper shadow formations, remember it *must* be replaced with a suitable heat-resistant plain glass screen to minimize stray UV.

Fig. 12.1 Density and transmission curves for colored light filters
These typical curves are based upon 'Cinemoid' manufactured by Strand. The clear area of the graph represents the spectral distribution of the colored light transmitted. The shaded area shows absorption.
 (6) Primary red
(39) Primary green
(20) Primary blue
(14) Ruby
(48) Bright rose
(13) Magenta
 (5) Orange
(33) Deep amber
 (1) Yellow
 (4) Medium amber
(23) Light green
(21) Pea green
(24) Dark green
(16) Blue-green
(15) Peacock blue
(18) Light blue
(19) Dark blue
(25) Purple
(26) Mauve
(42) Pale violet

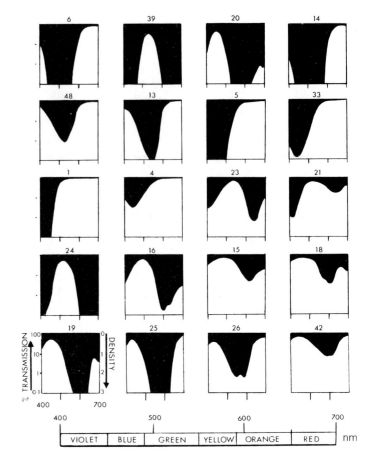

Color-correction filters

You have already met the idea of using a color-correction filter/conversion filter when the color quality of the prevailing light is different from the system's color balance. For example, daylight-balanced film stock (5500 K; 182 mireds) may be used with tungsten lighting (3200 K; 312 mireds) by fitting a Wratten 80A filter (−130 mireds) to the camera lens.

Another regular problem arises when the scene is lit by both daylight and tungsten lighting. Even although light is pouring through a nearby window and helping to illuminate an interior, we often need additional *tungsten* lighting for optimum picture quality. There is too little daylight; its direction or coverage is wrong; or there is excess contrast.

If you find that the differing color quality of the *mixed luminants* is unacceptable – because areas look too blue or orange-yellow – you have the option of color correcting the daylight coming through the windows or putting suitable corrective filters over the lamps.

1

2

3

Sky filters

1. When a sky is over-bright, it appears blank and detailless.
2. A graded neutral density sky filter can help to reduce the sky brightness, while leaving the rest of the shot correctly exposed.

Colour filters can be used for differential filtering in monochrome work, or where you want dramatic sky effects in color (e.g. 'pseudo-sunsets').

3. Stopping the lens down (smaller aperture) improves sky detail, but will reduce the brightness of the entire scene.

Daylight-to-tungsten conversion

Should you decide to color correct the *daylight*, an extensive range of filters is available to compensate for wide variations in its color quality. The most widely used *CT orange* daylight conversion filters include *full, ½, and ¼ correction* –

- 6500 K daylight to 3200, 3800, 4600 K (−159, −109, −64 mired shift).
- 5500 K daylight to 3200, 3800, 4500/4600 K (−131, −81, −35/40 mired shift).

The amber-orange *Wratten '85'* series of filters converts daylight to tungsten lighting (+112, +131, +81 mired shift).

Rolls of amber-orange gelatine or plastic film can be stretched over a window, attached to a portable support frame, or stuck to individual panes when wet. Unfortunately, sheets of media tend to distort the scene beyond, and any unevenness or buckling can catch the light, producing random reflections and hotspots. Stretched sheets are liable to rattle and shake in the slightest breeze.

Large rigid heavy-duty acrylic panels of '85' avoid these problems, and are preferable when a camera actually sees the window. They can be attached externally with gaffer tape or double-sided tape.

Alternatively, you can saw a panel to fit the window or other opening, or use it self-supported. One particular advantage of these filter sheets is that they can be tilted if necessary, to avoid light reflections.

Neutral-density (ND) filters

A daylit exterior may be as much as 4–5 stops higher than the interior lighting, and this disparity in light levels can be a real problem. Either the exterior is going to appear burned out in the picture or the interior underexposed.

Although film can accommodate up to 2 or 3 stops' difference, a video camera is likely to show overexposed exteriors if there is much more than a 1–1½ brightness difference. This is where *neutral-density* (*ND*) filters are particularly useful.

The neutral-density filter is a gray-toned medium which reduces light levels without affecting color quality. It can be used to reduce the intensity of daylight through a window, to subdue an overbright exterior scene, and help balance the brightness of the daylight to that of the interior lighting.

A range of ND filter densities is available, depending on the material you use. Gelatine, for example, ranges from 0.01–80% transmission (*N4 to N1 grades*). Polyester and acetate media typically have 12.5, 25, and 50% transmission (*N9, N6, N3 grades*). You can combine different grades for more accurate adjustment, altering them if necessary, as daylight levels change.

Some forms of *CT orange* daylight color-correction material incorporate a neutral-density filter of N3, N6, or N9, with corresponding reductions in light levels. These combined media are often identified as 85N3, 85N6, 85N9.

Table 12.1   Neutral-density filter factors

ND	0.1	0.2	0.3	0.4	0.5	0.6	0.7	0.8	0.9	1.0	2.0
N	N1	N2	N3	N4	N5	N6	N7	N8	N9	N10	N20
Light transmission	80%	63%	50%	40%	32%	25%	20%	16%	13%	10%	1%
Light reduction by	20%	37%	50%	60%	68%	75%	80%	84%	87%	90%	99%
Filter factor	1.3	1.6	2.0	2.5	3.1	4.0	5.0	6.3	7.7	10	100
Open lens aperture by	0.25	0.5	1.0	1.25	1.5	2.0	2.25	2.5	3.0	3.25	6.75

Tungsten-to-daylight conversion

There are many occasions when it is much simpler and quicker to convert the tungsten lighting to the color temperature of daylight by using blue gelatine, plastic film or dichroic filters over the lamps (so-called *boosters*). These considerably reduce the red end of the spectrum, so effectively raise the lamp's color temperature. In doing so, they can reduce the light from your lamps by up to 50% (1 stop).

Again, filter media offer a wide range of opportunities, converting tungsten light (usually 3200 K) to various daylight color temperatures. The most widely used *CT blue* tungsten conversion filters include *full, ¹/₂, and ¹/₄ correction* –

- 3200 K tungsten to 5700, 4300, 3600, 3400 K (137, 78, 35, 18 mired shift).

Some are available that incorporate diffuser material.

The blue '*80*' series Wratten filters converts lower color temperature tungsten lights (e.g. 3200, 3400, 3800, 4200 K) to daylight (5500 K). Internal reflector PAR lamps fitted with daylight filters (coated or clip-on) have a typical color temperature of 5000–5200 K.

Filtering fluorescent sources

A number of filters are available to suit most types of fluorescent lighting: 'daylight' (CCT-6800 K), 'warm white' (CCT-3600 K), 'white' (CCT-4300 K) or 'cool white' (CCT-5700 K). Some are used in conjunction with Wratten CC and '85' lens filters. There are differences between fluorescent tube phosphor specifications in the USA and Europe, so corrective filters vary accordingly.

Various shades of green (and orange) are used to match –

- Fluorescent lighting to daylight and tungsten illumination;
- Convert 5500 K dichroics or FAY lamps to 'daylight' fluorescent;
- Convert tungsten to fluorescent 'daylight'.

Where there is a deficiency of a particular hue in the illumination, or you want to considerably modify the overall color balance, a *color-compensating filter* may be desirable. These filters cover the primary hues (red, green, blue) and secondary hues (magenta, cyan, yellow). Some lighting cameramen prefer to use CC filters to correct certain fluorescent lighting (e.g. 30M/magenta + 20Y/yellow, to correct 'warm white' fluorescent lamps).

Filtering carbon arcs

The color temperature of light from carbon arcs depends on the types of carbons fitted (e.g. 'white flame', 6000 K; 'yellow flame', 3350 K) and the filters used with them. Apart from their use in limiting ultra-violet radiation, filters are used with arcs

- As conversion filters, to mix carbon arcs with tungsten light sources (3200 K). Pale to strong amber (yellow/orange) with 6000 K or light green with 3350 K arcs;
- To improve the color balance when arc lighting is mixed with daylight;

● To increase the color temperature of 'yellow flame' arcs to daylight (light green).

Diffusers

To simplify rigging, and avoid having to fit separate diffusers and color gels, combination filters are available. The most useful are the combined diffusers and blue color-correction 'boosters' that convert 3200 K tungsten sources to 'daylight'. When combined with color, there is inevitably a considerable light loss; and this may be as much as 75–90%.

Using filters

When you are using filter medium it is best to fit it into a proper *color frame/diffuser frame* which clips a short distance in front of the light fitting. If you simply wedge the filter into position the reduced ventilation is likely to overheat (and crack!) any fresnel lens, shorten lamp life, and distort and degrade even the toughest filter material.

For brief set-ups you can temporarily fasten thin color gel (or diffusers) onto the flaps of barndoors with spring-clips or even clothes-pins, but it can be difficult to avoid unwanted white-light leaking around its edges, and spilling over the scene. Wrapping the gel around the fitting to avoid light spill is liable to result in overheating.

When you sandwich two or more filters together their individual characteristics will add. So while, for example, a single sheet of quarter-booster blue raises the color temperature by 300 K, a double thickness provides a 600 K boost. By combining different filters you can increase the effective range of your filters. However, as we saw earlier, all filters introduce some light loss, and where you are using strongly colored medium (especially dark blues and greens) these losses can be considerable.

When a lamp is dimmed, or the supply voltage is below normal (because of voltage drops in long cables; or heavy current demands on available power), lighting will have a lower color temperature than usual. You can compensate for this by fitting a suitable booster blue filter. Similarly, a quarter- or half-booster blue filter can be fitted to an HMI lamp as it ages and falls below its original 5500 K ('daylight').

Filter effects

The effect of any filter will depend on several factors:

● The particular color shade you choose. Most color filters cover a spread of the spectrum.

- The color temperature of the light source. The effect of any color filter can change as you dim a lamp. 'Low K' red-orange light from dimmed lamps noticeably changes color mixtures (particularly blue-green).
- A filter material can produce varying results when used with fittings of differing power; even though their color temperatures are similar.
- The color of the surface being illuminated (Figure 2.15).
- The color characteristics and color balance of your film or TV system.

You can *identify* filters simply by holding them up to the light and comparing them with samples in a maker's color swatch. But you will get a better idea of what they will look like on camera by shining light through the gel onto a representative surface.

Filters in practice

Where rehearsals and recording take place within a day, using new filter material, there are few problems. It is as well, though, to check a rig to ensure that the right media have been fitted throughout. It is not unknown for an amber groundrow to include the odd gel that is a lighter or darker grade than its neighbors.

Re-used gels are another matter. Because blue and green filters absorb more radiant heat, they tend to deteriorate more rapidly than e.g. yellow, orange and red hues. Filters that are used over a long period – in a daily studio production – can pale out unnoticed.

When lighting a large area (e.g. a cyclorama) with a series of colors remember that they will additively mix where they overlap. So blues and greens, for instance, can mix at the edges of light beams and produce rogue cyan patches.

Any white light falling onto a surface lit with color will dilute it, or reveal the background color/tone instead. This frequently happens when a performer in a following spotlight moves near the background, or white light reflects from a shiny surface (floor, piano).

Colored light

You can use colored light for several different applications –

- To simulate realistic effects such as firelight, moonlight, sunsets;
- To provide color and pattern on neutral scenery (e.g. gray flats, cycloramas);
- To decoratively light action areas.

Where colored lighting is concerned, beauty is certainly in the eye of the beholder. What one person considers a subtle color relationship, another may well dismiss as wishy-washy and indefinite. One person's strong and vibrant effect is another's idea of brash and gaudy. While one person will happily use red or orange to simulate firelight, another will only consider using various amber shades. Whenever you use colored light it is important to bear in mind how strongly it can modify the apparent tone and hue of everything else in the shot; the appearance of scenery, costume, make-up.

Don't overlook what people will be wearing. Otherwise you may find that attractive costume becomes insignificant as it merges into a background of similar hue, or colors clash badly.

Take care if you introduce a variety of hues into your background treatment. Some bizarre changes can occur as cameras intercut. While one camera may see a green area in close shots, another at a different angle may see a yellow background instead!

Certain background colors are obtrusive, and badly affect face tones. 'Straw' and 'amber' shades are safe and attractive – but may be overused. Some Lighting Directors even use them to color back light in order to make it less obvious on shiny surfaces; but that is a matter of taste. The mid to light blues (e.g. 'steel blue') are usually effective as background lighting for neutral settings. Greens need to be used with care.

It is best to avoid strong vibrant colors – e.g. bright reds or yellows. In video pictures red shades generally appear noisy. In some systems magenta/violet colors may vary between cameras if color balance has drifted. For decorative lighting the richer blues and greens can form a foundation for other colors.

As a general rule, think twice before allowing colored light to fall onto performers. Colors that are extremely successful in stage lighting (e.g. 'surprise pink') can produce some very odd effects with film or video cameras. There are, of course, circumstances where even the most startling color treatment adds to the fun of the occasion and nothing can be too outrageous, but be warned!

13 Picture control

In this chapter we shall be looking at various ways in which the tonal quality of the picture can be adjusted and the contribution picture control can make to the visual impact of the final picture.

Tonal limits

Tonal response

As we saw earlier, all photographic and television systems can only handle a limited brightness range. Within that range, reproduced tones are generally proportional to those in the scene. When *film* receives insufficient light it registers nothing but an even 'fog' density which we accept as 'black'. In a *photographic* system, as highlights in the scene approach the upper limits they will be reproduced as increasingly smaller tonal changes, until the lightest tones crush to an even 'white'. This gradual roll-off, which is referred to as the *shoulder* of the exposure curve (characteristic curve), is a natural feature of the film medium that allows highlights to be reproduced convincingly.

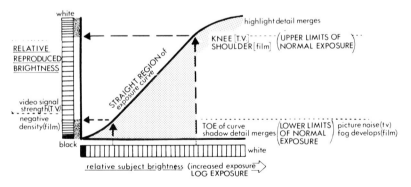

Fig. 13.1 Exposure curves
Typical photographic exposure curve. This shows the progressive response of a film or camera-tube to light. At the lower limits (the toe) the darkest tones become less differentiated, finally merging to a common black. At the upper limits (photographically 'the shoulder': in TV 'the knee') lightest tones similarly become less distinguishable. Between these limits there is a proportional change in the density of the film as the brightness of the subject tones (luminance) increases. The longer the straight region, the greater the medium's latitude.

The TV camera-tube has a linear exposure curve, so within its maximum and minimum limits the reproduced tones are proportional to those in the subject. However, on reaching its limits (clipper levels) all tones are reproduced as blank black or white. When a *knee* has been introduced electronically, extreme tones do not clip off in this way, but are progressively compressed at the limits.

369

Television systems behave rather differently. The *video signal* varies in strength with scenic tones. For dark tones it can be so weak that it becomes comparable with the inherent background noise in the system. So we see dark areas of the picture marred by the 'snow' or grain of video noise. In addition, because certain video defects develop in the pick-up device at low light levels we may see color smearing and trailing effects.

Unlike a photographic system, the *video camera* does not have this gradual tonal roll-off at the upper end of its exposure curve. Instead, the video signal increases progressively up to a point, then goes no further as it clips off to white. Instead of progressive crushing, this clipping (white level) results in blank white highlights. Look at TV pictures that include very light areas (e.g. white shirts, plates, paper and specular reflections) and you will probably see the effect. In addition, white overload can cause a camera tube to generate visual defects such as puddling (spread of highlights), black outlines around the edges of strongly contrasted areas, and comet-tails.

Thanks to technical developments such distractions have been considerably reduced in modern video systems, and cunning circuitry has *simulated* a 'knee' to provide the effect of the 'shoulder' of the film medium. This arrangement is used in 'Electronic cinematography', where video cameras are designed to produce typical 'film look' picture quality.

Tonal contrasts

The degree of contrast in a picture will depend on

● Subject tones;
● Lighting contrast;
● The contrast characteristics of the system.

While a lighting contrast of 1½:1 to 3:1 is typical for most portraiture and close subjects, stronger contrast is preferable for more dramatic scenes – say, 5:1 to 6:1.

As far as scenic lighting is concerned, contrast ratios of 5:1 up to even 8:1 have proved successful. But because so much depends on the inherent contrasts in the subjects themselves, as well as the angle and distribution of the light, one cannot be dogmatic.

Uneven lighting accentuates tonal contrast, as unlit dark tones become darker, and strongly lit light-toned surfaces become lighter still. If a scene has a low lighting contrast of only 10:1, and your lighting contrast is 3:1, this could result in extreme contrasts of 30:1 – the limit of the TV system.

The lens itself can modify the contrast in the reproduced scene. Despite surface coating of the lens elements ('blooming') which reduces light scatter and improves overall image contrast, there is still a certain amount of stray light interreflected within complex lens systems. This spills onto shadow areas in the lens-image, decreasing the overall contrast.

Thus a scene with 130:1 measured contrast may in fact be presented as a contrast of only about 30:1 to the film or pick-up tube within the camera. But, of course, you cannot rely on this form of 'contrast compression' as a method of accommodating high-contrast scenes.

Lens control

The lens aperture – *f*-stop

Within the camera lens system you can see a variable *diaphragm* or *iris*. The size of the aperture is calibrated in graduated *f-stops*, *f-numbers*, or *transmission numbers* (*T-numbers*). When you alter the size of this aperture, two things happen:

● You affect the overall brightness of the lens image falling on the film or video pick-up device (the camera tube or CCD sensor). This will adjust the *exposure* of the picture.
● You change the *depth of field* in the shot; i.e. the range of distances within which subjects will appear sharply focused. (Depth of field will also be influenced by the lens's focused distance and its focal length.)

When setting up the camera and lighting, therefore, you have to consider both

● The *depth of field* you want to work with; for that will determine the lens aperture needed; and
● The amount of light your system needs for correct exposure at that lens aperture.

The importance of depth of field

Some people assume that only the cameraman is really concerned with the technicalities of exposure and depth of field. If the depth of field is shallow, he finds it harder to focus accurately, especially in closer shots. In fact, the *f*-stop used also influences the work of other members in the team.

When the depth of field is limited the Director may have to rearrange subjects if he cannot get them all in focus. When there is considerable depth of field he may find things in the background far too sharply defined and distracting for his purpose. But more of that in a moment.

The effectiveness of lighting treatment, scenic design, make-up, costume, graphics can be directly affected by depth of field and the way tonal values are reproduced. When focused depth is shallow, slight tonal differences merge; modeling, texture and detail are lost.

Fig. 13.2 part 1   Lens aperture
When the lens aperture (*f*-stop) is altered it affects both the *depth of field* and the *exposure*.

1. *The effect of aperture on depth of field.* Keeping the focal length and focusing distance (FD) constant, the depth of field increases as lens aperture is reduced (stopped-down).

2. *The effect of subject distance on depth of field.* Keeping the focal length and aperture constant, the depth of field increases as the camera gets further from the subject: but image size decreases.

3. *The effect of focal length on depth of field.* Keeping the aperture and subject distances constant, the depth of field increases as focal length decreases (lens angle is widened), but image size decreases.

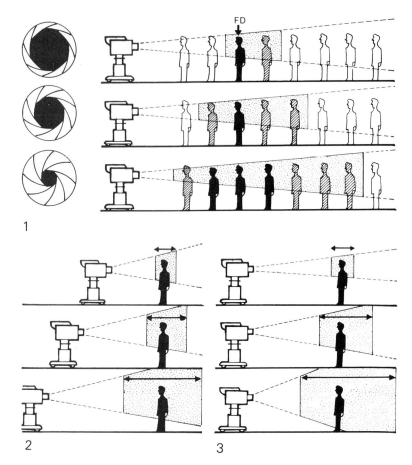

So, for instance, when the background is continually soft-focused a Set Designer may use a bolder treatment to avoid it appearing nondescript and formless. When there is considerable depth of field and everything is sharply focused a more subtle approach might be preferable, to avoid the background becoming obtrusive.

Selecting the depth of field

Shooting with available light

When you are shooting on location with whatever light is available you will usually have to adjust the lens aperture for the correct exposure and accept whatever depth of field results. Under low light conditions at maximum aperture, depth of field will be limited – especially on closer shots.

Where the lens-image is still insufficiently bright to correctly expose the shot the photographic solution is *forced processing*, and the electronic remedy to increase the *video gain*. Each will strengthen the overall picture brightness but will not increase the actual exposure.

Depth of field

The amount of the scene that is sharply focused will depend on the depth of field. This varies with the lens aperture (*f*-stop), the focal length of the lens, and the camera distance.

1. At *f*/22, the depth is considerable, so everything from the foreground to distant subjects is clearly focused. But a high light level is necessary for correct exposure.
2. At *f*/3.5, the depth is quite limited. You can focus selectively on a particular object (*differential focusing*)

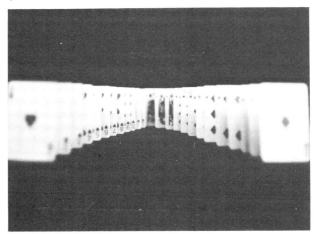

1

2

Background control

When as in (1) the background is obtrusive, or the subject does not stand out from it, you can reduce the depth of field (open up the lens), and (2) the surroundings become a blur.

1

2

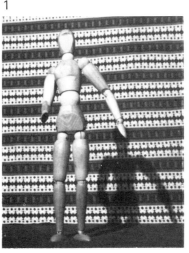

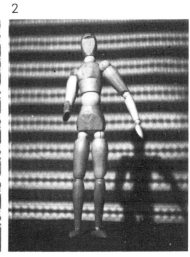

Adjusting depth of field

An inappropriate depth of field can prove embarrassing.

- Sometimes a Director will want to *isolate* subjects against a defocused background. A large lens aperture will achieve this. In this case, lower light levels are needed (or a neutral density filter over the lens).
- On the other hand, he may want to use *deep-focus* techniques, in which everything from foreground to far distance is in sharp focus. Then he will need considerable depth of field. That will involve stopping the lens down and using a great deal more light on the scene.

Exposure

What is 'exposure'?

When we adjust a shot's *exposure* we are altering the brightness of the lens image so that

- The scenic tones we are most interested in 'fit' within the system's tonal range;
- A given range of tones is reproduced convincingly;
- We achieve a certain overall tonal effect.

EXPOSURE CURVE

Fig. 13.2 part 2   Lens aperture and exposure

As the lens aperture is increased (e.g. f/1.9) the lens image falling onto the film or camera pick-up device brightens. Subject tones are effectively moved up the exposure curve. Lightest tones may be lost in the reproduced picture.

Less light is required on the scene; but the depth of field becomes shallower. Dim conditions can be accommodated by opening up the lens.

As the lens aperture is decreased (e.g. f/11) the lens image falling onto the film or camera pick-up device darkens. Subject tones are effectively moved down the exposure curve. Darkest tones may be lost in the reproduced picture.

More light is required on the scene; but the depth of field increases. Very bright conditions can be accommodated by stopping the lens down.

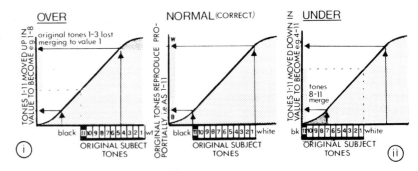

REPRODUCED TONES

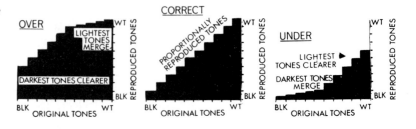

Clearly, we can alter exposure either by changing the lens aperture or the intensity of the lighting.

There is no precise figure that represents '*correct exposure*'. There *is* an optimum value for a particular purpose. What is an appropriate exposure on one occasion may not be exactly right for another.

Adjusting exposure

As the lens iris is *opened* –

- All tones in the picture become lighter.
- Light tones are progressively overexposed. Modeling is lost, and they merge to white as they crush out.
- Detail and modeling is seen in darker and darker areas.
- Color tends to appear less pronounced, more pastel (desaturated).

As the lens iris is *closed* –

- All tones in the picture become darker.
- Darker tones are progressively underexposed, modeling is lost, and they merge to black.
- Detail and modeling is seen in lighter and lighter areas.
- Color tends to appear more pronounced (saturated).

Even fairly limited changes in the lens aperture ($\pm\frac{1}{2}$ stop) can considerably affect picture tones.

The motion picture camera operates at a standard frame speed of 24 frames per second, a rotating shutter exposing each film frame for 1/50 second. This exposure rate affects both the amount of light falling onto the film and the sharpness with which movement is captured. In many cameras, this shutter has an adjustable opening (cut-out) that is variable between e.g. 50° and 200°.

While the *lens aperture* alters both the exposure and the depth of field, changes in the *shutter opening* allow the exposure to be adjusted without affecting the depth of field. Reducing the shutter-opening angle, however, also increases the sharpness of image-arrest. The spokes of a moving wheel may be a blur at 200° but sharply defined at 50°. The effect of the briefer shutter opening time is particularly evident with rapidly changing movements, such as flickering flames or fluttering leaves, and can produce unnatural movement break-up and stroboscopic effects. Rather than have this happen, a neutral-density filter may be preferable for light reduction. Many video cameras are now fitted with an adjustable shutter speed for improved detail of moving subjects.

When the shutter opening and the lens aperture are at maximum the only other expedient is to reduce the frame rate to increase the effective exposure. However, this increases the reproduced speed of movement, so has limited applications.

The effective speed of many film emulsions can be considerably increased by processes known as *flashing* or *latensification*. These involve exposing the raw stock to a very weak white light source before (or after) shooting. Such techniques can also be used to reduce picture contrast, particularly when shooting high-contrast subjects.

Auto-iris

This facility, fitted to many video and photographic cameras, measures the *average* light level received by the lens and automatically adjusts the lens aperture to suit. So if you move from a sunny exterior scene where the lens has stopped itself down to e.g. $f/16$ to a shadowy interior, the iris will automatically open up to, say, $f/2$ to compensate for the different conditions.

Although auto-iris keeps the lens-image tones within the system's exposure limits it does so very arbitrarily. It is influenced by *all* tones in the scene, whatever their importance. If an area of sky, a newspaper, or a white shirt happen to come into the shot the auto-iris will stop down. When they move out, or you pan away from them, the lens will open up. So instead of your *main subject* receiving *constant* exposure, its gray-scale value will fluctuate; lighter in some shots and darker in others.

Because the auto-iris system underexposes the shot whenever it sees a bright area behind a subject, or strong light reflections, some video cameras have a '*back light*' control that can be switched into 'compensate'. In fact, it arbitrarily opens up the aperture a stop or so above the auto setting, to improve the subject exposure – but, of course, overexposing highlights in the process.

So while self-adjustment of the exposure frees the cameraman from the need to compensate when working under difficult rapidly-changing conditions (e.g. when shooting off-the-cuff news items) it has important limitations.

Consistent exposure

If lighting is not reasonably *consistent* you will find yourself continually adjusting the lens aperture to correct exposure as the camera position changes. But that does not mean, as people sometimes assume, that lighting has to be *flat*! It is quite practicable to vary contrast and general level but keep the exposure of the main subject reasonably constant.

Just as the dynamics of *music* add to its appeal so those of light and shadow build visual interest and should not be 'ironed out'. Flat lighting may solve exposure problems but it is usually a poor artistic compromise. Crisp shadows add boldness and definition to a picture.

The odd underexposed shot as people pass through shadowy areas can be dramatically effective, but the poor tonal gradation and clogged shadows of dull-looking underexposure seldom sustain interest. Bad *over*exposure, on the other hand, is seldom acceptable, except as a special effect; e.g. explosions, lightning, blinding headlamps.

Measuring correct exposure

You cannot expect to look at a scene and accurately judge 'correct exposure' by eye. Our eyes adapt too easily to prevailing conditions.

'Guesstimates'

An experienced cameraman can become so familiar with the performance of his equipment, and the exposure needed under different conditions, that he can make remarkably accurate 'guesstimates'. He becomes accustomed, for instance, to working with a key light of 300 fc and knows precisely the lamp powers and distances that provide that lighting level. But what may look like a 'wet finger in the wind' technique is really based on known parameters. Processing readjustments compensate for errors.

Light tables

Similarly, knowing the sensitivity (speed) of the film you are using, it is possible to look up *light tables* and estimate probable exposure values for general subjects and conditions. Although this usually avoids gross errors it is only a general guide, but again processing can compensate in most cases. You can only expose *film* systematically by careful measurement.

Incident light

Most film cameramen, knowing the speed of the film they are using, use *incident light* measurements; either selecting the *f*-stop and adjusting the light levels to suit or assessing the prevailing light levels and adjusting the lens aperture to an appropriate *f*-stop.

The light meter responds to the amount of illumination falling onto its light-cell. A white plastic diffusion disc or hemisphere over the cell helps to reduce its directionality. You hold the meter at the subject's position pointing in turn towards the key light, fill light, and back light to check their intensities, and their relative proportions. This is in fact an *averaging* process, and does not take into account the tonal range of the scene or which tones *you* are primarily interested in. If the subject happens to be light-toned the exposure indicated will be too high; while for a dark subject it will be too low. For subjects of fairly limited tonal contrast, particularly people, this is probably the fastest and most reliable method. But you have to remember to make allowances if you want to expose more extreme scenic tones correctly.

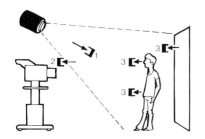

Fig. 13.3 Light measurement
Light measurements avoid gross over- or underexposure. Basic methods:
1. Incident light falling on subject.
2. Average light reflected from scene at camera
3. Measuring surface brightnesses (tonal values and contrast).

Other methods

There are several other methods of assessing exposure:

■ *Average reflected light method*, in which you hold the light meter near the camera position and measure the average amount of light reflected from the scene (integrated reflected light). This method is widely used for general photography.

■ *Brightness range method*, in which you separately measure the amount of light reflected from highlight areas (e.g. *f*/5.6) and then the shadows (e.g. *f*/3.5) and choosing a mid-point lens-stop.

A difference of	will indicate a contrast of
2 stops	4:1
3 stops	8:1
4 stops	16:1
5 stops	32:1

■ *Surface brightness method*, in which you use a spot-photometer to precisely measure the surface brightness of a specific area (e.g. a face) and expose for that. You can also check highlights and shadow areas to see whether they fall within the acceptable brightness range.

■ *Key tone (substitution) method*, which relies on measuring a standard gray card of 18% reflectance placed in the subject position. This ensures that mid-tones are located in the middle of the exposure curve. Where highlights are important an 'artificial highlight' in the form of a white card can be used similarly, to avoid overexposure.

Calculated light levels

To get a general assessment of the light levels you need, you can use the formula:

$$\text{Total light in foot candles} = \frac{25 \times f \times f}{0.02 \times \text{ASA}}$$

So if you want to shoot at *f*/8, with 100 ASA rated film stock at the normal 1/50-second (0.02-second) exposure the light levels from combined key *plus* fill lights would be:

$$\frac{25 \times 8 \times 8}{0.02 \times 100} = 800\,\text{fc}$$

If you want a lighting contrast of 3:1, the *key* will be around 520 fc and the *fill* around 280 fc.

Light levels

Knowing the light levels you need is one thing and assessing the lamps required to achieve it is another. Manufacturers provide data

Table 13.1   Methods of light measurement

Incident light method	Reflected light method	Surface brightness method
Meter positioned beside the subject, pointing at light sources.	*Meter positioned* beside the camera, pointing at the subject.	*Meter positioned* beside the camera, pointing at the subject.
Measures light intensity falling upon subject from each lamp direction in turn.	*Measures* average amount of light reflected from scene and received at camera lens.	*Measures* brightness of surface at which the instrument is directed.
Providing for (average) subjects of fairly restricted tonal range, typical incident light intensities and balance suitable to camera can be assessed. Base light, key light, fill light, and back light measured in turn.	*Providing* average, reflected light levels suitable to camera's sensitivity. Measure lightest and darkest tones separately, and use midway reading for guide exposure.	*Provides* readings by measuring surfaces of known reflectance (skin, 'standard' white, black). You can then deduce the suitability of light intensities falling upon them. *Also* allows scenic tonal contrasts to be measured to prevent overcontrasty lighting, overlit highlights, underlit shadows.
Ease of operation: method is simple and consistent. Does not require experienced interpolation. Widely used in motion-picture lighting.	*Ease of operation:* readings vary with meter angling, and experience is needed to make allowance for subject tones and contrast. Large dark areas cause readings to be falsely low, encouraging overexposure of highlights. Large light areas give high readings, which may cause underexposed shadows.	*Ease of operation:* method requires some experience in judging the *importance* of individual surfaces' brightness relative to overall exposure.
Advantages: when a show is to be repeated original levels can be duplicated readily. Balance between various light directions readily checked.	*Advantages:* method provides a quick rough check of average light levels. Can facilitate evenness of lighting.	*Advantages:* method is capable of assessing surface brightness and contrast very accurately.
Disadvantages: arbitrary allowance has to be made for subject tones.

The amount of light required depends upon the subject – which this method cannot assess.

Method only directly useful for 'average' subject tones.

Does not take into account tonal values, proportion of tones, and tonal contrast. | *Disadvantages:* meter readings are only of an 'average' nature; which varies considerably with tonal values and proportions.
Method does not indicate contrast range of subject or lighting.

Meter's 'angle-of-view' seldom identical with the camera-lens's.

Where a single surface (e.g. a face), is to be equally exposed in a variety of settings, measured exposure *should* be constant; but will vary as adjacent tones change. | *Disadvantages:* several separate readings are necessary to check evenness of lighting and contrast.

Method measures scenic tones, but does not distinguish their relative importance; and hence the desired exposure.
Tonal contrast measurements may not signify:
  if the tones measured do not appear together in picture;
  if their proportions are small and unimportant;
  if they *may* be acceptably 'crushed out' without impairing pictorial quality. |

Note: Where meter is held close to subject, measuring individual surface brightness, method becomes as for surface brightness method.

on their various fittings as a guide to their light output. For example, a typical lensless spotlight – 800 W 'Redhead' (3200 K) – may provide:

Distance				
meters	2	3	4	5
feet	6.5	9.8	13	16.4

Output				
Spot: lux	8750	3888	2187	1400
fc	813	361	203	130
Flood: lux	2500	1111	625	400
fc	232	103	58	37

(1 fc = 10.76 lux; 1 lux = 0.09 fc)

In practice, the age and condition of the lamp, its housing, the evenness of the light spread, adjustment, can all modify the quoted figures but they do give a general indication of what to expect. The safest thing is to measure typical units of the kind you use, to form a rule-of-thumb guide.

Sensitivity

The light levels that a film or TV camera system needs will depend on its inherent sensitivity, the lens aperture, and losses from any lens filters. If there is too much light for the required *f*-stop an appropriate neutral-density filter will reduce it. On balance, you are more likely to be embarrassed by a lack of light than by an excess.

Film

Color film systems require about two to three times the light of black-and-white stock. While many modern film emulsions now have greater sensitivity and enhanced performance, the price one has to pay for higher sensitivity is usually a more grainy image. The speed or rating of a film will determine the lens stop you can work at for a given light level: e.g.

ASA	25	64	100	160	200
f-stop	2.8	4	5.6	6.3	8

For a typical 100 ASA rated film stock an incident light level of around 300 fc (3230 lux) key light level would be needed when working at *f*/4.5. Larger lens apertures require correspondingly less light, and any existing light sources (daylight, practicals) will appear proportionally brighter.

Forced development by increasing the time and/or developer temperature effectively increases the film speed when the need arises. But this is not a desirable practice for tonal gradation (gamma), color quality and grain suffer. The typical latitude of negative color film stock is of the order of three stops for a 30:1 subject contrast range but less for reversal stock.

Video

The effective sensitivity of a video system will depend on the type of pick-up device used (i.e. the kind of camera tube or CCD image sensor fitted), the working lens aperture and electronic adjustment of the video system. Any filter or prompter attachment will cut down the light to the lens. If we take $f/4$ as a typical working lens aperture, then light levels may be anything from 75 to 250 fc (lumens per square foot) or 800–2700 lux (lumens per square meter). The relative amount of light needed at various lens apertures (taking $f/4$ as unity) varies:

$f/1.4$ 2   2.8   3.5   4   4.5   5   5.6   6.3   8

$\frac{1}{8}$ $\frac{1}{4}$ $\frac{1}{2}$ $\frac{3}{4}$ 1 $1\frac{1}{4}$ $1\frac{1}{2}$ 2 $2\frac{1}{2}$ 4

$\leftarrow$ Less light   More light $\rightarrow$

Table 13.2   Typical video camera sensitivities

Pick-up device	Typical light levels		f-stop
Plumbicon (lead oxide vidicon)			
1¼ in (32 mm)	116–150 fc	1250–1600 lux	[a]$f/4$
1 in (25 mm)	800 fc	75 lux	[a]$f/2.8$
⅔ in (18 mm)	600–750 fc	56– 70 lux	[a]$f/2$
CCD sensor			
⅔ in (18 mm)	325–500 fc	30– 47 lux	[a]$f/2$
½ in (13 mm)	600 fc	56 lux	$f/2$

[a] These stops provide similar depth of field.

Picture control in film

We have seen how the Lighting Cameraman/Director of Photography relies on measurements and experienced judgment for accurate consistent exposure and color quality. He cannot afford to be wrong in his assessments, for subsequent processing can only rescue him to a limited degree.

There are many variables, particularly when shooting on location. Despite stringent manufacturing limits, there can be slight deviations in the film material itself that may be revealed under certain conditions. However, ageing apart, these tend to remain constant for a given batch of stock.

By fitting a monochrome video camera to the film camera's reflex viewfinder the lens image can be checked on a nearby TV monitor picture (and videotape recorded for reference purposes). This check picture serves many valuable purposes, but it can only provide a general guide to the film's potential picture quality. The recorded image quality, color values, lighting balance and pictorial subtleties can only be seen in the final print. The review of the print from processing laboratories is the moment of truth.

Let's examine, then, the broad steps through which the film passes between camera and screen.

From negative to print

The bulk raw film stock is loaded into *magazines* which are attached to the motion-picture camera. After exposure this now precious commodity, the *camera master (camera stock)*, is unloaded and forwarded to the processing laboratories.

■ *Processing*   On receipt of the film the laboratories process it to the manufacturer's instructions. The actual steps involved depend on the type of film used. When *reversal stock* is used, the camera master is converted into a positive print. Alternatively, a *neg-pos* process is employed, in which the camera master provides the valuable *original master negative* from which *positive prints* are derived.

■ *Check prints*   As an initial check, *test strips (Cinexes)* are prepared by the laboratories, showing in a succession of frames from each scene the effect of printing exposure. These frames are progressively numbered with their *timing* or *printer lights* and enable the Lighting Cameraman to decide how light or dark a print he requires. His choice of density may be:

● *Corrective* – to compensate for over- or underexposure.
● *Matching* – to provide continuity of the average scene brightness.
● *Pictorial effect* – to enhance or create a low-key or high-key atmosphere.

■ *Rushes*   The Lighting Cameraman, Director, and others concerned then view the *rushes* or *dailies*. These are mute prints (positive) projected to give a general indication of the suitability of the material shot. Uncorrected, these prints (sometimes called *one-light prints*) are screened exactly as they came from the camera, including clapperboard marking, fluffs, retakes and all. For economy, check prints are usually provided in black and white, although selected sections may be printed in color as guide prints.

■ *Editing*   Taking these rushes, the Film Editor assembles a selection according to their required content and length to provide a *cutting copy (rough cut)*. This preliminary editing gives an indication of the form of the final film. The various soundtracks (dialogue, effects, music, etc.) are run in synchronism with it ('laying soundtracks') to complete the compilation.

■ *Identification*   Along the border of the original negative a series of *edge numbers* were stamped during manufacture. These indicate the film stock type, date and footage. (These are not to be confused with code numbers added during processing, for picture/soundtrack editing.) So when the 'cutting copy' print is forwarded to the laboratories for *negative cutting* the operative there can match the edited version exactly to the corresponding portions of the negative and introduce the transitions (cut, dissolve, wipe, etc.) indicated by the editor. A *time code* system is increasingly used, which records the precise moment at which each frame is recorded and serves to identify all picture and soundtrack.

■ *Grading*   The 'assembled master' is then graded ('timed'), each sequence being scrutinized to assess the brightness and color of printer light required for the most satisfactory picture quality. Both visual and automatic methods are used. There is less opportunity for such control where reversal stock was used in the camera.

According to these grading instructions, a *corrected print* (*answer, approval or grading print*) is produced. This print of the finished film is then examined by the production group and any further grading changes indicated. In such circumstances re-corrected versions (second or third answer prints) would be provided.

■ *Show print*   A final approved print (*show* or *release print*) is then prepared, which provides optimum color quality and continuity. This may be a *composite print* (*married, synchronized print*) in which a soundtrack has been physically printed alongside the picture (but displaced in time) or a separate magnetic track (*sep. mag.*), played separately but synchronously with the picture in another section of the projector.

Where copies of the *print* are needed (e.g. where a negative is not available, or reversal stock was used) a *dupe negative* may be printed from a corrected positive print. It thus becomes a *generation* away from the original. This duplication necessarily involves some deterioration in the quality of picture and sound, whether optical or electronic methods are used. These shortcomings are generally quite noticeable relative to the original by the time the third generation is reached. The situation may be unavoidable, however, where library shots (stock shots) are being copied.

Film in television

Where film is used solely for television certain steps may be omitted for economy. Even original negatives are transmitted (phase reversed) in some situations. Grading during processing may then be omitted and electronic correction introduced instead.

Picture control in television

The television system

The TV camera

Television/video cameras today take several forms; from the large studio broadcast cameras to self-contained lightweight units cabled to a portable videotape recorder (VTR). Increasingly used are the 'Camcorders'/'Combo' cameras/VCRs (videorecorder cameras) which incorporate a recorder taping both picture and sound and allow immediate playback.

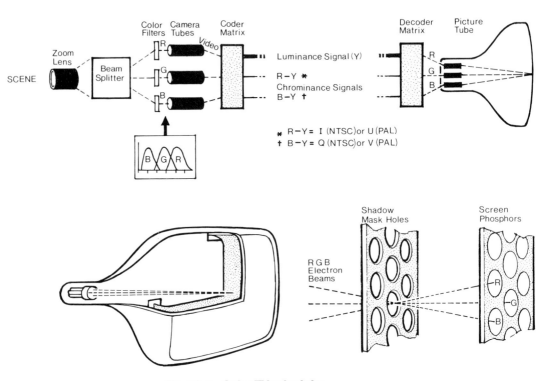

Fig. 13.4   Color TV principles
The focused image is split by a prismatic block (or diochroic filter-mirrors) into three light paths – covering the red, green, and blue regions of the spectrum. Camera tubes (3, 2, or 1) produce corresponding video signals. These signals could be transmitted in this form, and recombined to recreate the original color image (as in color printing processes). However, to provide a *compatible* color TV picture (that will reproduce on monochrome receivers) a complex intermediate system is necessary – NTSC, PAL or SECAM. This involves *coding* to derive special *luminance* (brightness) and *chrominance* (hue, color) video signals. (Monochrome receivers respond to the luminous component of the signal.)

In the color picture-monitor or receiver the separate RGB components are recovered by a *decoding matrix* (decoder), each video signal being used to control their respective electron-beam strengths in the picture tube.

The picture tube screen comprises three patterns of phosphorus (dots or stripes) that glow red, green, or blue when energized by their associated electron beam. The eye merges these tiny RGB patterns, seeing them as color mixtures; when proportionately activated they appear white to gray.

Then there is *electronic cinematography*, using video cameras specially designed to incorporate the characteristic features of a motion-picture camera. Used by film production units, material is shot in video form and photographic prints derived by an image-transfer process.

The television picture

Let us look briefly at the television process, for it will help you to appreciate the underlying principles of video control.

Behind the TV camera's zoom lens a glass prism-block or 'beam-splitter' produces identical versions of the scene on three pick-up devices – camera tubes or CCD solid-state image sensors. Each of these pick-up devices systematically scans the pattern of electrical charges on its surface, which corresponds to the light and shade in the image at each point. The resulting continuously fluctuating electrical signal is the *video*.

Fig. 13.5 The picture and its video
Showing a typical scanned line and its corresponding video signal.

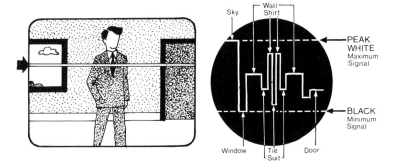

By fitting the pick-up devices with red, green and blue filters, respectively, they simultaneously analyze each part of the picture between them into its corresponding color proportions. The *luminance* (brightness) component of the picture is derived from the combined RGB outputs – since $R + G + B$ = white. This is the signal which gives us all the information about

● Light and shade;
● The overall picture brightness;
● The intensity (saturation) or dilution of colors;
● Subject details;
● Texture.

The actual process is complicated by several technical variants, the need to keep the entire system in synchronism, and to encode it for transmission (using NTSC, PAL, SECAM processes).

To reconstitute the color picture the video signals corresponding to the red, green and blue components of the scene are applied to the three electron guns of the television picture-tube. These regulate the strength of their respective electron beams, as they simultaneously excite their separate red, green and blue phosphor dots or stripes on the tube face. The phosphors glow in proportion to the strength of the video signals. Where three adjacent R, G, B phosphors are equally excited, their hues appear to the distant eye to combine as white in that part of the picture. With just red and green activated, the result is a small yellow area, and so on (additive color mixing).

Camera set-up/line-up

To ensure consistent high-quality pictures all video camera channels have to be adjusted to very close technical limits. All equipment drifts to some extent, and needs regular checks (daily, weekly, monthly). Among other things, these ensure that the gray-scale reproduction is neutral, the system is color-balanced to a standardized color temperature, that contrast range, resolution, and other parameters (e.g. geometry) are within specification. These checks may be carried out manually (using special charts and electronic tests) or by automatic circuitry.

Standard set-up

In the television studio the video equipment is generally adjusted during *set-up* (*line-up*) to produce correctly exposed pictures from a scene of 'average tones' lit to a standard lighting level of e.g. 75 fc/800 lux. Typically, the lens aperture may be around $f/4.5$. (See *Sensitivity.*) It is normally impracticable to work at much smaller lens apertures such as $f/8$, because of the high light levels needed.

The studio lighting is arranged to provide a general level (and color temperature) around that of the standard, and under these conditions most shots will be correctly exposed. Variations of up to $\pm\frac{1}{2}$ stop will be introduced wherever needed to improve the tonal quality of individual shots; to compensate for lighting or subjective variations.

Assessing exposure visually

Monitoring the picture

The television Lighting Director has a number of disadvantages, including the problems of multi-camera viewpoints, continuous shooting, and limited opportunities to correct or change the lighting set-up during production. He is less able to design lighting shot by shot. However, he has a *major advantage* – the opportunity to continually monitor results and adjust lighting levels and picture processing ('video control') at each stage as he watches. He can work empirically, can experiment, can vary lighting balance, and see the effects of alternative treatments as he is making them. So he develops experience rapidly, notwithstanding the pressing time scale involved. Light measurements are used only as a general or preliminary guide, for the picture monitor itself provides a continuous display showing him whether the exposure, the color temperature, the lighting balance are satisfactory.

Picture quality

The color quality of video pictures can be extremely high. Unlike film, they can be measured and corrected at any time, anywhere in the picture chain. But there are still opportunities for arbitrary judgments. Picture monitors drift, as do camera channels, and it may be arguable which is 'correct', particularly when working with one camera and color monitor. The camera itself invariably has a *monochrome* viewfinder tube.

When two or more television cameras are shooting a scene it is important that their pictures match closely, otherwise there will be variations in brightness, hue, contrast, on intercutting between them. So continual comparative assessments are essential to detect and correct any errors as quickly as possible. The eye soon becomes lulled into accepting inaccuracies. When using a single video camera it is important that pictures shot at different times match well, for there is usually no opportunity to color correct pictures at a later stage, during post-production editing.

Video control

As the processing laboratories are to the film camera, so you can regard the work of the *Shader/Video control operator/Video engineer* to the television picture.

Approaches to video control

'Picture quality' is a very subjective judgment. You can take a photographic or video camera, point it at the scene, and leave it to automated circuits to produce perfectly satisfactory routine pictures. Many do just that. But *are* they really satisfactory?

Clearly, this is a matter for artistic judgment. If you compare a run-of-the-mill picture with another that has been skillfully arranged and adjusted for maximum effect you will generally find that the latter

● Is more arresting,
● Arouses and holds the interest more closely,
● Is more pleasurable to look at,
● Engenders an emotional response.

The television system today allows program makers to follow three paths:

1. *Preset adjustments*, in which the system is set up to standard parameters (*f*-stop, color temperature, and light levels) and then

left alone. From then on it is up to the production team to adjust lighting, setting tones, costume and make-up to keep within the system's limits. Lighting is controlled to suit successive shots.

2. *Automatic control*, in which, after alignment, the video is switched to a self-adjusting mode. *Auto-iris* essentially avoids over- or underexposure. Auto-black circuits ensure that the darkest area in the shot is reproduced as 'picture black' (black level) – whether that is appropriate or not. Although this control approach avoids gross errors, automatic systems can be fooled, as we saw earlier. When intercutting, tonal values will change pretty arbitrarily from shot to shot. Dynamic contrast compression circuits can help to control highlights and to stretch blacks, so that we can see tonal gradation in shadows. But again, these changes are uncontrolled, and can affect other picture tones.

3. *Manual control*, which offers the greatest flexibility. Here, as we shall see, an operator monitors all picture sources contributing to a program (cameras, film, slides) and adjusts them to provide consistent quality.

Video control adjustments

There are two operational approaches to manually controlling picture quality: *CCU/RCU operation*, and *centralized video control*.

CCU/RCU operation

In the first system quality control is carried out by video engineers ('Shaders') sitting at *camera control units (CCUs)* or *remote-control units (RCUs)*. These house the main circuitry involved in generating and processing the picture video and the electronic adjustments for each camera. A long flexible cable connects each camera to its CCU. The main operational adjustments include:

- *Lens aperture/f-stop* – remotely controlled,
- *Black level (lift; sit: set-up)*,
- *Video gain*,
- *Color balance*,
- *Gamma*.

Centralized video control

In the second system these various adjustments are remotely controlled by a *vision control operator*, who monitors the picture quality of all sources at a console beside the lighting board. The video control operator has before him:

- A series of monochrome picture monitors, showing the continuous output of each video source (*previews*).

Fig. 13.6 Video control
The main operational controls for each camera channel can be centralized at a video control (vision control) position. A single control knob is used to adjust:

- *Exposure* (to and fro). The iris/diaphragm varies ± ½ stop.
- *Black level/lift* (rotate knob). This raises/lowers all picture tones simultaneously.
- *Preview.* Pressing the knob allows the channel's picture to be compared with the shot currently on line from the switcher (to match picture quality).

- A color *master monitor* displaying the studio switcher's output (line, main channel, studio out, transmission) that is passing to the videotape recorders or on to the transmitters for live transmission.
- A second color monitor providing a *switchable preview*, allowing any picture channel to be instantly examined and checked against the master picture.
- Where *waveform monitors* are installed, these display a graphical image of the video signal, allowing the operator to check white and black crushing, the strength of the video signal, etc.

Each camera channel has a grouped panel of controls. The main facility is the *channel controller*; a single combination knob with several different functions:

- *Forwards/backwards*, it adjusts the lens aperture, e.g. ±½-stop. This provides a fine adjustment of the camera lens iris to adjust exposure. (A further thumbwheel control gives full control from maximum to minimum aperture when needed.)
- *Twisted*, it adjusts the camera channel's black level – 'sitting the picture' up or down as necessary. This determines which of the lowest tonal levels in the picture is taken to black.
- When *pressed*, the picture on the monitor switches to show the shot on-line at the moment (i.e. selected for 'transmission' by the production switcher). This allows the quality of the two sources to be compared instantly, to ensure good visual continuity.

In addition to the channel control knobs, there may also be:

- *Camera channel gain adjustment.* This is usually switched in stages (e.g. +3 dB, 6 dB, 12 dB), and increases the overall amplification of the camera channel. Normally, extra gain is only introduced when operating under low light conditions, for it results in increased picture noise.
- *Color gain adjustment controls.* These adjust the video amplification of the red, green and blue components of each camera's picture. This provides rapid compensation for color temperature changes; e.g. on remote telecasts (outside broadcasts, OBs). And you can modify color bias for special effects. Instead of individual knobs, a

Fig. 13.7 Video gain
Adjusting amplification of the video signal moves it up and down the camera-channel's transfer characteristic, between the white and black clipping limits.
(*Left*) Insufficient amplification (undermodulation) results in pictures lacking snap; lightest tones too dim; lowest tones merge towards black clipper limit.
(*Right*) Excess amplification (overmodulation) takes lightest tones towards the white clipper limit; lightest tones merge, light areas look overbright (but no lower subject tones become visible; picture noise increases).

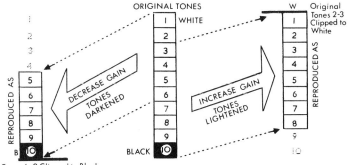

Original Tones 6-9 Clipped to Black

vertical joystick or 'paintpot' may be used, that will pivot hemispherically to any degree within a color circle. If, for instance, you detect a blue bias, the joystick is moved away from blue towards yellow to correct.

- *Color black trim adjustment controls.* These adjust the black level of each color channel; the neutrality of lower tones. Again, separate knobs or a vertical joystick may be used.

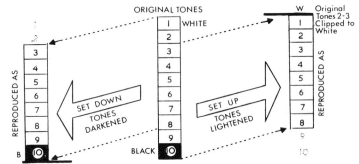

Fig. 13.8 Black level
(Set-up; sit). Adjusting camera-channel black level moves the picture tones down or up the tonal scale, the effect being most noticeable in darker tones.
(*Left*) Sat down (batting down): all dark tones clipped to black, light tones down-scaled (unless compensated for by video gain increase). No lighter subject tones become visible.
(*Center*) Normal sit.
(*Right*) Sat up. No dark greys or black in the picture; lightest tones may be merged, cut off by the white clipper limit. No lower subject tones become visible through sit up.

- *Gamma control.* This adjusts the coarseness or subtlety of tonal reproduction. It allows lightest and/or darkest tones to be expanded or compressed to improve their clarity or adjust the overall effect. Gamma correction invariably results in increased picture noise.
- *Filter adjustment.* A filter wheel located just behind the lens of the studio camera contains color correction filters, neutral-density filters, and effects filters (e.g. star effects). It may be operated by the cameraman or remotely by the video operator.
- *Detail enhancement.* Some systems include a control that allows the picture sharpness of each camera's picture to be adjusted electronically. Soft-focus pictures may be required for special effects. (Usually achieved by lens filters.) If over-sharpened, the effect looks very artificial. Edges, fine detail and texture are emphasized. Facial blemishes are exaggerated and skin is harshly modeled. Picture noise increases noticeably.

Video control techniques

Skillful video control, like film grading, is an unsung art. It is only when one has the opportunity to compare the pictures before and after treatment that the full effect is obvious. The difference is a qualitative judgment; rather like comparing the results of automatically processed color film with those achievable by a specialist film laboratory.

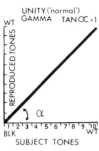

ORIGINAL TONAL PROPORTIONS

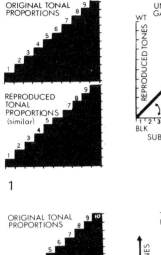

REPRODUCED TONAL PROPORTIONS (similar)

UNITY ('normal') GAMMA   TAN ∝ = 1

1

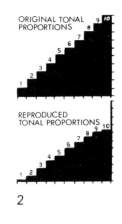

ORIGINAL TONAL PROPORTIONS

REPRODUCED TONAL PROPORTIONS

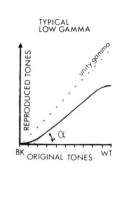

TYPICAL LOW GAMMA

2

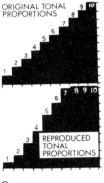

ORIGINAL TONAL PROPORTIONS

REPRODUCED TONAL PROPORTIONS

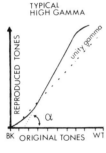

TYPICAL HIGH GAMMA

3

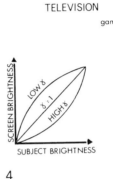

TELEVISION

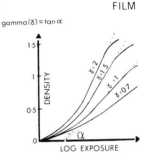

FILM

$gamma(\gamma) = tan\,\alpha$

4

Fig. 13.9 Gamma

1. Gamma influences the subtlety or coarseness of tonal reproduction. Where reproduced tonal values are directly proportional to those of the original scene, the transfer characteristic of the system is said to have a gamma of unity (tan α = 1). A film emulsion, camera-tube or video system usually has a gamma that deviates from unity.

2. The low gamma device accepts a wide contrast range, but compresses it to fit reproduction limits. The result is reduced tonal contrast (thin).

3. A high gamma device accepts only limited subject contrast, expanding to fit reproduction limits. The result is exaggerated tonal values.

4. The effect of a series of processes is multiplicative. Thus the high gamma of a picture tube (2.2) may be compensated for by low gamma corrective circuits. In monochrome, a gamma of around 1.5 compensates subjectively for the absence of color. In color systems, gamma variations can lead to color distortions.

Tonal accommodation

The basic functions of *video control* are to adjust exposure, match successive pictures from all sources, and complement the lighting treatment by electronic adjustments. Where a scene has a relatively limited tonal range there should be no difficulty in reproducing all its tones effectively. But where tonal extremes exceed the system's range you have to decide which are the most important for your purpose. You cannot expect to record *all* the subtle half-tones in the highlights and in the shadows. Are you going to allow the darkest tones to crush to black in order to reproduce the lightest tones well (the usual choice), or to reproduce details in shadows at the expense of other tones in the shot? Over- or underexposure inevitably creates false

Table 13.3   Relative exposure values for different lens apertures

f-no. (standard)	f-no. (Continental)	Relative exposure
1		1
	1.2	1½
1.4		2
	1.6	3
2		4
	2.2	6
2.8		8
	3.2	10
4		15
	4.5	20
5.6		30
	6.3	40
8		60
	9	90
11		120
	12.5	180
16		240
	18	320
22		510
	25	640
32		1 040

A change of one stop in a series doubles or halves light passed by the lens. Thus:

$$f/4 \text{ to } f/8 = \frac{f/4^2}{f/8^2} = \frac{16}{64} = \frac{1}{4}$$

∴ From $f/4$ to $f/8$ requires four times the light level ($f/8$ to $f/4$ requires ¼).

From the table, $f/4 = 15$, $f/8 = 60$. Thus:

$$\frac{15}{60} = \frac{1}{4}$$

tonal values in some part of the scene, even when there is a relatively restricted contrast range.

Sometimes you may deliberately merge shadow detail to create a dramatic effect. Alternatively, in a high-key scene, you might deliberately clip off the lightest tones to white, to lose all modeling in white surroundings so that the action seems to be taking place in endless space. But these are *effects* rather than normal exposure adjustments.

Key tones

In shots showing people from the waist up or closer, *face tones* usually guide the exact exposure. Where people are shown full-length or further away, we are more aware of the overall impression of their *surroundings*.

● You can adjust the lens aperture to expose the shot, so that face tones appear natural (neither too light, nor too dark), and allow other tones in the scene to fall as they will. (Extreme tones will be crushed.)
● Or you can select the exposure to suit the most important tones in the scene, and adjust the amount of light falling on faces until their tone is satisfactory.

If, after adjusting the exposure (and/or lighting) for the best overall effect, you decide that a particular area is *too dark* you have the options of:

● Increasing the local illumination on that area (using lamps or reflectors).
● Adjusting the gamma of the system to improve shadow detail. (This will modify other picture tones and increase picture noise.)
● Keeping the area out of shot.
● Accepting the situation.

Where you have exposed for the optimum effect, but find that a particular area is *too bright*, you can:

● Reduce the local illumination on that area by dimming lamps lighting it. Shield light off it with scrims.
● Adjust the gamma of the system to improve highlight detail. (This will modify other picture tones.)
● Keep the area out of shot.
● Accept the situation.

Corrective control

Video control corrects and enhances pictorial quality in a number of ways. Some are obvious, others are extremely subtle. Let's look at typical examples:

■ *Tonal range*   Video is usually adjusted so that the peak whites in the picture reach the *reference white level*, while the darkest tones in the shot coincide with the *reference black level*. This makes full use of the system's tonal range. Beyond these limits any tones are 'clipped off' electronically and produce as solid white or black.

■ *Uneven lighting*   The light and shade that produce interesting pictures, and create an illusion of depth and solidity, can emphasize tonal contrast and may produce uneven light levels. Skillful correction can compensate for these variations.

■ *Varying surface brightness*   The brightness of surfaces can vary significantly with the camera angle. When intercutting between cameras these differences could be very noticeable if not corrected by slight exposure or black level changes.

■ *Improve clarity*   When you want to see detail or tonal gradation clearly in very light or dark-toned subjects adjustments to the exposure and black level can help considerably. A regular example is a close shot of a book or letter which, left uncontrolled, could block-off and reproduce as a blank surface!

■ *Unsuitable tones*   When the tones of a hired costume, properties, furniture, drapes, prove unpredictably overbright and exceed the system's limits, or look drab and featureless, adjustments to exposure, black level, gamma, may compensate. It is a lot easier and cheaper than changing the problem subject.

■ *Off marks*   During continuous production an actor may move from his rehearsed position ('off his marks') so that he is out of his key, or even in shadow. Increased exposure may correct or improve the shot.

■ *Subjective effects*   Adjustments can compensate for subjective effects. We saw earlier how a person can appear lighter or darker when in front of a very dark or light background. If successive shots contain different tonal proportions they may visually jar when intercut. Tonal adjustments may make the transition less distracting. Faces may appear warmed or paled against strongly colored backgrounds. Slight adjustments to the camera's color balance can help to compensate for unwanted subjective color effects of this kind.

■ *Shot length*   Slight adjustment (e.g. black level) can increase the contrast of a *long shot* and improve its definition. Conversely, it can reduce contrast in close shots. During multi-camera production it may not be possible to adjust the lighting balance as shots are intercut. Face tones that appear appropriate in a longer shot may look too bright in a close shot. Selective exposure adjustment can improve the matching when such shots are intercut. Similarly, a dark area within a long shot may provide an effective compositional accent. But in a closer shot it could be overpowering. If the black level is raised slightly, darker tones may appear less heavy.

■ *Supplementing lighting*   By lightening or strengthening shadows, and by emphasizing or holding back highlights, video control can assist the lighting treatment so that it virtually becomes an extension of the lighting balance itself.

■ *Visual effects*   Video control can create or assist visual effects:

● Deliberate *overexposure* – e.g. to simulate dazzling sunshine.
● Deliberate underexposure – e.g. to enhance low-key scenes; to prevent lighter tones becoming too dominating; to emphasize an overcast day or poor weather conditions.
● Increasing *tonal contrast* – e.g. to improve dramatic impact.
● Decreasing *tonal contrast* – e.g. to improve misty, ethereal, or high-key treatment.
● *White-crushing* the lightest tones – e.g. to make a light surface appear blank white.
● *Black-crushing* the darkest tones – e.g. to merge darkest tones to an even overall black; a regular practice with black-background drapes, and black title cards.
● Introducing a deliberate *color bias* – to simulate moonlight, firelight, gaslight, etc.

Color and picture control

Color continuity

The eye is not particularly good at assessing the *accuracy* of color. In fact we do not always like accurate color reproduction when we see it. Leafing through a glossy magazine, we can happily accept the colors before us, until we happen to notice an identical picture in another publication – and see how different these two versions are. A girl may be a redhead in one picture yet have chestnut hair in the other. Suntanned in one version, she has a paler complexion in the other. The grass is green-yellow, or green-blue . . .

The high standards of color reproduction we see around us daily are a miracle we all take for granted. Variations pass unnoticed, and it is only when we have the opportunity to directly compare two versions that inconsistencies become obvious. The probability is that neither is accurate. But both may be pleasing to look at. Few will query the color quality of a single TV receiver; but place several together, and we are immediately aware of differences.

A regular dilemma of motion pictures and television is the problem of maintaining consistent color quality. Color variations can arise from a number of causes:

● Technical changes – e.g. a video camera's color balance may have drifted. Inherent film errors;
● Changes in natural light quality;
● Color reflected from nearby surfaces causing color casts;
● Varying color temperatures of lamps.

Whenever the picture cuts to another view of the same subject we make an instant comparison. If the sun shines as boy speaks to girl, and the sky is overcast when we take the reverse shot of the girl replying, we have two shots with differing brightness, contrasts and color temperatures. The differences in picture quality can be quite noticeable on a direct cut between them. When there is an unavoidable mismatch between two shots one solution is to introduce a brief *buffer shot* of another subject between them. Then, because there is no direct comparison, we can overlook even quite large variations.

Because eye and brain are unable to perceive *gradual changes* color faults can develop unnoticed over a period, and only become obvious during subsequent editing. This is a potential problem in all film and single-camera video production. In multi-camera TV production, pictures are continually intercut, and this soon reveals any discrepancies. Fortunately, both film processing and video control can compensate for differences to a surprising degree.

TV color monitors

To ensure consistent color quality all video equipment from camera, switcher, picture monitors, through to videotape recorders is carefully aligned to stringent regular performance standards. From then on, one relies on picture monitors (and, to some extent, waveform monitors) to assess results.

In order to standardize performance, picture monitors are usually set to peak-white, and adjusted to a color temperature of

Luminant C (6774 K) – NTSC   *Luminant D* (6504 K) – EBU

Tontal step-wedges are displayed to check gray-scale tracking; i.e. the neutrality of successive tonal values. Without such care you might find that while you are looking at pictures of a warm welcoming firelit interior scene, the version that others are seeing is brightly lit, cold, cheerless . . . !

The performance of the *home receiver* can be very arbitrary. It may be poorly color balanced, oversaturated, set up quite empirically, and watched under varying ambient light conditions. Checks on home receivers have revealed color temperatures of 9000–12 000 K. It is a reminder, though, of how readily one comes to accept color quality that bears little resemblance to the original subject!

Color discrepancies

All color systems have inherent limitations and modify color fidelity to some extent. Dyes, pigments, printing inks, all absorb some of the light that they should transmit. All introduce luminance (brightness) and saturation errors. Some colors are reproduced darker than normal, others become changed in hue. Most produce attractive pictures within their particular restrictions.

Color systems are normally balanced to produce effective flesh tones, for we are more critical about how skin is reproduced than any other color subjects. This may cause gray values to have a slight color bias. Subjectively, we generally prefer flesh to look warmer and somewhat yellower than in life.

In Table 13.4 you can see typical ways in which color reproduction can vary. At times, these discrepancies can be very obvious. A costume reproduced as violet by a video camera may appear as plum-colored in a film insert.

Table 13.4 Typical color discrepancies

	Color film		*Color TV*[a]
Red	May shift towards orange	*Deep red*	Reproduced too dark
Pink	May reproduce too bluish	*Green*	Can become desaturated, darker
Red, orange	May reproduce too light	*Blue*	Reproduced too saturated
Yellow	Ditto. Desaturated	*Blue-green (cyan)*	Reproduced too blue
Blue, cyan, green	Tend to be too dark, desaturated	*Purple*	Reproduced too blue
Cyan, greens	May shift towards blue		
Magenta	May shift towards red		
Purple	May reproduce too red		

[a] Improved with 'extended red' camera tubes

Very pale colors tend to lose chrominance, and appear monochrome.
Very pure (saturated) hues may not reproduce with equal sensation of saturation to the original (e.g. in red, orange, yellow). Particularly avoid saturated blue-greens, jade greens, purples where possible.

The television system itself can introduce some unexpected side effects. In the NTSC system, hues can change with subject brightness (due to differential-phase distortion). A fairly dim pure color may look lighter than normal. That is because highly saturated, low luminance colors have inherently restricted detail (low bandwidth) and electronic noise increases its effective brightness.

Color into monochrome

There are no direct relationships between color and gray-scale values, so when a colored scene is reproduced in black-and-white the transformation can be quite striking:
- Colors of differing hue can have similar brightness, and merge together in monochrome.
- The gray scale values of saturated hues, especially red and blue, will often reproduce as too dark.

● Colors that convey emotional overtones, or have physiological effects (*lateral color adaptation*), lose their visual impact in monochrome. A startling color is just another gray tone in black-and-white!

Looking at the scene through a *monochrome* video viewfinder it is easy to overlook the impact that color will be making in the final picture.

Color film on TV

Watching a film on the large screen of a darkened cinema, any shortcomings in its color quality are not particularly obvious. But see the same print on TV in a lighted room, and any color inaccuracies or color cast are clearly visible. Viewing conditions affect our assessment of picture quality; color fidelity and tonal values. In near-darkness the eye adapts, the brain interprets, and we accept the color quality we see. But in well-illuminated surroundings this does not happen, and we are much more liable to become aware of hue discrepancies and any color bias (color cast). Some film review theaters have a broad white border around the screen to help prevent such adaptation when assessing film for television transmission.

Ambient light falling onto the TV screen grays-out the darkest tones, generally thinning out the picture and desaturating colors. Although a contrast range of some 160:1 may be achieved on the cinema screen, 30:1 is more typical of the home TV screen. Although TV picture-tubes have become much brighter, a high contrast-range (e.g. 100:1) with good half-tone reproduction is only really achievable in dark surroundings.

Film is reproduced on television either by using a TV camera coupled to a film projector or through special *telecine* equipment. The latter scans the film electronically ('flying spot' or solid-state scanner) and not only allows the televised image to be corrected in various ways but suppresses blemishes (marks, scratches).

Correction may lead to side-effects. Additional gain or gamma correction increases picture noise. If the color film has high contrast, gamma adjustment may noticeably affect the saturation and hue of certain colors – particularly greens and blues.

Reversal stock is less tolerant of exposure errors and correction during laboratory processing is limited. So to save the expense, time and money of preparing color corrected answer prints, final quality adjustment may be made electronically. Electronic corrections are based on the known color characteristics of a particular type of film, and a standard gray-scale in the film leader provides a tonal reference. Any further corrections are made empirically while viewing the film.

Video correction – 'TARIF'

Grading during processing ('timing') becomes more critical as luminance and saturation are increased, for the eye is less aware of errors at low luminance (brightness). During grading, the color mixture of the printer-light is adjusted to improve any *overall* color cast.

Variable video correction (TARIF) can be more specific, and supplement or replace laboratory grading. The red, green, and blue components of the picture can be altered *independently*:

- *Gain* (amplification) – will vary the proportions of the primaries.
- *Black level* – will affect shadow density.
- *Gamma* controls (contrast laws) – will modify gray-scale tracking.

So overall color cast and the hue and density of highlights and shadows can be adjusted over a wide range. A video operator can make these corrective adjustments while watching the film 'on-air'; or experiment during rehearsal and use these stored file settings to control the transmitted film.

Film errors

Matching errors in the film medium can arise from density and gamma differences between its three built-in color layers. These can often be reduced by video adjustments. So, for example, pink highlights may be neutralized by increasing the video gain of the blue and green channels (cyan). Over-blue shadows may be improved by gamma changes, and by adding yellow ('minus blue').

A film dye intended to control one color can be spuriously affected by another. Then, despite *dye-masking* in the film stock, certain purer hues can gray-off (lose luminance). Switchable *electronic masking* can modify gamma and gain of the color channels to compensate effectively for red and blue errors. It is less successful with cyan (blue-green) errors, but, fortunately, scenes seldom include saturated hues in this part of the color spectrum. Electronic masking can also be used to compensate for *analysis errors*. Film stock is designed to be optically projected, and this is a *subtractive* synthesis process.

When color film is analyzed electronically in telecine equipment and displayed on a color TV picture tube the process involves *additive* synthesis. Consequently gray-scale errors develop – e.g. a magenta cast – particularly in areas of lower brightness (low-luminance). Masking of this sort will correct gray-scale but not luminance or saturation errors.

14 Scenery

The art of scenic design has developed over the years from the theatrical to the sophisticated staging techniques we meet in today's film and television studios. Here we shall be examining various scenic features and their influence on lighting techniques.

The role of scenery

Film and TV scenic design

The form and functions of scenic design largely depend on the type of production involved. Today, motion pictures are mainly based on the everyday scene, and that is reflected in the realistic forms of scenic design. An increasing number of films are made on locations, which are selected, modified and augmented to suit the requirements of a particular production. The abstract and decorative settings found in large musicals have largely become a memory.

Generally speaking, television/video scenic design covers a wider range of functions which we can summarize as:

- *Neutral settings* – non-associative, non-representational backgrounds, suitable for talks, interviews, demonstrations, newscasts.
- *Decorative settings* – stylized display in which design is predominantly arranged to delight the eye and stimulate the imagination; suitable for games shows, childrens' shows, dance, music, etc.
- *Realistic settings* – built to resemble typical environments and intended to appear as actual places. Typically, these are found in soap operas, and traditional drama.
- *Video effects* – staging in which most or all of the environment is inserted electronically, from graphics, photographs, etc.
- *Location settings* – modified and augmented to suit the production.

Scenic function

Occasionally, there is no scenery of any kind, and one has to light the studio walls 'attractively'! But normally, the most basic scenic treatment is something like a gray flat and a stool for a guitarist, or plain cyc for a dance group. On the other hand, scenic treatment may be so extensive that it entirely fills the studio!

Scenery is usually designed to provide a *total environment* for the action, e.g.

- *Drama* – a complete, fully furnished and decorated three-walled room (the fourth wall is usually omitted).
- *Games show* – grouped furniture or desks in front of a scenic background.
- *Dance* – a performance area bordered by scenic units.

In most cases the Set Designer creates a setting that will give the Director a series of visual opportunities. The Director selects shots to suit the action and to make full use of the set's features. Alternatively, the Director may have prepared a *storyboard* showing the kinds of shots he wants at each point in the show, and the Designer has this in mind when designing the staging. Where the Director is only taking close shots, instead of building a complete set, the Designer may provide just a limited *section*; e.g. the front door set in a brick wall to suggest the entire house. If video effects such as chroma key are being used the Designer will selectively arrange any scenery to supplement the inserted image.

Scenic prominence

Even the most brilliantly designed scenery relies for its effect on the shots the Director takes. The longer the shot, the greater the overall visual impact of the scene and its associated lighting treatment. The closer the shot, the less we see of the surroundings and their lighting. Features may even be reduced to random areas of line and tone in the picture, as the scene is fragmented and depth of field becomes increasingly restricted. A magnificent colonnade can become just a pole growing through the actor's head! At the same time, however, portraiture becomes increasingly prominent.

The role of lighting

The precise role of lighting can vary with the kind of production. In most, but not all, appropriate *portraiture* is important.

■ *Primary role*   Sometimes the pictures' visual appeal rests almost entirely on the lighting treatment, and the scenic contribution is nominal when

- Action is against a completely black or 'white' background;
- Plain surroundings are decorated by light – color, projected patterns, cast shadows, etc.;
- A rudimentary setting is transformed by the interplay of light and shade – e.g. back-lit columns standing in front of an open cyc.

■ *Enhancing role*   In both 'neutral' and 'decorative' staging, carefully positioned lighting is used to complement the scenic design, so that together they achieve an attractive visual effect. Through shading, varying emphasis, structural lighting, the set is given form, dimension and perspective.

■ *Supplementary role* Well-controlled light breathes life into a realistic scene. When lighting is arranged to appear completely natural, the studio setting becomes real to the audience.

Basic scenic units

Most set design is based on a series of standard scenic units. These are arranged and decorated to suit the needs of individual productions. Although we do not have room here to discuss scenic design and construction in any detail, the subject is of considerable importance to the Lighting Director, for, as you will see, it directly affects his own contribution to the production.*

For our purposes here, scenic units can be divided as follows:

■ *Standard flat* A flat sheet of plywood or prepared board on a wooden framework. Joined side-by-side to form the walls of 'rooms'. Flat heights range from 2.5–3 m (8–10 ft) in smaller studios to 3.0–3.6 m (10–12 ft). Occasionally, very high flats become necessary in larger settings (e.g. up to 4.6 m/15 ft).

Fig. 14.1 Scenic flats
Scenic flats are lashed together to form 'walls', and supported by scenic braces or jacks (hinged brace).

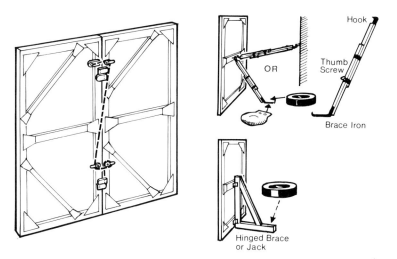

■ *Profiled flat* A flat with an irregularly shaped edge. Used vertically to suggest contoured surfaces (e.g. a rock face). Set horizontally to form a scenic 'groundrow', simulating an horizon (e.g. hills, city skyline).

■ *Set pieces*
● *Architectural unit* – a flat including a door, window or fireplace.
● *Built pieces/solid pieces* – structural units such as pillars, arches, stair units.

*The subject is covered in detail in *TV Scenic Design Handbook* by Gerald Millerson, Focal Press.

Fig. 14.2 Profile pieces/cut-outs

1. Profiled groundrows suggest progressive planes. Floor lamps can be hidden behind them.
2. Profiled flats are used to simulate irregularly shaped areas; e.g. rock face, foliage, broken masonry.

Fig. 14.3 Architectural units

Stock architectural features can be combined.

1. Contoured flat or frame (an example of a single-sided unit – viewable on one face only).
2. Door plug.
3. Fireplace plug.
4. Window unit (an example of a double-clad unit – viewable on both faces).

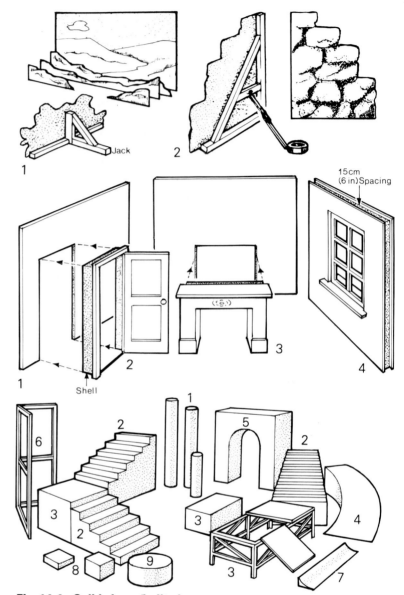

Fig. 14.4 Solid pieces/built pieces

1. *Pillars:* cylinders or half-shells of 0.15–0.6 m (0.5–2 ft) diameter and up to 4.5 m (15 ft) high.
2. *Staircases (stair units):* groups of two or more treads, usually 0.15 m (6 in) risers, matching heights of stock parallels (rostra).
3. *Parallels (platforms, rostra, risers)* which are variously shaped level platforms on folding/dismantling frames with boarded-in sides.
4. *Ramp:* sloping plane surfaces.
5. *Arch.*
6. *Drape frame:* light framework in single or hinged units carrying draperies (otherwise suspended from batten or bar).
7. *Cove:* shallow sloping surface used to merge horizontal/vertical planes and hide cyc units.
8. *Step-box (riser block):* wooden shells from e.g. 0.15 m (6 in) to 0.6 m (2 ft) square. All-purpose unit that provides half-steps, display tables, for raising furniture, etc.
9. *Podium (drum).*

● *Raised areas* – demountable 'platform/rostrum/parallel/riser' used to provide a horizontal level surface, e.g. 150 mm to 1.8 m (6 in to 6 ft) above floor level.

■ *Drapes*
● *Cyclorama* – a large suspended plain cloth, stretched to form a detailless background.
● *Set drapes* – material usually hung in folds; at windows, or to decorate a setting.

■ *Decorative screen or panel* Ornamental units (hung or self-standing), with surface decoration, translucent panels, tracery, on a timber or metal frame.

■ *Backgrounds*
● *Plain backdrop* – detailless vertical canvas sheet suspended outside windows and other openings. Usually introduced to imply space, and prevent the studio beyond from being seen.
● *Decorative backdrop* – vertical canvas sheet with pictorial, scenic or ornamental design.
■ *Studio floor* The floor surface may be painted or decorated for scenic effect.

Practicality

No matter how sophisticated and imaginative scenic design may be, in the end one has also to judge it in terms of its *practicality*. In order to create particular atmospheric effects you need to position lighting fixtures at appropriate angles. The surfaces and structures of settings can assist or inhibit your lighting treatment, according to how they are selected and arranged.

Surface tones

It is good scenic design practice to limit the overall tonal range of scenery wherever possible. In television, the lightest background tones are usually restricted to a reflectance of no more than 60% (peak white), while the lowest tones are not darker than about 3% (black) for areas in which one wants to detect modeling.

Under the light and shade of lighting treatment, scenic tones can easily exceed the 30:1 contrast range of the system. Simply by adjusting the amount of light falling on a surface, you can alter its effective tone in the picture. A surface with a reflectance of 30% can be lit to the equivalent brightness of a 90% reflectance surface, or shaded until its brightness is similar to a 10% reflectance surface.

Theoretically at least, you should be able to reduce tonal contrasts in the settings by lighting darker subjects more strongly, and keeping light off the lightest areas, to bring it all within the system's tonal range. But, in practice, one's opportunities are limited. On balance, therefore, it is far better to control subject tones wherever possible by careful selection, or by surface treatment.

Surface contrast

Our main subject is *people*, and because, generally speaking, we want them to stand out from their backgrounds, the reflectance value of the human face is a *key tone* to which all others should be related. Typical tonal reflectance values of skin range from Caucasian 30–40%, to mid-brown skin 20%, and black skin 10%. Although we can adjust the effective tonal values of backgrounds through lighting, it is best wherever possible to keep them within reasonable limits in the first place.

Fig. 14.5 Face and background contrast
1. Face tones should contrast with their background to provide subject isolation.
2. In static portraiture, contrasting edge tones are often used to achieve subject separation.
3. Sometimes outline rimming is effective; but as in part 2, can become mannered with over-use.

Color contrast may successfully isolate subjects, although its monochromatic separation may prove negligible.

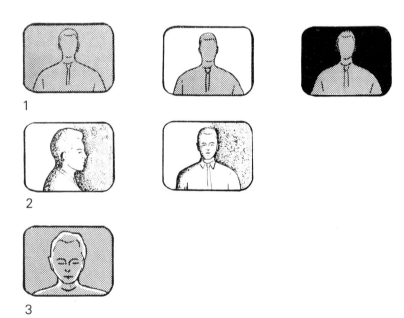

Under normal conditions, flesh tones are typically arranged to be about one and half to two times as bright as the background. (Or, put the other way round, a background of two thirds to half the face tones.) For highly dramatic situations, there may be a 3:1 contrast or greater. Where backgrounds are much lighter than faces, as in high-key 'limbo' treatment, the result can prove tiring to watch for any length of time, and subjectively, face tones will appear darker than usual.

In a black-and-white film or TV system you need to maintain reasonable tonal contrast to avoid faces merging with their backgrounds. In a color system, where differences in hue help faces to stand out from backgrounds, contrasts are still necessary to enhance the illusion of depth.

Surface finish

While a matte surface finish is technically 'safer' and relatively problem-free, it does lack the dynamic appearance of a surface with a slight sheen. An object with a dull, dead surface finish looks less well defined and undynamic by comparison. The light reflections and sparkling highlights in glossy surfaces give them a livelier, more interesting appearance.

However, the trouble with shiny surfaces is that they do reflect unwanted light all too easily. As a camera moves, specular reflections of studio lighting and any lamp on the camera will shift around distractingly, and in TV may cause picture defects such as 'comet-tails' and 'streaking'.

In a static black-and-white picture a large bright area on a polished panel looks acceptable – even attractive. But in a color system, a strong light reflection is reproduced as a *white* glare on a *colored* surface, so is much more noticeable. Dimming the light until the glare is innocuous only makes it ineffective as a light source.

Light bounces off a shiny surface, so that it invariably appears darker than a matte equivalent. A mirror looks dark – until we see something reflected in it. Cover a wall with a gold foil, and it will look black and insignificant from some angles and from others there will be strong light reflections.

Curing the problem

To reduce flare, specular reflections, and glare from a shiny surface, the quickest solution is often to angle it slightly. (A small wad of paper behind a wall picture or mirror will often clear a distracting reflection.) Slightly adjusting the camera position can help. But occasionally you may even have to relight the area.

A regular remedy is to dull the offending surface with wax spray ('dulling spray, anti-flare'), paste, latex spray, water-based dye, flock powder, adhesive nylon gauze, modeling clay or putty. But any of these treatments will alter the characteristic appearance of the surface to some extent; and even then, may not be effective from all camera positions. If there is a particularly distracting hotspot (on a table, for instance) it may be better to hide it in some way. A table cover of low-reflectance matte material might be the quickest remedy.

Reflected speculars

Light bouncing from metal objects or shiny materials can produce some very embarrassing reflections that can be quite tricky to track down. Regular culprits are:

● Jewellery reflecting glittering 'pock marks' onto the wearer's neck and chin.
● Cutlery, tableware, silver, chromium, glassware.
● A shiny piano top reflecting a frontal spotlight up onto a darkened cyc behind the piano.

Materials

Fabrics and textiles have their own associated problems, whether in clothing or as drapes in a setting. It is important here to distinguish between the *tone* of a surface and its *finish*.

Any light-toned clothing is easily overlit (particularly by back light) and, unless heavily textured, looks pale and lacks modeling. But where the material has a very smooth, glossy, or shiny surface, such as satins, plastics and glazed finishes, the situation becomes more difficult. Even the dark-toned materials may reflect bright detailless patches that reach the system's upper limits and crush out ('blooming').

Under intense lighting any light-colored clothing may reflect light of that color onto the face and neck of the wearer, but the situation is worsened if the garment has a very smooth or shiny finish. Some materials, such as velvet and darker velours, are very light-absorbent, and usually appear much darker and less saturated than smoother surfaces of identical hue. It can be quite difficult at times to reveal draping and detail without overlighting the surroundings in the process. To reduce the high contrast of deep red velours and white lace nets, for instance, can be time consuming as one tries to increase light on the former while shielding light off the latter!

Certain weaves will cause a fabric to change color with the light direction, so that a green material, for example, may appear green, gray or blue from different viewpoints. Moiré taffeta and shot-silk are well-known extreme examples of this phenomenon.

Horizontal surfaces

Horizontal surfaces such as table-tops, work-benches, script paper are all liable to appear excessively bright as back light bounces off them straight into the camera lens. Again, the regular remedies are re-angling the offending surface, darkening or dulling it down, or covering it over. In the case of table-cloths a color dip (light tan, light blue) or spraying down with black washable water-color may help. But where a smooth surface texture is the root of the problem the only answer may be to drape it with black net or replace the troublesome material. Relighting is obviously the last resource.

Surface color

There can be noticeable disparities between colors in the scene before the camera and the reproduced colors on the screen. Some are due to the processes involved. Others, as we saw earlier, are caused by physical and psychological factors that affect our interpretation of the screen version.

When the camera isolates a segment from the scene, places a border around it, and transforms it into a flat image, our normal responses toward what we see there alter. We now 'see' relationships in line, tone and form where they do not really exist – 'pictorial composition'. But our interpretation of color also becomes modified.

The choice of background hues is all-important. Desaturated colors can emphasize foreground subjects. However, while flesh appears prominent against a pale blue or green background, it merges all too easily into a background of beige, yellow, or desaturated orange.

If your color pictures are also going to be seen in *monochrome* (as in a compatible TV system) there are certain issues you cannot ignore. It is easy to overlook that there is no direct relationship between what something looks like in color and the way it appears in monochrome. Areas of quite different hue can have exactly the same luminance and translate into identical gray-scale values! So while a bunch of daffodils, for instance, would be strongly contrasted against a blue-gray sky, they could merge with it in monochrome. As one graphic artist found, with bad luck even a detailed multi-color map can be reproduced in monochrome as a detailless gray surface if all its features have exactly the same gray-scale values!

Although the situation arises less frequently, you will sometimes meet extremely poor color pictures, in which color relationships are garish, inaccurate, or ill chosen, yet look very satisfactory in monochrome. In a *monochrome* picture a surface simply looks lighter or darker as the amount of light falling onto it is varied. Its gray-scale value changes. But in a *color* picture our impressions of its saturation and even its hue can also vary with the surface brightness. As more light falls on the surface it appears to become desaturated, eventually crushing to white, whatever its actual hue. If something in the background of a monochrome picture is defocused we tend to ignore it. But in a color picture an attractive hue will still take the eye, even when we can no longer discern details; particularly if it is strongly saturated and moving. Whether we are frustrated or just distracted as a result depends on the circumstances.

Set features

There are various aspects of the ways settings are designed and used that directly affect lighting opportunities and problems. Here you will find a summary of typical situations to look out for.

Scenery usage

Television scenery tends to be used in several different ways.

■ *Total studio set* In most TV shows all the settings for the production are normally erected at the same time. After camera rehearsals and videotaping they are dismantled ('struck') and the studio cleared for the next show. Consequently, you usually have to rig and set lighting for the entire studio, prior to camera rehearsal.

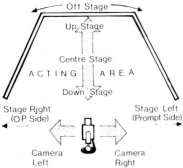

Fig. 14.6 The staging area
Locations are normally described in relation to a given camera's viewpoint.

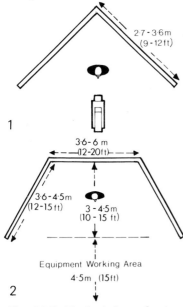

Fig. 14.7 Typical sizes of set
1. *Two-fold* – bookwing, book flat.
2. *Three-fold* – 'box set'.

Fig. 14.8 Scenic background
A single plain or decorated flat, a photo blow-up, of a painted backdrop (cloth) can provide an effective background to limited action. But it may be difficult to avoid shadows of the performer falling onto it.

(For a filming session, only the current set may need to be lit.) This can make heavy demands on facilities, manpower, scheduling, etc.

■ *Permanent set/standing set* Here, a studio setting remains in position for regular use. This is a typical arrangement for a daily newscast or talks show, or a regular weekly cooking show. The rest of the studio staging area may be used for other productions. With such settings one usually develops a 'standard' lighting plot tailored to suit the normal action and augmented slightly where necessary.

The main disadvantage of this staging arrangement from a lighting point of view is that it can permanently tie up a lot of lighting fixtures. Unless you have plenty of spare fixtures available for other shows in the studio, you are liable going to find yourself continually removing, replacing, repatching and resetting them.

■ *Major resets* This method is normally used when there is insufficient room in the studio for all sets at the same time. Scenery is erected throughout the studio ('total studio set'). Then, after general camera rehearsal, all the show's action in a *selected setting* is taped. It is later removed (overnight, perhaps) and another set erected in its place. This has then to be lit before camera rehearsal continues.

■ *Minor resets* Sometimes, when a production involves a series of short sequences (e.g. 'one-liner' sketches), a succession of single settings are placed in the same area of the studio in order to save space and simplify production mechanics. You may be able to use the same lighting treatment for them all (i.e. only the scenery changes), or it may be necessary to have a group of interswitched lighting fixtures, arranged to suit individual action.

Set layout and proportions

The general shape and layout of a setting can determine possible lighting treatment.

■ *Scenic background* A single plain or decorated flat, a photo blow-up, or a painted scene can provide a background for limited action. Its size usually depends on the shot required. For a longer shot it needs to be correspondingly larger. Shadows are the main problem here, particularly where people stand too close to the background. When it is plain, any shadow immediately beside the subject will be distracting. If the scene is intended to look realistic (e.g. distant countryside), not only must you avoid any shadows destroying the illusion but the light directions and quality of foreground and background subjects should match reasonably well. Two regular hazards here are

● A *key light* that is too steeply angled in an attempt to throw a person's shadow downwards onto the floor; and
● An oversteep *back light* where the person is too near to the background.

Fig. 14.9 Open sets
Economical and flexible, open
settings are adaptable for stylized and
realistic situations.

■ *Open sets*   In this stylized treatment a series of scenic units are
arranged in front of an open cyclorama; free-standing or suspended,
columns, screens, skeletal units, etc. The treatment can be used
decoratively or semi-realistically, with furniture drapes and architectural elements (door, window) creating a surprisingly convincing
'room' for a discussion group. An economical yet effective staging
style for a wide range of subjects, it is not without its limitations. The
main difficulties here are in lighting these isolated scenic units
without casting accidental shadows or spill light onto nearby areas;
particularly if a light-toned cyc is used. Where a spotlight follows
someone walking around among them, you have a problem!

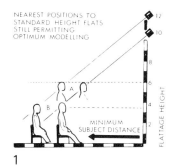

Fig. 14.10 Set proportions
1. *Optimum lamp height.* If the
   maximum vertical lighting angle
   for good portraiture is around 40°
   (allowing for typical 10–20° head
   tilting), to prevent oversteep
   lighting, a lamp should not be
   closer than line (A) for a standing
   position, (B) for a seated person.
2. *High sets. (a)* Tall settings can
   result in steep lighting (A), unless
   lamps can be hung low within the
   setting area (B), and raised for
   up-angled shots. (*b*) High flats may
   be confined to essential regions,
   and reduced elsewhere to permit
   lower elevation lighting.
3. *Narrow sets.* In lighting narrow
   settings (hallways) steep lighting
   on walls (and people) may be
   unavoidable (A). Frontal key lights
   may produce flat unatmospheric
   results, and create camera and
   microphone shadows. Maximum
   light break-up may improve
   lighting treatment.
4. *Deep sets.* Balanced downstage
   soft light (A) usually proves of
   inadequate brightness to provide
   filler for upstage areas (B). Mid-
   stage take-over filler may be
   unavoidable, despite its height,
   and potentially to excess top
   lighting at (C).

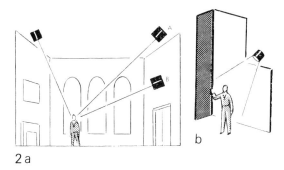

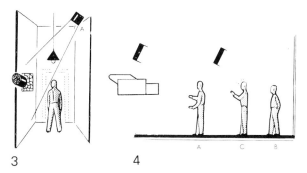

■ *Shallow sets*   Where a set is shallow it is not usually possible to
light people and the background separately. Although limiting, that
may prove unimportant where your key light can light both
effectively.

■ *Narrow sets*   Where narrow settings such as corridors and
hallways have general overhead lighting this can be used or
simulated. People are lit from around the camera position at an
intensity that overcomes the inherently toppy modeling. But where
overhead lighting would be entirely inappropriate for the location
shown (e.g. a long corridor 'in darkness') you have a dilemma;
especially in long shots, where a ceiling is visible! If there are doors
in the side walls, these may be left ajar, and light introduced through
these openings.

In a ceilingless setting, closely shuttered spotlights on top of the side flats may light the opposite walls. On location, lamps may be hidden behind gobos, on cross-poles between the side walls. If a strongly dappled key from around the camera position is heavily diffused, it will usually provide very acceptable results.

■ *Deep sets* When there is a considerable distance between the front and back of a setting (e.g. over 9 m/30 ft) the usual fill light at the front of the setting is largely ineffective for action upstage. Although you could increase its intensity to compensate, this would overlight downstage action. Instead, it may be preferable to provide a second 'layer' of filler mid-stage that can be faded in for upstage action (Figure 14.10).

Structure

Certain design features in a setting can badly impede lighting treatment. You may be able to light round them or under them, but they prevent lighting fixtures being placed in the best positions. One has to light to accommodate the structure rather than suit the action or the needs of the production! The most regular examples are ceilings, high walls, and large overhanging structures.

A large decorative canopy over an orchestra, for instance, may look great when seen in the odd long shot but if it results in the lighting for the orchestra itself being poorly positioned, to avoid casting shadows over the players, its value is debatable.

Sometimes through coordinated and sympathetic set design the effect the Set Designer is seeking can be achieved some other way. A subterfuge may help. A large chandelier, for example, that looks so impressive in the ball scene may be 'dropped in' for the couple of long shots in which it is seen, and 'cleared' for the rest of the time to allow optimum lighting.

Camera viewpoints

Sometimes, in order to allow a wider variety of camera positions, parts of a setting are removed. These 'wild walls' may be slid aside, hinged open, or hoisted ('flown'). For the Lighting Director, the change often imposes a number of lighting compromises under typical television production conditions:

● The original functions of all lamps in the area have altered. The soft filler at the front of the setting, for example, is now to the side or back of the new viewpoint.
● Extra lamps have to be introduced for the new camera position(s). This can further stretch limited resources.

- There is usually little time to relight and check the new situation.
- Where a camera peeks through a side or rear wall of a set using a pull-aside drape, a 'wall trap' (small hinged wall panel), or a fireplace opening it may not be possible to provide appropriately positioned lighting to suit the new viewpoint.

Walls

You will seldom want to light walls *evenly*, except perhaps in exterior settings. More often, they will appear more attractive or more compatible, if they are shaded, blobbed, dappled, or have an appropriate shadow pattern cast across them. This has a practical as well as a decorative purpose. Top-shaded walls help to throw faces into greater prominence, and to give a feeling of enclosure – a ceiling to the room. They give an overall illusion of depth and solidity. In addition, shading can help to avoid accidental shadows from the sound boom. Wall decorations and fixtures (practicals, mirrors, pictures, etc.) influence how the wall can be lit, to avoid unattractive shadows, hotspots, reflections, etc.

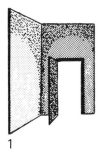

Fig. 14.11 Wall shading
1. Careful air-brush spraying (blowing down) of tops and corners of walls can supplement shading by lighting. Such shading must not conflict with lighting treatment.
2. As walls are progressively shaded the background is transformed from a blank, open, spacious surface to one implying the presence of a ceiling, and giving a sense of enclosure. Extensive shading conveys intimacy, cosiness, compactness and a confined space, even oppressiveness, where top shading is heavy.
3. (*a*) A naturally confined space (boat, aircraft, train, tent) is imitated most convincingly with (*b*) top and base shading.

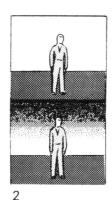

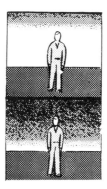

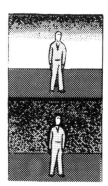

2

a

b

3

Flat heights

Wherever performers face downstage, their keys will be placed somewhere at the front of the set. But a high proportion of action is cross-shot with cameras either side of the set, facing inwards. People are angled, looking across the set. Consequently, the optimum positions for their keys and back lights are directly over the side walls of the set. If the walls are high (e.g. 3.6–4.5 m (12–15 ft)) this must increase the height of these lamps and result in much steeper lighting, particularly if people are positioned fairly close to the walls. You can, of course, hang lighting fixtures within the setting lower than the top of the flats, provided they do not appear in shot, cast unwanted shadows, or create lens flares. But then, it might be argued, there was no need for the flats to have been so high in the first place.

In smaller studios a low ceiling height and an overhead lighting grid will usually determine the maximum height of flattage. Most studios use a 3 m (10 ft) flat height wherever possible, resorting to 3.5 m (12 ft) only where necessary.

The maximum height of flats is normally chosen to avoid cameras *overshooting* (*shooting-off*) the tops of the setting walls, seeing lights and the studio beyond. So the Set Designer takes into account –

- The longest shot (widest view) cameras are taking.
- Whether cameras are working a long way from the back of the setting. (Even a slight upward tilt will overshoot the set.)
- The maximum lens angle being used. (Wide-angle lenses cover a greater vertical angle.)
- Whether cameras are taking upward-tilting shots (particularly from low viewpoints).
- Whether action is taking place on raised areas (e.g. on stairs or platforms/rostra).

Instead of increasing the height of the flats, the Designer can often introduce an intermediate scenic feature that prevents the camera seeing the top of the flats, such as a border, cutting piece, or an architectural feature such as a beam or an arch.

Windows

Windows are an important feature of any interior setting:

- They help us to locate the setting – a rooftop view of a town, perhaps, or countryside. Even a skycloth and a little foliage outside the window seems to make the setting look more realistic and 'authentic'.
- Windows can reveal the time of day, the weather, even the season.

■ *Window size and shape*   The size and shape of a window can determine the effectiveness of both the interior and the exterior scene beyond it. Light coming through the window has a strong influence on the atmosphere and mood of the interior.

Small windows, particularly in a deep embrasure, reveal little of the exterior, and the amount of light you can project into the room is strictly limited. A large window, on the other hand, reveals a great deal of the exterior scene. To project a sharp complete light pattern from such a window onto a wall of the room requires a strong spotlight on a lighting stand some distance away (e.g. a 10 kW quartz fresnel spot or 250-A arc at least 6 m/20 ft away). The closer the lamp is to the window, the larger and less distinct the light pattern. In most studios there is seldom the space available to do this directly.

■ *Lighting through*   Any 'sunshine' or 'street lighting' shining into a room is normally provided by a spotlight on a lighting stand outside a window. This needs to be positioned so that it cannot be seen by cameras within the setting. Consequently, for a window in a side wall

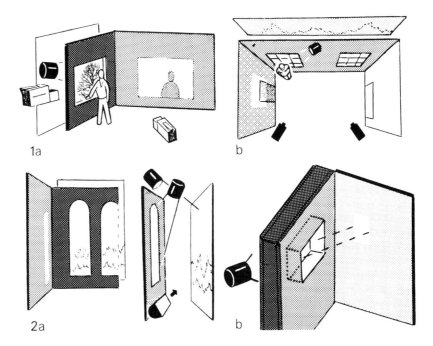

Fig. 14.12   Window design
1. *Window position*
    (a) A window in the side wall of a setting has considerable advantages: people can look out, cameras can look in, the backing is clearly visible, time of day (day/night) and location are quickly established. 'Sunlight' or 'moonlight' can be projected through the window onto people and the background.
    (b) When a window is located in the back wall of a setting any through-light is usually angled obliquely, with little opportunity to project directly onto people or the background.

2. *Window size*
    (a) *Large windows* pose certain difficulties.
        Unless the backing is well-spaced from the windows it may not be possible to light the entire visible backing evenly.
        It is seldom practicable to cast proportionally large light patterns from big windows.
        If the interior scene is bright, and the exterior backing dark, spill light from the interior is liable to fall onto the backing (spurious light patches and shadows).
    (b) Disappointingly little light can be projected through small windows, deep windows, embrasures (e.g. in castles, prison cells) and light patterns are usually too small to have any artistic value.

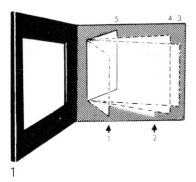

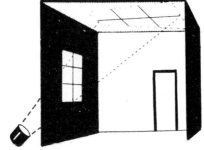

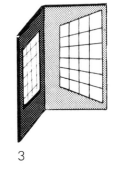

1

2

3

Fig. 14.13 Pattern shape

1. The size, shape and proportions of a window pattern can influence our impressions of an environment.
   (1) A limited wall pattern can be ineffectual, appear mean and restrictive; very suitable for confined or sordid interiors.
   (2) Strong daylight, sunlight.
   (3) A broad, bright, delicate, airy effect.
   (4) Expansive effect from close lamp.
   (5) Lamp too low for sunlight.
2. *Ceiling light pattern.* The ceiling pattern (from a lamp on the floor) suggests light from a street lamp below. (In practice, the effect can be distracting.)
3. *Diverging light pattern.* If a lamp on a lighting stand is close to the window, the cast shadows will spread unnaturally.

Fig. 14.14 Window patterns

For sharp, clearly defined, undistorted window patterns, a distant point source is essential – particularly when casting 'rain-on-window', window lettering, billowing curtains, and similar shadow effects. The closer the fixture is to the window, the larger the light pattern and the less distinct the shadow effect.

1. *Restricted window pattern*
   The more oblique the lighting angle (1), the more restricted will the light pattern be. Although the lamp position (2) may provide a more effective light pattern on the wall the fixture itself may be visible in cross shots.
2. *Limited space*
   When the scenic backing is close to the window opening there may be too little room to successfully position the lamp providing the 'sunlight', *and* it may not be possible to light the backing evenly.
3. *Disrupted window pattern*
   Window light patterns will usually be ineffectual when projected onto surfaces that are patterned, dark, or shiny. When light patterns are broken up by furniture (a), or distorted by the shape of the setting (b), they lose their impact. Similarly, light patterns can be interrupted by wall decorations, drapes, mirrors, etc., or by architectural features such as doors, arches, etc.

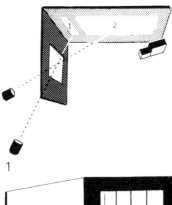

1

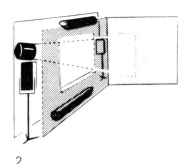

2

3 a

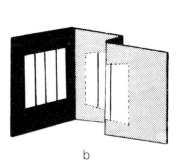

b

it is invariably placed at its downstage edge. Because it is at an oblique angle, the effective width of the light pattern from the window is considerably reduced. For small windows little useful light may get through.

■ *Window dressing* The way in which window openings are dressed will often affect your lighting opportunities:

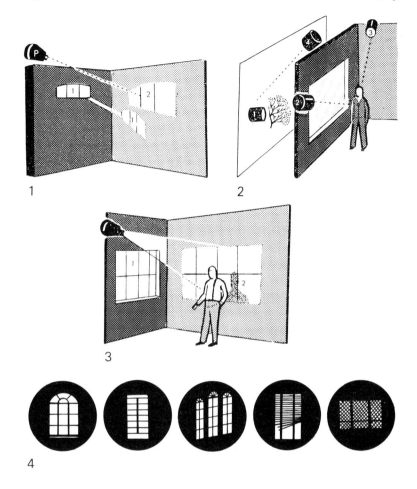

Fig. 14.15 part 1 Cheated window effects
1. *Poor window pattern*
   Whereas in (1) the wall pattern produced by shining light through a window opening is unsuccessful (e.g. too small, wrong shape, unattractive), it may be preferable to project an imitation instead (2).
2. *Covered window*
   When the window opening is covered with sheers, nets, scrim, etc., the best solution may be to:
   (1) Light the translucent material separately; perhaps introducing a leaf dapple pattern.
   (2) Key the action with an interior lamp.
   (3) Add a rear kicker.
   (4) Intensify the lighting on the backing.
   Try to avoid light from within the room falling onto the surface, or it will produce a 'blind window' effect, and obscure the backing beyond.
3. *Projected window pattern*
   When a window pattern is projected there can be various giveaway signs in the picture:
   (1) The light direction on the subject, and cast shadows, do not correspond to the actual window position.
   (2) The subject's shadow on the background may be unconvincing (wrong angle, wrong size).
4. *Typical projected window patterns*
   The projector must be very stable, and not sway. (Projectors supported on a rope or short bar may move in air currents.)

Fig. 14.15 part 2 Artificial shadows
Where space is limited (e.g. insufficient throw, poor angle), a painted imitation background shadow may be more effective than an indifferent projected version.

- Full drapes on either side of a window opening restrict the amount of light that can be introduced as a key or effect.
- Light-toned blinds or drapes are easily overlit by any through-lighting, even when the latter are drawn back ('swagged').
- Materials such as sheers, lace curtains, net curtains, voiles, stretched across the window opening, generally preclude effective light patterns. Rather than introduce through-lighting, it is often better to light the backdrop outside the window more intensely to emphasize the scene beyond, and give it greater definition.

■ *Night hazards* Night interiors usually present two hazards when there are large uncurtained windows:

- Lighting *within* the room can throw shadows of the window frame onto the backdrop beyond; a particular problem when keying someone standing beside an uncurtained window.
- Any strong light from outside the window, intended to simulate moonlight or street lighting, can hit the back of the window flat and bounce onto the backing, mysteriously overlighting it! Careful barndooring and a black drape over the back of the flat usually solves that dilemma.

Backings

Any scenic plane placed on the far side of an opening in the setting (window, door, arch) is, strictly speaking, a *backing*. It may take various forms, including a canvas backdrop, brick-wall flat, a scenic cloth, a photo blow-up, a cyclorama. (The term *background* is a more general one, for any kind of surface appearing behind a subject.)

If we simply want to prevent the camera seeing through the opening, closed drapes or obscure glass (frosted, or heavily patterned) would achieve that. But when there is a backing outside a window or door, it seems to extend the setting, and connect it to the 'external world'.

■ *Lighting the backing* While a splash of light may be sufficient on the backing outside an interior door, just to reveal that it is there, exterior backings must be lit carefully if they are to be convincing. In most cases you will need to light the backing evenly with soft light from above, below, perhaps either edge. To achieve this, it must be 2 m (6 ft) or more from the window opening. Less than this, and its edges will be hot, while its center remains darker. Lit from above only, any sky in the backing scene is likely to be overlit. Lit from below, the shaded result is usually bizarre.

■ *Re-using the backing* For economy, the same scenic backing is sometimes used for both day and night scenes – either by throwing a black scrim (gauze) over it or leaving it unlit for night. The result is seldom convincing. Sometimes a photo backing can be treated, with rear-lit window openings, or frontally lit reflective patches to simulate a night scene.

There are limits to how far you can hope to alter the appearance of a backing by selective lighting. With luck, a monochrome photograph can be lit with white light to imply leaden skies and a snowy landscape, and with colored light to provide blue skies and green meadows, but this is exceptional.

The cyclorama

The *cyclorama*, or *cyc* as it is more usually called (pronounced 'sike'), is a remarkably versatile arrangement. It can be used –

● As the main background to action;
● In conjunction with other scenic units;
● As a backing to a built setting.

■ *Design* The cyclorama is simply a suspended cloth hung from a straight or curved track to form a background along one or two walls of a studio. The cyc may hang free but it is usually stretched taut, and arranged in either a 'curved' or 'wraparound' form. It may be anything from 3–6 m (9–20 ft) high to some 6–18 m (20–60 ft long), depending on the size of the studio.

Fig. 14.16 The cyclorama
1. *Curved cyclorama.* A shallow C-shaped hung cloth background.
2. *Wraparound cyclorama.* A widely used form of cyc. Corners are gently curved, e.g. 2.7 m (9 ft) radius.
3. *Paper cyclorama.* A roll of *background paper* may be stapled to a run of flats, to form a continuous background.
4. *Merging coves/ground coves* can hide the join where the cyc meets the floor, and create a continuous plane.
5. *Cyc lighting* can be hidden behind a *groundrow*.

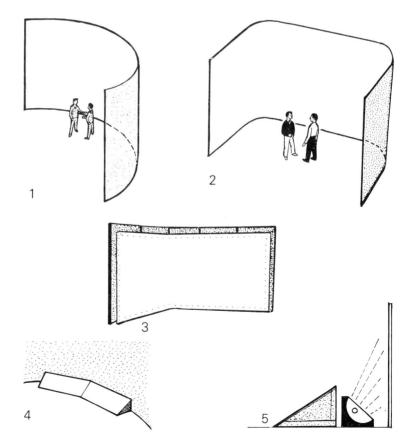

■ *Materials*   Various materials are used, including canvas, muslin, duck (linen or cotton cloth), filled gauze, shark's-tooth scrim, felt, and even velours. In a small studio, *seamless paper*, either suspended or attached to a run of flats, may form a cyc. Cycs are available in a range of colors, including off-white (60% reflectance), light gray, mid-gray, dark gray, black, light blue, dark blue, and cobalt-blue or green for chroma key effects.

■ *Lighting the cyc*   As you saw earlier (*'Decorative lighting'*) one can light the cyc in an almost endless number of ways. Formerly, cycs were lit by large open-reflector lighting fixtures such as 'skypans', 'scoops' or 'five-light' troughs, or with blended fresnel spotlights, and some people still favor these methods. But problems with spill light and/or unevenness led to the development of the present cyclorama lighting units. These have specially designed reflectors which provide an asymmetrical pattern of light, which is remarkably even over the cyc's entire height.

Fig. 14.17   Lighting the cyc
1. *Soft overall lighting.* Suspended soft lights and soft groundrow lighting combine to provide even overall lighting.
2. *Merged spotlights.* A series of spotlights with soft-edged beams merge to provide reasonably even illumination on a background. Fitted with *cookies/dapple-plates* (Figure 10.14), they can produce an overall dappled effect. You can 'spot' (focus) lamps to obtain pools or blobs of light. Barndoors can restrict light into columns, squares, etc.

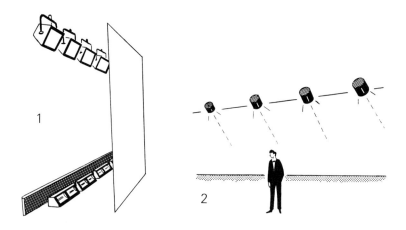

When cyc-units are used as a *groundrow* you can light the background evenly or, by moving them nearer the background or adjusting the reflectors, produce a top-shaded effect. Hidden behind a scenic 'cove', this upward lighting can provide a very attractive white or colored border of light to backgrounds. Some studios have a special pit in the studio floor at the regular cyc position, which houses hidden cyc units, thus avoiding the need for scenic coves.

Used decoratively, the cyc gives you the opportunity to project light patterns, cast shadow formations, selectively illuminate with shafts or blobs of light, color-mix, and so on. Used as a 'sky', you can run the gamut from dawn to sunset, with clouds, sun, moon, stars at the touch of a button . . . given sufficient time and facilities!

When lighting a cyclorama with the denser colors it is worth remembering that –

● Only a fraction of the light emerges (e.g. 10–20%), so you will need more powerful sources, and perhaps more lighting fixtures.

● The light falls off rapidly, leaving more restricted coverage than normal and perhaps localized hotspots.

Scrim/scenic gauze

The term 'scrim' is another of those multi-meaning terms. It refers to both:
● The *diffuser* used to soften light and reduce light levels, and
● *Scenic gauze*, a thin cotton or synthetic netting with a mesh of around 1.5–3 mm ($\frac{1}{16}$–$\frac{1}{8}$ in); available in black, gray, or white.

Scrim (scenic gauze) can be used hung loose or stretched in several ways:

● Over a cyc cloth.
● As a cyclorama against black cyc drapes.
● Over scenic backings – to soften outlines, lessen tonal contrast, and to lighten color. It reduces the artificiality of painted backgrounds.

1

2

Scrim – scenic gauze
1. When the scrim is lit from the front, it appears solid.
2. If we leave the scrim unlit, and light the subject behind it, the scrim almost disappears. The overall image is slightly softened.
3. Here the subject is lit as before, but the scrim has been illuminated from behind. The result is an overall haziness.

3

- Over windows, it diffuses the scene on the far side, and suggests aerial perspective.
- As a substitute for glass, it is non-reflective, and allows lettering to be attached or painted on its surface.
- To simulate a misty or foggy atmosphere.
- For scenic effects, where background detail is painted on the scrim (see Figure 15.14).
- As scenic drapes.

Scrim is often used very effectively, stretched taut over a light-blue (or gray) cyclorama, to help hide storage creases or wrinkles. Sometimes the stretched scrim is pitched (sloped away) from the cyc at the bottom, so that cyc-lights can be placed between the scrim and the cyc. It is arguable how useful the idea is in practice, for it positions these lamps too close to the cyc, and emphasizes irregularities through edge-lighting.

If the scrim is spaced a short distance from the cyc cloth you can produce blurred or double images of shadows or light patterns – one on the scrim and another on the cyclorama. This will soften a shadow, and often gives it 'depth'. The multi-shadows of bare tree branches can suggest a wood. Similarly, you can produce multi-clouds and 'watery' moons. However, if you want a *single* clear-cut image the scrim must be stretched in direct contact with the cyc. When you light anything painted or attached to the surface of *black scrim* the material itself is unseen, but the added details are clearly visible. You can light through black scrim with quite acceptable light losses.

White scrim looks 'solid' wherever light falls on it, and is often stretched on a timber framework to form a 'ceiling' to a setting. When lit from above, this arrangement resembles an illuminated panel rather than a normal ceiling. (Any structural cross-pieces will cast shadows.) Only a limited amount of light passes through it; insufficient to illuminate subjects below effectively. If you increase the amount of light the panel will be overlit ('burn up'). Lit from below, the ceiling piece appears normal and effective.

Studio floors

■ *Film studio floors* In the *film studio*, scenery is frequently nailed to the wooden floor for greater stability ('spiked'). Scenic structures are often more extensive and elaborate, and more robust. To ensure smooth movement, camera dollies may run on specially laid tracks/rails attached to the floor. The Designer in a film studio can lay coverings of boarding, canvas or prepared surfaces on the floor to decorate its surface. Where the film camera is static, or running on tracks, floor panels of profiled cobblestones, paving, grass, boards, etc. can be introduced to provide a variety of finishes.

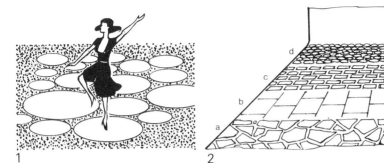

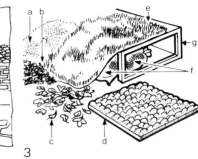

Fig. 14.18   Floor treatment
1. The studio floor may be treated decoratively (painted, stenciled, stuck-on designs), or lit with a series of light patterns.
2. Painting simulates surfaces economically, leaving them flat for camera moves: (a) crazy paving; (b) paving slabs; (c) pavé bricks; (d) cobbles.
3. The floor surface can be covered by: (a) sawdust; (b) scattered peat; (c) dead leaves; (d) sheets of rubber/plastic cobbles, brickwork, etc.; (e) grass matting (surface contours changed by (f) sawdust sacks, or (g) sack-filled platforms).

■ *TV studio floors*   In the *television studio* the floor is a smooth, specially levelled surface, which allows camera dollies to move around freely without picture-judder. Scenery is supported by wire ropes, weighted braces and struts ('jacks'), and never fastened directly to the floor. The Designer has a smooth, clean floor surface, which can readily be decorated with waterpaint, adhesive patterns, sheeting, etc.

However, because TV camera dollies cannot move without judder over floor coverings such as carpets, matting, turf, any built-up floor treatment must be limited, leaving the floor area clear in order that cameras can move around freely. Decorated floor cloths and panels are used to some extent, and their appearance will vary with the materials. An effective way of overcoming this difficulty is to decorate the floor to imitate these features. Using a water-soluble paint, a scenic artist or a special machine with interchangeable rollers paints cobbles, stones, pavé, paving stones, floor planking, parquet flooring, earth, by the yard, wherever needed.

■ *Lighting problems*   Although simple, cheap, and convincing enough at a distance, painted treatments have disadvantages as far as lighting is concerned. However much you paint it, the floor itself remains smooth! It presents no actual *texture*. Back light bounces off it, producing bright detailless patches, and the painted treatment disappears! And even if the surface is disguised by scattering tanbark, peat, or sawdust, the camera will still pick up the general sheen of the smooth floor surface beneath. (The materials themselves are liable to become spread around the studio and stuck to feet and wheels of equipment.)

The shiny black floors that look great in a dance routine as they reflect scenery and dancers can also reflect every lamp in the studio!

(They can also show every footmark, and require frequent cleaning or refurbishing to maintain that spectacular appearance!) When lit from the side with cross-keys, one does at least avoid speculars from back light; but you may find yourself lighting to avoid hotspots and flares rather than to suit the subject itself. Although some Lighting Directors tackle the situation by using total soft light, so that any speculars are at least of larger area, this is, at best, only a palliative.

Foliage

Foliage is used in the studio in many forms: as suspended tree branches, small trees, bushes, undergrowth, hedges, etc.

■ *Problems*   It can be notoriously difficult at times to light foliage to obtain a completely natural effect. Because incandescent lighting is somewhat deficient in the green-blue region of the spectrum it often needs a considerable amount of light to create any visible impact on-camera. Much depends on the type of foliage, of course, but with darker leaves, coniferous shrubs or trees, you may well find that the odd fifty kilowatts or so have been used to light the trees and bushes in a studio 'exterior' scene in order to give them any sort of vitality. The glossy leaves of rhododendrons, laurels, hollies, and similar foliage can reflect strong speculars, while still remaining dark on-camera – an effect that is quite unnatural, and totally frustrating.

■ *Position*   If you have to pour light onto dense foliage so that it will register successfully, take care to ensure that it is not going to overlight the sky cloth or a wall nearby. Where there is foliage just outside the window of an interior set it should usually be sparse; otherwise it will tend to look dense and poorly defined from within.

Practical lamps

A wide range of *practical* light fittings (*pracs*, for short) is used in settings both as decorative features and implied light sources. They make a valuable contribution to both the realism and the general ambiance, so it is worth taking particular care in the way they are presented. It is essential, therefore, to know the position and type of all practicals when designing the lighting treatment; particularly where they are to appear as the main light sources illuminating the scene.

■ *Variety*   Pause for a moment, and consider the enormous variety here: table lamps, wall brackets, pole lamps, standard lamps. Overhead light fittings such as chandeliers, globes, central shades, fluorescent troughs, ceiling panels, etc. Then, of course, there are candles, oil lamps, lanterns, flashlights, torches, flambeaux, Chinese lanterns, gas lamps, sconces . . . Each lights its surroundings in a

Fig. 14.19 Practical lamps

1. Unattractive light pattern. (A) Upward streak suppressed by an inserted piece of diffuser. (B) Bulb seen as localized hotspot (see part 3).
2. Ineffectual light output. From small low-power fittings. This requires localized simulated wall-blobs from e.g. projector spotlights. (a) Incorrect imitation. (b) Correct imitation.
3. Overbright bulb. Paint down or shield front half of bulb, leaving full light to illuminate the wall. (a) Before treatment. (b) After treatment.
4. Incompatible light direction. Subject lighting direction does not agree with the position of the practical.
5. Shadows of a practical lamp falling onto nearby surfaces can look (a) unnatural, and/or (b) unattractive.
6. A hanging practical fitting will often cast obtrusive shadows onto people or the setting. Assuming that the practical itself cannot be repositioned, or a barndoor used to restrict the light, it may be necessary to re-angle the key; offsetting it as in (a) or raising/lowering it as in (b).

- Hanging practical lamps are usually positioned in upstage or non-action areas to avoid spurious shadowing, or obstructing the sound boom.
- Long suspension cables for overhead practicals can look unrealistic in low-angle long shots. Unless brail lines are used to steady them, they are liable to sway under strong studio ventilation.
- Shadows of hanging practicals can sometimes be hidden by casting them onto the floor, throwing them out of shot, or casting them onto dark or broken-up surfaces.

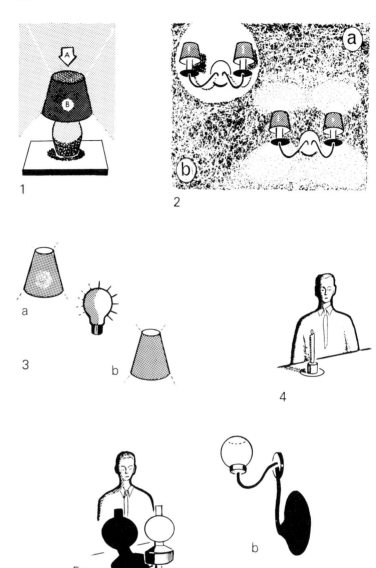

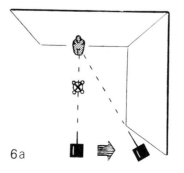

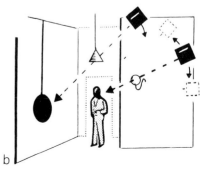

particular way; each creates a certain characteristic atmospheric effect.

■ *Design* The effectiveness and performance of any practical light fitting varies with its design.

- In some, the light source is visible.
- In others it is shielded.
- Some are purely decorative and give out little light.
- Others strongly illuminate their surroundings.

Your lighting treatment will be affected by such design features.

For each practical light fitting we usually need to ask ourselves the basic questions:

- How effective will this practical fitting be? Will it be bright enough to actually use as a luminant (completely or partly), or will I have to ensure that it *appears lit*, then *imitate* the effect on its surroundings?
- Will the fitting be overbright on-camera, or insignificant?
- Will the design of the lamp pose particular problems? If, for instance, a decorative wall practical contains a hidden bulb that casts a glow of light around it, will its effect be visible on-camera under studio lighting? Can you simulate the light around an elaborate wall bracket without casting multiple shadows of it onto the nearby wall?

■ *Source brightness* The apparent brightness of any practical source will depend not only on its design but also on the general lighting level. Under e.g. 300 fc, a practical fitting will appear quite dim – even unlit. Under e.g. 30 fc, it may be distractingly bright.

You will generally find that by the time a visible light source such as a gas mantle or a bare bulb is bright enough to illuminate its surroundings effectively, it is very distracting. On a video camera, it is likely to be causing electronic problems. Reduce its intensity until the source is suitably bright, and it will not appear to be lighting its surroundings!

■ *Uprated lamps* When a large, heavily shaded practical fitting appears insufficiently bright, you may be able to solve the problem by uprating its lamp, or using a Photoflood lamp instead of the regular bulb. The increased light from the overrun lamp may look exactly right as it spills onto a nearby wall. But there is the danger of overheating the lamp holder and its shade, and many practical fittings are too small, or have too little room for overrun lamps. Just occasionally, you may be able to overrun the regular lamps by supplying them with a slightly higher voltage (+5–10%), but this is not a reliable method.

■ *Cheating* Some fittings can be modified to provide more light. For instance, you may be able to cut away part of a table-lamp shade, so that more light emerges in a certain direction. But practical lamps

Fig. 14.20 part 1 Ceilings

Omitting ceilings on studio settings gives greater freedom to lighting and sound treatment. Sometimes, however, they are essential to prevent low cameras shooting over the walls, for atmospheric effect, or to add realism.

1. *Ceilings can degrade lighting in several ways*
(a) (1) Spill light on vertical planes.
    (2) The ceiling may remain unlit, where any lamps used would be in shot.
    (3) Where the ceiling cuts off frontal lighting, its shadow may fall onto the background. Part of the background may remain unlit.
    (4) Portrait lighting may be limited to frontal and cross-lighting.
(b) (1) The ceiling piece can prevent your using the best lamp position for good portrait lighting.
    (2) If you place the lamp at the ceiling edge it is then too close, and too steep.
    (3) Although the lamp's height would be more suitable at (3), its direction would be inappropriate.

2. *Solutions*
(a) Where there is a ceiling (or partial ceiling) you may be able to
    (1) Bring action forward of the ceiling area.
    (2) Use hidden lamps to light walls and ceiling.
    (3) Occasionally, practical lamps may help to light the setting.
(b) You may light action and setting through openings – windows, doors, skylight.

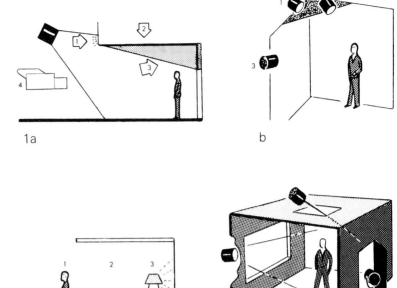

1a b

2a b

that appear to be lit yet have no effect on their surroundings are incongruous.

To make many practicals look convincing you will often need to 'cheat' by

● Uprating the source where possible.
● Imitating the way its light would fall on people. (Using additional localized spotlights.)
● Arranging light patches on the wall to suggest the emergent light falling on the surroundings (e.g. simulated by pattern projectors).
● Shielding the source to prevent its being overbright from the camera viewpoint. (Using a clip-on diffuser, light-shield, or half-painted lamp.)

You can even cheat so that the actual light source itself does not need to be lit at all! For instance, a small hidden spotlight may illuminate a gas mantle (or even a card replica!) so that it appears alight. Easier still is to modify a gas fitting, so that it can be powered electrically, its mantle being replaced by a low-power frosted 15 W lamp ('pygmy').

■ *Common problems*
● Some translucent shades and white opal globes have the frustrating property of appearing to be alight, even when not powered, as studio lighting falls on them.

Fig. 14.20 part 2   Set design
The setting may be designed to give
the impression of a ceiling on
camera, yet keep obstructions to a
minimum.

1. Restricting the ceiling to just the
    part seen on camera.
2. By sloping the ceiling piece.
3. By localizing the ceiling (corner
    pieces).
4. By introducing false returns in the
    walls, so that an apparently solid
    wall has a gap for a concealed
    lamp.
5. By using cutting pieces (false
    soffits, beams) to conceal lamps.
6. Hanging a vertical plane (e.g. a
    scenic cloth) in the position of
    camera shoot-off.
7. Using a translucent ceiling allows
    light to penetrate – but:

    The ceiling itself will look
    illuminated (not always
    appropriate).
    Little light gets through to
    illuminate people or the setting.
    The translucent ceiling's
    supporting structure may be
    visible.

Other methods of simulating the
ceiling include camera mattes, optical
printing, electronic insertion.

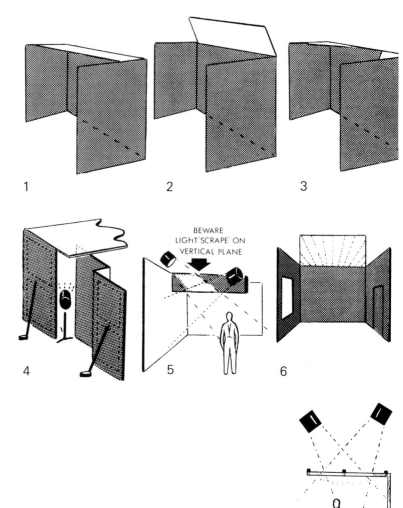

- Although very dark lamp shades shield direct light from the camera, their interior linings may appear extremely bright.
- The emergent light pattern from some light fittings is unattractive or distracting. A judiciously placed piece of diffuser will usually lessen or suppress the effect.

15 Visual effects

Visual effects are part of the magic of film and television production. They enable us to create an illusion which, if used skillfully, can appear completely natural, and will be accepted as 'the real thing' by the most critical audience. The various effects we shall be concentrating on here are widely used in studio production, and directly affect lighting techniques.

Light effects

Firelight

Firelight has a strongly evocative quality for us all, ranging as it does, from quiet restfulness and romatic allure to the frightening excitement of a fierce conflagration. Firelight has become symbolic, and can make a persuasive contribution to the atmospheric quality of a scene. But it needs to be imitated well if it is to be successful. Some attempts strain one's credulity.

If you look at real firelight closely you will find that it has two elements: a constant glow, and a flicker. The fiercer the fire, the more agitated the flicker. When imitating firelight, it has usually to be exaggerated to some extent. Real domestic fires tend to contribute little to the illumination of their surroundings unless the room is in darkness.

■ *Constant glow* You can simulate the low-intensity glow from inside a stove with a piece of orange-red color medium; randomly daubed with black paint, perhaps, to break up its evenness. 'Coals' of solid glass (cullet) lit with colored light can be effective too. To imitate the light from such a fire, fit a spotlight with a red-orange filter and a diffuser, and place it on the floor or a low stand. Its constant low-intensity light should be softened and fairly localized.

Colors for firelight are controversial. Orange and deep amber can be successful if the light beam is broken up but are less realistic as an unrelieved flood of light. At all costs, avoid strong reds.

■ *Random flicker* Everyone seems to have their own favorite way of creating this type of fire effect. There is nothing to excel a 'flicker stick' (Figure 15.1) fitted with a succession of narrow strips of translucent cloth. This is held across the front of a lamp fitted with an orange-red filter. Slide the stick slowly back and forth across the lamp in a series of short, irregular jerks. More like an unsteady hand than a regular rhythm. If you jiggle or wave the stick, the effect will become too violent. In the right hands, coupled with varying brightness by dimmer adjustments, the effect is totally convincing.

427

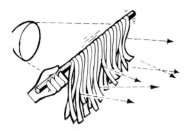

Fig. 15.1   Fire flicker stick
Narrow strips of rag attached to the
stick are gently shaken in front of a
ground lamp to simulate flickering
firelight.

Although some enthusiasts advocate thin multi-color strips of medium (e.g. red, orange, yellow, etc.) for the flicker stick, they tend to sway rather than 'dance', and may be heard rustling under quiet conditions.

There are other methods of creating 'fire flicker', including:

- Counter-rotating discs, clear drums with undulating patterns, projected smoke or flame effects. Generally speaking, the results are rather mechanical-looking.
- Smoke from burning incense or rope, or burning gas jets in front of a lamp, may be used to cast shadows and/or distort the light rays. Given sufficient smoke, the effect can work well enough, but appears as moving superimposed shadows rather than flickering flames.
- Colored light reflected from silvered plastic sheeting produces an effect more reminiscent of water ripple than firelight.

Shafts of light

There is something highly emotional about shafts of light. We see them shining down from the heavens, from church windows, cutting through the gloom of a cellar . . . They are impressively dramatic, and they are understandably requested by Directors and Designers. Unfortunately, because light itself is invisible one can only achieve 'light beams' if the atmosphere is *smoky* or *dust-laden*! If there is too much smoke, mist or fog about, the light is dispersed, and does not produce a clear-cut beam. As most studios need to have efficient air-conditioning systems to dissipate the heat from lamps any smoke is quickly whisked away. And then there are the smoke-alarm systems!

On balance, the optimum methods of producing light beams are:

- Arrange the appropriate lamp to produce the light beam *and* ensure that no other lighting in the area is going to result in unwanted beams. Use no soft light. Switch off the air-conditioning, and introduce a minimal amount of well-distributed smoke from a carefully controlled smoke-generator.
- Have light streak along the surface of stretched scrim (scenic gauze), and superimpose this on the required area.
- Superimpose a defocused graphic showing the required beam.
- Use video effects.

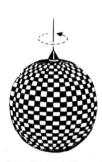

Fig. 15.2   Mirror ball
A strong spotlight illuminates the
multi-mirror-faced ball. Revolved by a
motor the resultant pattern of light
dots moves across the scene.

Passing lights

■ *Mirror ball (Figure 15.2)*   The mirror ball, familiar in dance and disco, reflects a traveling pattern of tiny light-dots across the scene as it rotates. It is lit with one or more very localized spotlights from a roughly similar height, taking care that they do not accidentally

Fig. 15.3　Mirror drum
A polygon of plane mirrors reflects a regular sweeping pattern of rectangular lights across the scene as it rotates (e.g. train windows).

overspill past the ball onto the surroundings. The motor-drive is usually adjustable, and a slow speed is most effective.

■ *Rotating drum (Figure 15.3)*　When a rotating drum faced with a series of mirrors, pieces of metal foil or similar materials is lit with a spotlight it reflects moving light patterns. In a *mirror drum* a rectangular pattern of lights sweeps across the scene to suggest the windows of a passing train. A *cyclodrum* consists of large, rotating, circular framework. This may be used to provide

● Passing shadows from an internal lamp, or
● Passing lights from reflective patches.

Its main application is for vehicle shots, where passing shadows (day) or lights (night) help to create the illusion that a car is moving (Figure 15.4).

Fig. 15.4　Passing shadows and lights
1. A rotating framework fitted over a lamp, and hung with a stencil cut-out, branches, etc., to create passing shadows. Patterns must not be obviously repetitive.
2. A spotlight focused onto *mirror drum* faced with plastic mirror, produces 'passing-lights' effect for vehicle interior shots.

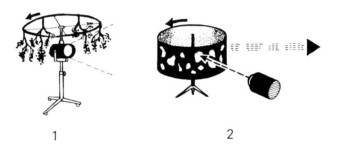

1　　　　　　　　　　　　　2

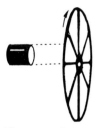

Fig. 15.5　Flicker disc
As its spokes cover and uncover, the beam of the lamp flickers. Multi-color segments may be used for decorative effect.

■ *Flicker disc (Figure 15.5)*　Although extremely simple in concept, a large, vertical, rotating flicker disc has its applications. Made in clear plastic with painted-on shapes, or from a circular plate with holes, the device can be set up in front of a lamp to provide pulsating light or vari-colored light effects. When the disc is turned at slow speeds, and the light switched/dimmed, you can illuminate a car interior to suggest moving lights or shadows, from passing vehicles, shop lighting, etc.

Swinging boomlight (Figure 15.6)

Fig. 15.6　Swinging boomlight
A counterbalanced spotlight provides a swinging key light, to convey movement (e.g. rolling ship). A motorized telescope can also be used for continual height changes.

A spotlight fixed onto a counterbalanced boom can key through the porthole of a 'ship', rising and falling to suggest that it is rolling. Shadow movement creates a very convincing effect, especially when coupled with corresponding camera movements (rhythmical tilting) and a swinging practical lantern (fishline operated). Alternatively, a motor-controlled hanger (telescope) can provide automatic height variations.

Strobe lights

Using the vari-speed flashes from an electronic flash-tube, stroboscopic lighting can create speed changes in movement, arrest movement, hold moving machinery 'still', or show dance with 'momentary blink' illumination. The technique is not, however, withouts dangers for the performers, for at lower speeds the flashing light can give rise to physiological spasms (epileptic fits, headaches, dizziness). And where 'freeze motion' is used with *rotating machinery*, people have been known to have accidents through mistakenly believing that it has stopped!

Lasers

A laser produces a narrow virtually parallel beam of monochromatic high-intensity light (0.3–30 minutes of arc is typical). A combination of argon and krypton lasers will achieve red, blue, green, cyan, magenta, gold, and white light.

Lasers are occasionally hired for decorative effects in pop presentations, dance, and dramatic visual effects. As with normal light rays, a smoke-filled atmosphere is needed to reveal the laser beams. The beams have vertical movement (Y-axis), horizontal movement (Z-axis), and intensity control, and by combined scanning, a wide range of mobile multi-color images can be created and automatically animated.

Car headlights

Headlights are usually simulated by turning a single spotlight on a lighting stand slowly across the scene. Slightly closed top/bottom barndoor flaps limit the beam-spread. An alternative arrangement uses a centrally pivoted bar with small spotlights at either end ('double head') to provide a more realistic effect.

Practical lamps (see Chapter 14)

As you saw earlier, practical lamps have their associated problems. By the time a practical is strong enough to illuminate its surroundings convincingly its light source is often too bright on-camera. Other sources such as matches and candles are normally not bright enough to illuminate the subject.

A cunning solution to this dilemma is to use a small, powerful pea lamp/peanut lamp as a hidden light source. It can be cupped in the hands as a match-flare. Concealed within a dummy candle or a candlestick, or a low-intensity lantern, it will light the person holding it quite realistically. Although it is often contended that candlelight is 'soft', in fact it is clearly a *point source* throwing hard shadows, as you can check for yourself. However, the *subjective effect* is more attractive if you diffuse the pea lamp a little. If you want a mystical glow, then a diffusion disc over the camera lens is the answer.

Real candles are largely uncontrollable, even with short wicks. Because they burn away quite rapidly (especially in a draft) it is necessary to keep an eye on visual continuity during shooting. Otherwise it may be short in one shot and long in the next. When shooting with video cameras under low-light conditions the flame may comet-tail or streak.

Lightning

Forked lightning is often imitated with animation, electrical discharges, or video effects, but little equates to the real thing on library film (stock film) clips. Sheet lightning can be imitated effectively enough, provided you take care to prevent accidental shadows being cast onto the 'sky' background, or gross overexposure during the flash.

The regular methods of producing sheet lightning effects are:

● A momentarily struck xenon arc or carbon arc.
● A rapidly operated 'venetian-blind' shutter.
● A rapidly switched group of overrun lamps (Photofloods).

If a tungsten lamp rated above about 2 kW is flashed, the filament heating time reduces the sharpness of the flash. Holding a card over a lamp and flapping it open for the flash is a very basic approach, but even that sometimes works.

Whether several rapid flashes or a more prolonged burst proves most effective depends on the application. The best *direction* for the flash will vary with dramatic movement:

● Frontal – an open effect, sudden revelation, exposure to the storm.
● Back light – the flash reveals outline detail. Threatening.
● Seen through a window – a sense of security; safe from the storm.

For blue-illuminated 'night exteriors', high color temperature white light is most successful, but when seen within a room at night, light blue (e.g. steel blue) makes a convincing contrast.

Projected patterns

The simplest way to create a light pattern over a large background area is to place a large cut-out stencil, foliage, etc. in front of a point light-source. The smaller the area of the luminant, the sharper the shadow will be. If you use a regular fresnel spotlight to project the shadow it will be less distinct than a lighting fixture with a small open lamp. Removing the fresnel lens will improve sharpness, but the light output from the spotlight will fall. To provide fairly soft-edged shapes – e.g. white clouds in a blue sky – try cutting holes in blue color-medium, clipped in front of fresnel spotlights.

You can make some very interesting multi-color patterns and 'stained glass' effects by taking two or three sheets of different colored media and cutting out shapes in each, before stapling them together. A combination of blue and yellow gels will produce blue, yellow, green patterns. Even the cut-out pieces of color medium can be stapled onto another sheet for a further effect!

Ellipsoidal spotlights (profile spots; projector spotlights) can project a sharp pattern of a metal stencil (*gobo*). These may be stamped or etched in stainless steel sheet, or hand-made in aluminum foil or sheet. Some extremely attractive gobo patterns are commercially available, including many types of foliage and tree patterns, geometrical forms, streaks, cloud formations, stars, landscapes, town skylines, maps, windows, window blinds, emblems, symbols, abstract shapes, etc. For more complex gobo designs a fine mesh is laid over the stencil to support details, and to make it more robust. Heat-proof glass gobos are used for photographic half-tone and negative images, and for intricately detailed subjects.

Composite gobo effects combine the patterns from several projectors to form the complete display. Using a different color medium before each projector, you can build up a multi-color image to provide anything from street signs to fireworks displays; from stained-glass windows to Christmas-tree lighting. Using an add-on motor-drive, you can rotate specially prepared gobos to simulate rain, snow, flames, rippling water, waves, moving branches, as well as providing flicker and moving pattern effects. Some are more realistic than others, but they can often help to supplement other environmental effects.

The area covered by a pattern-projector will depend on the angle of its lens, and its distance from the background. A wide-angle projector lens (short focal length) provides a larger image, even at shorter distances, but the light output is usually lower, and the image more likely to be distorted. The coverage of a projector with a narrow-angle lens (long focal length) is more restricted, for it is intended to be used some distance from the background. Its image is likely to be more accurate and sharper overall.

The main problems with projected patterns are:

- The effect is weakened by other light falling onto the background.
- The effects may be insufficiently bright.
- Small detail may be disappointing due to light losses and diffraction effects.
- If the projector is steeply angled the image will be distorted, and may not be sharp overall.

For shadows of water running down windows, curtains or foliage blowing in the wind the only reliable method is to use the shadow of a real scenic unit projected by a powerful distant point source.

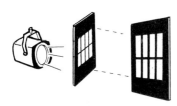

Fig. 15.7 Cast shadows
Objects can be positioned in front of a spotlight to cast shadows on the background. The shadow is sharpest when the lamp area is small (lamp fully flooded), when the subject is far from the lamp, and is distant from the background.

Fig. 15.8   Pattern distortion
If the image is projected at an angle to a surface it becomes distorted ('keystoning'). However, this effect may appear quite natural. 'Pre-distorted' gobo patterns in the reverse shape, can compensate: e.g. pattern E projected from position B would provide an undistorted pattern similar to C.

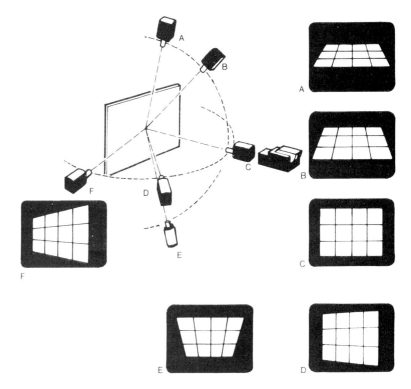

Flashing signs

A flashing electric sign can be actuated by bi-metal contact switches or electronic timing circuits – often fitted to lighting control boards. If the sign would normally illuminate its surroundings then any additional lamps lighting the surroundings will be fed from the same circuit.

A *dummy* flashing sign can be made from white or reflective lettering on a black background. Lit by a flashing studio lighting fixture (with a color filter, perhaps), the graphic will appear to flash convincingly.

Image distortion

■ *Lens flare*   When the light from a bright source enters the camera's lens system it may become internally reflected and scattered. The result is a *lens flare*, which takes the form of a multi-edged disc (from the lens diaphragm) or streaks of light. In another form the flare will gray-out black areas, reducing picture contrast, and causing desaturation. Modern lens systems are designed to minimize lens flares (lens coatings; lens hoods). In video cameras, electronic *flare correction circuits* appreciably reduce such effects.

If lens flares happen to arise during production it is usually due to low-angle back light, specular reflections, or when shooting into light sources. The normal remedy for back-light spill is to barndoor it off the camera, raise its height, or suspend a gobo (border) to shield light off the lens.

When light flares are requested by a Director as an atmospheric effect the best one can do is to use low back light pointed straight at the camera, remove any lens hood (sun shade) and hope! But, frankly, such effects are not easily repeatable, and a 'successful' flare during rehearsal may not recur during the take.

■ *Wavering image*   Where a lamp beam is projected through flames or smoke it can become diffracted into a strangely undulating light – a projected heat-haze shimmer. It is most pronounced on flat graphics, where the surface can appear to writhe in a weird, dramatic fashion. Burning wax, firelighters, oil-soaked rag, and incense have all proved successful for this purpose.

■ *Image diffusion*   There are several regular methods of softening detail throughout a picture. Clip over the front of the camera lens –

- Plain glass thinly smeared with oil or grease.
- A *diffusion disc* – Plain glass with fine concentric ribbing.
- Gauze, chiffon, net, tights.
- Fog filters – These come in several densities; giving slight diffusion, reduced saturation, and an overall reduction in contrast.

■ *Color filters*   Unless you want gross color distortions for dream sequences or abstract effects only a very restricted range of color filters can be placed *over the camera lens* for effects. Then it may be more economical to filter the lens than to place color medium over all the studio lights. The blue light simulating 'moonlight' is a typical example.

■ *Reflected shapes*   Reflective sheets in various forms offer some very interesting visual opportunities. With a little experimentation you can produce all manner of static or moving effects, including animated titling, reflected shapes, nebulous light patterns, and a host of magical and supernatural illusions; from apparitions to aurora borealis; from abstractions for dream sequences to mental turmoil.

Paint designs or lettering onto sheets of metal foil, aluminum, flexible plastic mirrors, or thin plastic, and reflect a light-beam onto a background (front or rear projected). When the sheet is flexed, the image will distort, writhe, compress, elongate. Patterns at either end of the same reflector can be manipulated separately or superimposed. You can treat the surface with grease or varnish to vary the brightness and texture of parts of the reflection.

To decorate a translucent screen, or a plain background, take slightly crinkled or folded metal foil and reflect the light from several differently colored spotlights. The multi-colored random patterns will merge to form attractive abstract shapes.

Fig. 15.9   Reflected shapes
1. Reflective material (metal foil, plastic mirror-sheet) held in a sharply focused light beam projects its image onto nearby surfaces. Bending the reflector distorts the reflected shape.
2. *Reflected patterns.* Patterns on mirror, foil or metalized sheet can be reflected onto backgrounds. Creased/crumpled foil provides random reflections.

Water ripple

The effect of rippling water is a regular requirement for shots of people sitting beside a river, leaning over the rail of a ship, looking through a ship's porthole. The most effective way of imitating water ripple uses pieces of mirror or mirrored plastic in a shallow pan of water (e.g. 1.2×0.6 m (4 × 2 ft)), water 50 mm (2 in) deep. Angle a fresnel spotlight onto the pan, and the reflected light will shimmer convincingly. For the best results:

- Use a spotlight on a lighting stand, about 1 m (3 ft) above the water. This will give a better spread of ripples than a hung lighting fixture. It is more easily adjusted, and quickly cleared after the shot.
- Reflective areas about the size of your hand are better than large mirror sheets, or small fragments.
- All you need to do to produce the ripples is to move fingers slowly in the water, taking care not to over-agitate it. Rocking the pan is less controllable, and liable to spill the water.

Fig. 15.10   Water ripple
Floor tray. Light is reflected from mirror fragments in a tray of water. *Stand tray.* A spotlight shines through a water-filled glass tray. The rippling light is reflected by an adjustable mirror onto the scene.

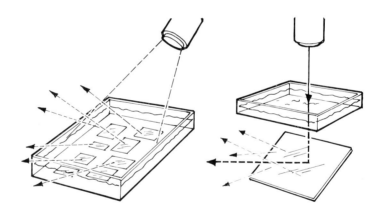

A water-filled pan is heavy, and not easily moved. So an alternative compact arrangement uses a glass-bottomed tray fixed to a wheeled lighting stand. A spotlight shines through the clear water-filled tray and its light is reflected by a lower mirror. The device has less coverage, but it is quickly adjusted by angling the mirror. Slightly tapping the tray or dipping a finger in the water is sufficient to create ripples.

Rain, mist, fog, smoke

■ *Rain and mist* Whether water is dripping, running down window panes, a steady drizzle, or a heavy downpour, it is best lit from side keys, or offset back light. Take care not to overlight rain, or the water drops will appear unrealistically bright. Where the exposure time is fairly long, rain can appear somewhat elongated.

'Rain' is achieved by adjustable overhead sprays or garden hoses. When the water particles are very small, the effect is of 'Mist'.

Any water provides an electrical hazard, so take care that sprayed water does not go anywhere near hung lighting fixtures, cyc lighting (groundrows), lamps on lighting stands, cables, plugs, etc. Apart from electric shocks, any water falling onto globes/bulbs is likely to cause them to explode. While wind machines may give rain a certain realism, blown spray is an even greater hazard, and is more difficult to contain.

■ *Smoke and fog* Mist, haze, smoke and fog are all variations on the same theme. They are generated by *smoke machines/fog machines* of various designs. One form uses a special non-toxic water-based fluid (odorless or with various flavors, including freesia, rum and strawberry!). Versions for heavy and light smoke are available. This pressurized 'smoke fluid' flows into a heated chamber, where it is converted into a thick white vapor. The controlled output is adjusted to suit the density required. The second type of equipment uses 'dry ice' (solid carbon dioxide). When heated in hot water, this produces dense white clouds of heavy vapor, which are mostly used for low-lying (waist-high) swirling clouds.

The denser the effect, the more care is needed in the positioning of lighting:

● Spotlight beams become very distinct in mist.
● In dense mist, light forms a bright area near lamps, and falls off rapidly with distance.
● Back light may be used for night scenes, but is incompatible with daylight, where illumination would naturally be flat and unmodeling.
● Side light is successful for night street scenes with fog in lit surroundings.
● Soft frontal lighting is very effective, but does not penetrate well.
● All frontal lighting tends to hit the mist and make it appear 'solid' (as when lighting scrim).

Mist/fog effects gray-out the picture; desaturating color, and reducing contrast as darker tones are lightened. For *daylight* the lighting must be flat, and not appear overbright. *Night* scenes in heavy mist can be difficult. Close shots easily look like 'poor daylight', while in more distant shots one can be very conscious of light beams.

Sometimes it is better to create the illusion by using fog 'filters' over the camera lens (image diffusion) together with a very modest amount of mist rather than filling the studio with impenetrable fog. Occasionally, just a scrim (scenic gauze) in front of the action will do the trick instead.

Photographic backgrounds

Instead of shooting a scene on location, or building a setting in the studio, it is often much cheaper and more convenient to combine the studio action with a *photographed* background. A number of systems are used in film and television production that enable this to be done. Although each has its own advantages and limitations, they are all able, at best, to create a totally convincing illusion, suggesting that the actor is actually there within the photographed scene:

- Photographic backdrop – photo-mural/photo blow-up
- Rear projection/back projection
- Reflex projection/front projection
- Traveling matte
- Keyed insertion/electronic matting – external key; chroma key

To be successful, the foreground and background images must be compatible in scale, proportions, perspective, viewpoints, color quality, etc. But your main concern when lighting is to ensure that the *direction, intensity*, and *contrast* of your lighting treatment in the studio reasonably matches that in the background photograph. If they differ substantially, the combined image will be unconvincing.

Photographic backgrounds can be used:

- As a *total background* behind the action,
- To provide a *backing* outside a window,
- To create a *scenic extension* to a setting; showing the environment beyond an arch or door.

Some regular TV settings (e.g. newscasts) use a *translucent* to suggest the city scene beyond a window. This is a large color photographic transparency mounted on a rear-lit translucent panel. Its overall brightness can be controlled accurately, and the general effect can be very convincing. Apart from the possibility of any studio lighting reflecting in its surface, it poses no lighting problems.

Photographic backdrop

Photographic enlargements stuck to canvas sheets ('drops') or scenic flats are a well-tried method of providing a realistic scenic background. Many designers prefer to use 'black and white' *photo-murals/photo blow-ups* and to color them lightly with aniline dyes rather than more expensive color photographs, which may appear rather oversaturated. Apart from *matching*, the main lighting considerations are to light the background photograph *evenly*, without emphasizing any surface irregularities, and to avoid any *shadows* of performers or scenery falling onto it.

Rear projection/back projection

Here the photographic film or slide is projected onto the rear of a large translucent secreen of 'frosted' plastic sheeting to provide a background for action taking place on the other side. Once used extensively in motion pictures and monochrome TV, rear projection (back projection, BP) has been largely replaced in film by process work (matting during optical printing), or reflex projection; and in color television, by chroma key (color separation overlay, CSO).

Fig. 15.11 Rear projection
1. A slide or film image is projected onto one side of a matte translucent screen. On the reverse side, a camera shoots action for which the screen image provides a background.
2. The rear-projected image can be used as a complete scenic background with foreground set-pieces, furniture, etc.; or (3) used to supplement a built set (e.g. window backing); or (4) to extend a built set.

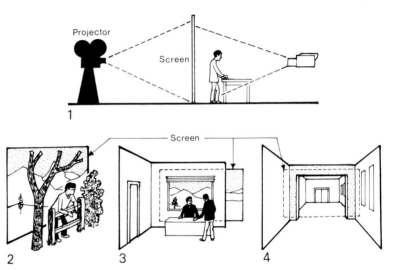

One of the system's weaknesses lies in the way any stray light ('spill light') from the foreground action dilutes the background image, graying-out shadows, reducing contrast and desaturating color. A black scrim over the face of the screen, or surface tinting, may reduce the effect to some extent.

To avoid light or shadows falling onto the screen foreground lighting must be carefully controlled:

● Subjects should not usually be closer to it than e.g. 2–2.5 m (6–9 ft).
● Soft light must be kept to an absolute minimum.
● Frontal lighting tends to be comparatively steep. Spotlights must be horizontally barndoored, and any bounce-light from the floor limited.
● Dead back light will usually be too steep, and angled back light is generally more suitable.
● For many situations the predominant lighting is usually from vertically barndoored side keys.

It is not easy to obtain a rear-projection image of even overall sharpness and intensity. There is a tendency to a central hotspot, which moves with the camera's position. The image brightness falls if the camera is angled to the screen axis. Consequently, if two cameras shoot the action the relative background brightness changes when intercutting.

Rear projection is mainly used in TV today for limited areas such as small-screen display panels, backings to car windows, and slide-display. Translucent screens are also used to display silhouettes, shadows and decorative pattern effects.

Front projection

Any image projected onto a surface seen directly by the camera is '*front projected*'. We have discussed how front-projected patterns can be used to imitate sunlight patterns on walls, decorate the open cyclorama, and so on.

Although direct front projection has many applications, and can be applied to a wide range of productions, it does have disadvantages:

- Keystone distortion due to the projector being angled to the background. (The effect may appear natural.)
- Parts of the image may be unsharp (limited depth of field).
- Diffraction effects may degrade small details in the image (projector optics).
- Tones and texture in the background may glare through the projected image.
- Spill light will dilute the image.

Reflex projection

The heart of this system (which many confusingly call 'front projection'!) is a special highly directional beaded screen of Scotchlite sheeting. Identical in principle to reflective traffic signs and night-safety clothing strips, the screen reflects nearly 92% of the projected background image back to the camera.

In a *fixed* camera set-up the background scene is projected onto the screen via a semi-silvered mirror (pellicle) set at 45° to its path. The film or TV camera shoots through the same mirror straight onto the beaded screen, and sees both the subjects placed in front of the screen *and* the strongly reflected background picture.

Because the screen's beads reflect all light directionally, straight back to its source, the normal subject lighting cannot dilute the background image unless placed beside the camera. Although the projected background image falls onto the subject or foreground objects it is swamped by the action lighting and so is not visible on-camera.

The system itself requires very accurate alignment, and can only be used with a static camera, but the results can be near-miraculous!

Fig. 15.12 Reflex projection
1. In Reflex (axial) Front Projection, the background scene is projected along the lens axis, via a half-silvered mirror, onto a highly directional glass-beaded screen. Actors' lighting swamps the image falling on them.
2. People appear *in front* of the background scene.
3. If people move behind a surface covered with beaded material, they will appear *behind* corresponding areas of the background scene (walls, trees, etc.).

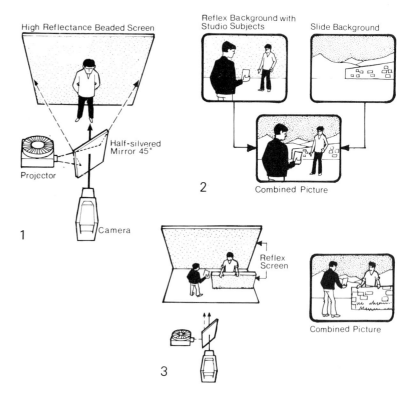

Lighting restrictions are few, and the main consideration is the usual one of matching the foreground lighting treatment with that of the background.

Electronic picture insertion

In these processes the background image does not appear behind the studio action but is inserted by an optical or electronic process from a separate source. Unlike the previous systems, where the general sharpness of the background image varies with the studio camera's focusing, the inserted scene remains in focus throughout. One is more likely, therefore, to be aware of any incompatibility.

Keyed insertion/electronic matting/inlay

You can think of this electronic system as an area-blanking process. A *selected section* of the background picture is blanked out and an exactly corresponding area of the foreground action in the studio is inserted in its place. Any color can be included in the foreground and background pictures.

This matted-out area can be derived from a hand-drawn silhouette graphic placed in front of a camera (camera matte), or more often by

Fig. 15.13 Electronic picture insertion

1. *Special effects generator (SEG) – inlay.* We select the specific area of the *background scene* to be treated and arrange a mask shape to match it. The equipment then automatically blanks out just this section and inserts the corresponding part of the *subject shot.*
2. *Camera matte – external key.* Here a separate silhouette on a graphic is set up in front of a studio camera or a video rostrum camera. This shape is made to correspond with a chosen part of the background picture. The equipment, as before, inserts that part of the subject instead.
3. *Chroma key – color separation overlay (CSO).* The subject is arranged before a plain backdrop of the keying color (usually a blue cyc. or flat). The subject itself must not contain or reflect blue, or false background break-through results. Wherever blue appears in the subject shot, the equipment switches instead to the background scene (which can contain any hues). The combined effect is the subject (less its blue surroundings) within the background scene. (You can use blue, yellow, green hues etc. as convenient for the subject.) Both sharp and soft-edge transitions are available.

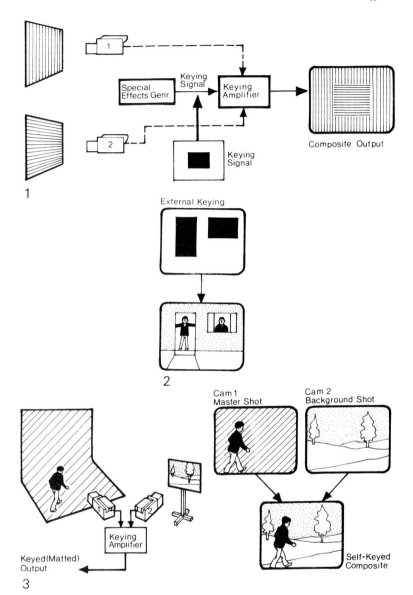

a *special effects generator (SEG)* which 'wipes in' a range of basic shapes. The subject only appears within the matted (masked) area of the composite picture. Elsewhere it disappears. This system imposes no limitations on lighting, except, of course, the usual need for matching.

External key/chroma key

This is an *automatic* self-matting process broadly termed *chroma key* or *CSO (color separation overlay)*. It is a color-selective method of picture insertion that requires the subject to be set up in front of a totally blue background (e.g. cobalt blue). Yellow or green are sometimes used instead. Wherever the system 'sees' blue in the *foreground* shot, it switches to the *background* picture source, and inserts that area into the composite shot instead. The inserted subject appears 'in front of' anything in the background picture, which can contain any hues.

Because the system is switched by *any* blue area, look out for blue items or reflections anywhere in the foreground scene. Otherwise the system will be fooled, and there will be background breakthrough. Some systems are more color-specific than others.

Modern chroma key systems, including allied techniques such as *Ultimatte*, are more flexible and able to handle transparent surfaces, shadows, and fine outlines (e.g. feathers, mesh, smoke) that cause spurious switching on simpler systems. In a typical chroma key set-up a blue backdrop or cyclorama is used – and perhaps a blue floor cloth. Although it is possible to use a neutral background to the action lit with blue light instead, there is always the chance that some may accidentally spill onto the subject, and cause inaccurate matting. As always, you should arrange the direction, contrast and intensity of the subject lighting to suit the background picture.

The main hazards when lighting for chroma key are:

● Many chroma key systems do not reproduce shadows. Here you will see that subjects appear shadowless and 'suspended' in the composite picture.

● Any shadows or hotspots on the blue keying surface are liable to cause uneven clipping, with spurious edge-ragging or background breakthrough. So you need to take particular care that the keying surface is evenly illuminated. It may even be necessary at times to add extra lamps to light-out heavy shadows.

● Where a chroma key system allows 'shadow insertion' look out for any inappropriately inserted shadows in the composite, caused by folds in the blue keying material or scenic shadows.

Traveling matte in film

The wonders achieved by motion-picture makers using traveling matte systems are legendary. Again, this technique relies on a selected background hue to prepare a matte. The major traveling matte system uses the 'sodium-vapor' process. Here the action is lit with normal 3250 K white lighting in which a very narrow color band has been suppressed (monochromatic yellow – 589 nanometers) by didymium-coated filters. The surface behind the action is illuminated with monochromatic yellow light from sodium-vapor lamps of this wavelength. A beam-splitting film camera is used, which simultaneously shoots both

● A color record of the foreground action. Because the color film stock used is relatively insensitive to monochromatic yellow, the action reproduces in full color while its background appears black.

● A monochrome film record. Shot through a yellow-selective filter, the action is reproduced as silhouettes against a light background.

During process printing, the black matte from the monochrome film is used to obscure parts of the new background scene and the foreground action is inserted in its place.

Scrims/scenic gauzes

As we saw in Chapter 14, *scrim* or *scenic gauze* is a very versatile material:

White scrim

When *white* scrim is lit from the front it appears 'solid'. Anything painted on its surface is clearly visible against a light background. Insufficient light passes through it to illuminate whatever is behind it. Even if subjects behind the scrim are lit quite strongly, they are still not seen on-camera until the light on the front of the scrim has been reduced. Then the scrim produces a misty diffusion overall.

Fig. 15.14   Scrims/scenic gauzes
Lift from the front alone (1), the scrim appears a solid plane. Surface painting or decoration shows up brightly against a plain white or black background. Unlit subjects behind the scrim are invisible.

By reducing brightness of front illumination (1) and lighting the subject (2) (behind scrim), the subject and setting are revealed with outlines and contrast softened, surface painting having almost disappeared.

A third lamp (3) added to rear-light the scrim, increases the mistiness over the scene, while silhouetting details on the surface of the scrim.

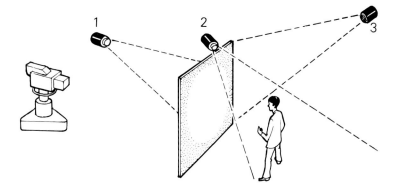

Rear-lit without frontal lighting, any surface painting is silhouetted, and the scrim adds diffusion to the lit scene beyond. If you light only the subjects behind the scrim, while leaving the surface itself unlit, it will 'disappear'. So by adjusting the relative light intensities you could begin, for instance, with the painted exterior of a house, and dissolve to show an interior scene beyond its walls.

Black scrim

When *black* scrim is lit from the front it is barely visible, but anything painted on its surface in mid to light tones stands out clearly against a black background. Surface painting on black scrim may be used to create decorative and scenic effects, to simulate illuminated street signs, etc. Although reduced in intensity, sufficient light passes through the scrim to illuminate whatever is behind it. When subjects behind the scrim are lit they are slightly diffused, but clearly visible on-camera.

The usual problems when lighting scrims are

● Action and scenery are too close to the scrim to allow it to be illuminated evenly over its entire surface. Any hotspots or underlit patches spoil the effect.
● Light may pass through the scrim and inadvertently reveal whatever is hidden behind.

Semi-silvered mirrors/one-way mirrors

Although you will only meet the 'see-through' mirror occasionally, it is a gimmick that can be troublesome. When glass is surface-coated with a very thin layer of silver or aluminum it behaves as a normal mirror when seen against a dark background yet appears transparent when viewed from the reverse side. So while one camera shows a make-up artist at work in front of the mirror, another can provide head-on shots through the mirror of the demonstration.

The trick lies in the relative light levels on either side of the mirror. While the rear side is in comparative darkness, the other must be brightly lit, to compensate for light losses through the mirror. Only a quarter or less of the light reaches the rear camera, and its glass may give the image a slight color-cast. So for optimum results you will need to light the action area to a correspondingly higher level, the front-side using a neutral-density lens filter or much-reduced lens aperture to avoid overexposure.

As with all mirror shots, you will usually find that if you attempt to light the person from around the front camera position they will be unattractively side lit, while lighting from above only results in ugly top lighting. Instead, a fairly low spotlight shooting straight into the mirror from within the set will serve as both a reflected key light and a back light.

Fig. 15.15 One-way mirrors
The thin reflective coating on this one-way mirror makes it appear transparent (with a considerable light loss) from the rear unlit side (Cam 1); but as a normal mirror in the brightly lit scene (Cam 2).

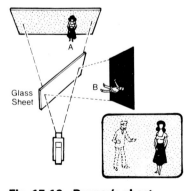

Fig. 15.16 Pepper's ghost
The out-of-shot person (B) is reflected into the lens as a transparent ghost within the main subject picture. Depending on the relative illumination, you can see A or B, or both (B ghostly).

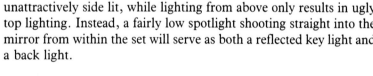

Pepper's ghost

This is a very simple way of lighting any subject *dead frontal along-the-lens axis* in order to entirely suppress texture, surface modeling and shadows. You will recognize it as the basic principle used in *reflex projection*.

The camera shoots through a sheet of glass or clear plastic angled at 45°, which is also reflecting the beam of a spotlight onto the subject. Although there is a considerable light loss, this is an effective way of illuminating multi-layer graphics where movable sections provide animation effects. A similar principle is used to intermix images (ghostly or solid) for trick effects.

Follow spot

The familiar spotlight is an excellent, if theatrical, way of isolating someone from their surroundings. And it is an excellent expedient where full overall lighting is impracticable. Widely used in arena spectacles, ice shows and stage presentations, the follow spot must be operated skillfully if it is to be really effective. It might raise a laugh if the performer moves and the spot has to chase after them . . . but it may not be the right occasion!

There are a number of points to look out for when using a follow spot:

● Adjust the spotlight mounting so that its horizontal and vertical movements have a certain amount of *drag*. If it moves too freely, it will be difficult to control accurately. Remember, most spotlights are used with a long throw (lamp distance) of 9 m (30 ft) or more, and even a slight jerk is greatly emphasized.

● Check all adjustments beforehand (iris, douser/blackout, focus, etc.) for smooth operation.

● The operator should be able to provide a spot instantly and accurately on a subject anywhere in the acting area. If the spotlight does not have a sighting device, and one cannot be improvized (a wire sight), the only solution is to use a 'pinpoint' spot to locate its position and bring it up on cue by removing the douser.

● Ensure that the beam will be uninterrupted wherever it moves. Check that no hung lamps or scenery will be casting shadows as it follows action.

● The operator should know how much of the subject is to be covered – full-length, waist, head, hand. The maximum beam spread will depend on the focal length of its lens. In a well-organized show the spot-operator has an opportunity to learn the action moves, to know when to broaden or narrow the beam or which person to follow if a pair of skaters move apart, and so on.

● In a full-length spot there should be some headroom visible between the person's floor shadow and the top of the spot. Where a sound boom is being used, watch out for the microphone dipping into the top of the beam.

● Where the follow spotlight is used to light people near shiny surfaces (e.g. polished floor, piano) there may well be reflections onto the cyclorama or nearby scenery. There is nothing you can do about it, except move the performer to a place where the reflection (often inverted) is less noticeable.

The spotlight circle is only clearly seen on a light-toned floor. On a black, shiny, or patterned surface it is ineffectual.

Close shots of people lit by a spotlight tend to be flat and over-contrasty. Strong back light can enhance the effect; fill light can reduce its harshness, but supplementary lighting tends to reduce the dramatic value of the spotlit effect.

Moving light sources

In drama productions characters sometimes move around in the dark, carrying a candle, lantern or flashlight as their sole source of illumination. As we saw earlier, you cannot rely on most practical lamps to illuminate the actor or the surroundings. So we need a convincing technique to simulate the effect.

Wherever possible, the light source should look appropriately bright, and seem to be lighting the person carrying it; even if this involves cheating with concealed pea (peanut) lamps in the practical fitting. You can tackle this situation in several ways. The simplest is to have someone holding a small lighting fixture, walking ahead of the actor, keeping just out of shot. Alternatively, you can have a localized lamp on a lighting stand, turning to follow the actor as a traveling key.

Fig. 15.17 Follow shots
Where a subject moves through a darkened setting, it may be lit by:
1. A traveling (following) key.
2. The follow shot subject may otherwise be lit by a series of interfaded localized areas.

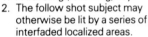
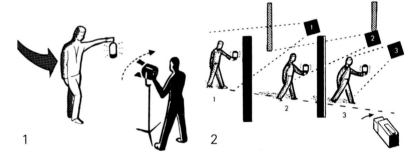

These methods are simple enough, and may work sufficiently well at times but –

- The direction of the illumination will often be inaccurate,
- Unless the practical lamp is held at arm's length, its shadow is liable to fall onto the actor,
- The appearance of the surroundings does not alter appropriately as the actor moves about.
- If anything comes between the actor and the operator (e.g. he moves *behind* a pillar), then unless the lamp-operator douses the effects lamp with a hand or card at the right moment we shall see the light unrealistically pass over it and throw a spurious shadow.

The most effective methods use a series of lamps along the actor's route. Each lamp is faded up as the actor approaches and down as he passes by, to imitate the changing light on the background. This can all be done quite convincingly, as long as the lamps can be positioned at appropriate angles; but in restricted spaces this may not be possible.

Finally, if there is sufficient room, and you have the manpower, you can have a variant which requires three lamp operators: one to provide the constant illumination falling on the actor's head and chest, another lighting just in front of him, and a third lighting just behind him. Results can be excellent, but three operators plus a camera dolly can need a lot of unrestricted space to maneuver!

16 Safety!

At first sight, lists of *safety precautions* are a bore! They appear so restrictive, overcautious, and downright pessimistic. Extra care takes extra time. For most of us the turning point comes when we see for ourselves (or experience!) the terrifying effects of electric shock – an exploding lamp showering hot, razor-sharp glass onto the people below – a rapidly growing fire starting from cloth draped over a groundrow – a falling wrench injuring an unsuspecting stagehand below . . . Then these tiresome rules and regulations take on a different significance!

Many of these important everyday safety issues, including some you have already met, are largely a matter of common sense. There are a lot of *dos and don'ts*, perhaps. But remember, the life you save may be *your own*!

Load

Add the power ratings of lamps you plug into any circuit, and ensure that their total consumption does not exceed the circuit maximum. It is particularly easy to overload circuits when plugging several lamps into one power outlet on the patchboard.

On location, remember that there may already be equipment elsewhere, using the power circuit you want to plug into. Your added load may exceed its capacity. It is always advisable to enquire before connecting to other people's supplies.

Fuses

Each lamp should preferably have its own associated *fuse*. Power supply circuits should have separate fuses, cut-outs, or circuit-breakers. Make it a routine, when rigging, to find where fuses are located, and to check out the supply of spares. Make sure that they are of the appropriate rating and type for the lighting fixture, and never use a higher or lower rating.

Most fuses blow (rupture) due to lamp failure, or cable faults (frayed ends, damaged cable). Wherever possible, track down the cause immediately rather than simply replace the fuse.

Live equipment

Switch off the power and unplug equipment before tightening connections, changing lamps, or checking faults. Don't rely on gloves or rubber-soled shoes for total protection. Always replace

447

frayed cables, split or burned insulation, and shield open conductors.

Electric current flows (conventionally speaking) from the *live* (L) side of the supply, through the lamp (the *load*), to the return common lead (*N* – *neutral*). If you touch either supply lead (L or N) current will flow through you instead, as it takes an easy path through the body to ground (earth). 'Ground' can be anything from the floor itself to a conductive surface nearby that is in contact with the earth – e.g. a metal pipe, fencing, scaffold, plumbing, etc. A higher current will flow if the skin is damp.

Even if an electric shock is minimal, it can be enough to cause you to drop equipment, or jerk away and injure yourself against something nearby. Muscular reactions can cause the hands to grip onto the live surface. At worst, a victim is unconscious, requiring artificial respiration.

Certain equipment contains high voltages that are potentially extremely dangerous and need particular care – e.g. ballast units for discharge lamps, TV monitors.

Phases

If you are using public utility AC power on location you should be aware of a further potential hazard – the danger of mixing electricity supplies of differing *phases*. AC electric supplies, including those from some mobile power units, are generated using a *three-phase* (*three-wire*) system. Three different supplies are produced simultaneously which are of identical voltage but differ in their relative timing or '*phase*'. Some heavy-duty equipment (e.g. three-phase electric motors) needs to be connected to all three phases.

However, the lighting and heating equipment we are concerned with requires only a *single-phase* supply. The original three separate phases are distributed independently to different consumers. So, for instance, while a house or apartment is fed from one phase its neighbors may be supplied from the other two phases. This standard practice is totally safe. *But* it is essential that equipment fed from one phase does not come within touching distance of equipment being fed from another. Anyone simultaneously touching equipment connected to different phases could receive a *double-voltage electric shock*!

When shooting on location with lightweight equipment you can often plug a lamp straight into a convenient power outlet and use household current. If to avoid overloading nearby outlets you also cable lamps into supplies from neighboring areas, this matter of differing *phase* can arise, so be warned! It has happened!

Grounding/earthing

Ensure that *all* equipment is *grounded* (*earthed*). Then if, for any reason, the casing becomes live (frayed or loose wire, or damp equipment) current will flow to earth and blow a fuse, rather than await an unsuspecting victim! All metal structures and equipment should be grounded, including lighting bars, scenic framework, practical light fittings, scaffolding.

Connections

Make sure that all lamps (bulbs) fit firmly in their holders (sockets). Do not overtighten a lamp or exert too much pressure. You could break the base-to-envelope seal, or make removal difficult after use. There are a number of different types of lamp bases, including forms that screw, clip, push and turn, have screw terminals, or have base adaptors. Loose lamps lead to burned and corroded contacts, and poor connections. Some sockets become distorted or lose pressure in use as sprung contacts deteriorate in the intense heat. If contact surfaces oxidize and become pitted, they form high-resistance points and local overheating worsens the condition. It is good practice, therefore, to check the condition of the old lamp base and the socket before fitting a new lamp.

Local overheating and burning leads to problems with cable connectors too. Make a point of double-checking connectors when plugging cables into outlets to ensure that they are secure. Some types of connectors can be locked into position; others are liable to be dislodged if a cable is inadvertently pulled.

Hot surfaces

Light fixtures become extremely hot in use! Obvious enough, but one often overlooks the fact. A lamp that happens to rest against a drape or cyclorama cloth can, in a short while, become hot enough to ruin it. Although most scenery is flameproof or fireproofed, it takes very little time to damage a costly item.

Even when correctly fitted, new color medium and scrims can heat and give off fumes. But if placed too close to a lamp, they will deteriorate rapidly, may reduce lamp life, and crack any fresnel lens.

Heat is a particular problem when using overrun lamps (e.g. Photoflood) fitted into domestic light fittings such as wall-brackets, ceiling lights, table lamps, desk fittings. They may overheat and damage the fixture and nearby surfaces.

It is inadvisable to move lighting fixtures around when they are lit or very hot, and lamp filaments are particularly fragile.

Heavy weights

Back injuries and wrenched muscles are regular industrial hazards. Sometimes they arise because people try to carry heavy weights instead of getting assistance or using appropriate equipment. More often it is because they use poor lifting techniques, and strain muscles unnecessarily. In Figure 16.1 you will see recommended methods.

Cables

A range of different electric cables are used in television and film lighting: from thin flex carrying a few amps (used to wire up practical light fittings) to cables nearly as thick as one's wrist. The cable's size depends on the current-carrying capacity required, and the length of time it is to be used. Thinner lightweight cable may be acceptable if used only for brief periods. But in continuous use (e.g. +20 minutes) it would overheat, and a heavier cable (thicker conductor) is necessary.

The conductors in cables may be made of aluminium or copper. For the same current capacity, aluminum cables are thicker, lighter, less costly than their copper equivalent. Any wire offers resistance to electricity, causing losses in the form of heat. So it is always advisable to keep cable lengths to a minimum where possible, and avoid unecessary losses (*voltage drop*).

When current flows through any *coiled* wire it effectively becomes an 'inductor' and offers additional resistance ('impedance') to the flow of AC electricity. As a result, a coiled cable becomes hotter when large currents flow: hot enough perhaps to overheat and damage the conductors and the insulation. At the least, it can give plastic-covered cable an inbuilt curl, making it brittle and reducing its flexibility. So although all cables should be stored in coils or on drums, and never left in random heaps, they should always be fully unwound when used. If a lighting cable is longer than you actually need, don't neatly coil up the surplus. A figure-of-eight pile would be preferable.

Check the condition of cables regularly. In use they become cut, squashed, bent, knotted – conditions that damage the conductors, and may remove insulation.

It is worth taking care to route your lighting cables safely. People trip over them, or tug them out of their way. Cables may accidentally be draped over hot surfaces (heating conduits, cyc lighting, floor lamps, etc.). The cables feeding lamps on lighting stands are best secured to avoid the fixture being pulled out of position. Even the cables hanging beneath suspended lighting fixtures are vulnerable, for they can catch in any scenery that is being raised or lowered.

It is good working practice to tuck floor cables away neatly at the bottom of walls or flats; support them in hangers or rope slings; tape lightweight cables to walls with gaffer tape. If there are people walking around, use a mat or board over critical places. Where vehicles are likely to drive over cables, cover them with low ramps or double boards to avoid damage. If you tie a string of cables together

Fig. 16.1 part 1   Lifting and carrying techniques
1. Check object's weight. Check your route.
2. Stand close to object, facing route. Bending hips and knees, crouch down to the load. Always keep your back straight. Elbows tucked into thighs, grip load either side if possible, above and below. (Use all of your hand, not the fingers alone.) If there are no handholds, place a rope round the object.
3. Raised head and chin in. Equally balanced on either leg, begin to lift the object (keeping a straight back!) with arms close to body and leaning forward slightly. Smoothly straighten legs, keeping object close to body. Carry object close to body, with your lower arm straight.
4. To put the object down, reverse the lifting process with a straight back and bending at the hips and knees. Do not bend over!
5. If lifting with one hand, place the other on a nearby bench, table, chair; or kneel on one knee, using it as a support.

Fig. 16.1 part 2   Carrying weights
Carry weights on your shoulders where possible, not in front of you with extended arms. Always avoid twisting or bending your back when lifting or carrying. Wherever possible, do not carry heavy weights, but use a wagon, trolley, castered skids, etc.

tightly, or bury them, remember that this reduces their current-carrying capacity, compared with that in free air.

To summarize, use the right cable capacity for the job. Use it fully unwound. Keep it out of the way and protect it. Check out its condition.

Safety bonds

Wherever possible, use some form of safety bond, such as a wire loop with a snap-hook, between all suspended fixtures and accessories to the nearest secure point (e.g. overhead pipe). Then if a barndoor should fall, or a fixture become loose for any reason, it cannot fall to the ground. Particularly where lighting fixtures are suspended at 20 ft (6 m) or more, or hung over an audience, such precautions should be obligatory.

Many lighting fixtures are designed to be rigged by hand without tools. But if you need to use a spanner, wrench or screwdriver, have it fitted with a wrist-cord loop so that it cannot be dropped. A convenience for you and a safeguard for a potential victim below!

Step-ladders/lighting ladders

Even in studios where lighting fixtures are suspended on adjustable battens, barrels, monopoles and similar devices, there are times when one needs to climb a step-ladder (treads) to attach or adjust lamp accessories (e.g. a flag, scrim, gobo). Because regular ladders are not self-supporting they are unsafe to use for lighting or scenic work; even in studios with a low overhead pipe-grid (e.g. at 9 ft (3 m)). Instead, either folding step-ladders or wheeled lighting ladders are used.

Folding step-ladders are safe enough if used correctly:

- It is most important to secure them properly: fully open, with locking stays in position, and based on a horizontal firm surface.
- Have someone stand on the bottom rung to prevent the steps from shifting.
- Do not move or carry heavy lighting fixtures while standing on steps. (They can be hoisted alongside by rope.)
- Never lean over sideways to unhook or remove a fixture. This will not only strain you but can throw you off-balance. Position the steps just in front of the work.
- Never be tempted to stand on the top step to reach an overhead fixture (e.g. to remove it from a grid or bar)!
- A safety chain from high steps to an overhead grid has saved many a disaster.

Some studios use a special 'lighting ladder', which consists of a permanent step-ladder structure fixed to a castered platform.

Fig. 16.2 Lighting ladder
Wheeled lighting ladders are available in several forms; including four-wheeled and tricycle base with a lockable steering wheel. Somewhat cumbersome and requiring space to maneuver, they are very stable and safe when handling heavy lighting fixtures.

Although more stable than a folding step-ladder it is a bulky contraption that can be difficult to maneuver in a crowded studio.

A step-ladder has the advantages that it is easily carried (when folded) and can be used in confined spaces, even on platforms (rostra, parallels). It is readily stored – e.g. on a bracket fastened to the studio wall.

The wheeled version, on the other hand, takes up a lot of room, even when stored. It can only be used on the flat studio floor, and is often difficult to position. If rolled over floor cables, the wheeled casters will damage them; yct the structure is too heavy to lift easily. Because of its height, the ladder may hit hung lamps and scenery.

Scenery

In film studios, settings are invariably built as extremely rigid structures. Scenic units may be nailed or screwed together, and fastened to the floor to ensure maximum stability.

Television scenery, which is designed to the highest standards, must usually be constructed and erected to suit a much quicker production turnround time. A higher proportion of the scenery is derived from stock units, which are stored for re-use. In the TV studio, scenic units are temporarily lashed or clamped together, and supported by scenic jacks or weighted braces. This difference in staging techniques influences the extent to which lighting fixtures can be attached to scenery in the film and television production.

In both media, lamps are often clamped or screwed to scenery in order to light inaccessible places. Most flats will support smaller lamps (500–1000 W) on a top bracket, but before considering attaching any lighting fixture to the wall of a setting, confirm that it is safe to do so. Will it remain stable? A ceiling piece, for example, may sway or unbalance if a lamp is clamped onto it. A lamp on a door-flat will vibrate and may become displaced if the door is slammed.

Take particular care when leaning heavy lighting fixtures against scenery. Lighting fixtures on a hung bar can push flats out of position, upsetting their alignment. Large clamp-on hangers (e.g. wall sleds) which allow heavy lighting fixtures to be suspended against vertical set walls are an excellent facility, provided the scenery is suitably robust.

Always ensure that there is plenty of space around lighting fixtures. Close liaison between the Scenic Designer, the Lighting Director, and their respective studio crews is a must at all times, but particularly where safety is concerned. The fact that some lamps are unlit, apparently unused, when the scenery is rigged does not mean that they will not be switched on later.

It is not unknown for drapes, cloths, or scrim (gauze) to be lowered during the show, just in front of hot lamps! The Lighting Director finds it frustrating, for light is cut off from the scene. The Set Designer finds it frustrating when an expensive hired drape has a hole burned in it!

Lighting stands

Always assume that floor lighting stands are vulnerable, and liable to overbalance, become displaced as someone squeezes past, or have their cables pulled (tipping or twisting the lamp). Apart from having someone standing by to protect a vulnerable lamp, you can weigh the bottom of the stand to make it less mobile (using sandbags, water bags, or stage weights), or better still, lock off casters or use floor wedges or weights to prevent them moving.

Handling lamps

Make it a rule never to handle the bulb/envelope of any lamp but to use gloves, cloth or some similar protection. Some tubular lamps are fitted with a tear-off paper sleeve to simplify fitting. Tungsten lamps with envelopes of silica glass or borosilicate can develop blisters (see '*Bulb blister*') through surface contamination if held in bare hands. The fused quartz envelopes of tungsten halogen (TH) or HMI lamps will be blackened or fail prematurely if handled, as body chemicals (acids, sodium) can attack the surface and cause devitrification. If you handle a lamp inadvertently, wash it off immediately with alcohol.

When a lamp fails in a lighting fixture that has been alight for some time, think twice before attempting to remove the hot lamp, even with thick gloves! The lighting fixture itself can cause third-degree burns. It is sometimes more prudent to rig a new lighting fixture alongside, to replace the one that has failed.

Some people remove lamps from lighting fixtures before transporting them. But, on balance, it seems that frequent handling is more likely to increase the chance of damage and cause undue wear and tear on lamp sockets and bulb-to-base seals.

Protective mesh

There is a lot to be said for fitting lighting fixtures with a protective wire-mesh or scrim over open-fronted lamps, to restrict flying glass if a lamp should happen to explode. Fresnel lenses have been known to crack and fall out of spotlights too. Always take special care with lamps hung over an audience. Scrims and safety bonds are cheaper than legal claims.

Water

Water and electricity don't mix! Don't underestimate the hazards when working in damp or wet surroundings. Moisture on plugs and lighting fixtures not only leads to electric leakage and shocks but can encourage carbon tracking across insulators, which progressively worsens in use and short-circuits supplies.

Some lighting fixtures designed for external use can withstand any amount of water, but most are very vulnerable. Allow rain or a water spray to fall on open lamps (e.g. scoops, broads) and they will explode!

Lamp inspection

All lamps deteriorate in use, so that the light intensity and quality changes:

● *Bulbs blacken.* Less of a problem in tungsten-halogen lamps (quartz-lights), but liable to develop if they are run dimmed-down for long periods. Large studio tungsten lamps contain tungsten granules with which you can periodically remove inner blackening of the bulb (envelope, globe).
● *Filaments sag.* Lamp filaments expand when hot. Filament sag may become excessive: through age, due to a reflected hotspot (misaligned reflector), or excessively steep burning angles.
● *Bulb blister.* The bulb of a lamp can become locally distorted to form a blister, which eventually bursts and destroys the lamp. Some types of lamp and bulb materials are more susceptible to this than others.

The usual causes of blistering are localized overheating due to high supply voltage, burning the lamp at too steep an angle, poor lamp ventilation, a misaligned reflector. Much depends on the design of the lighting fixture.

Appendix

Conversion table (footcandles and lux)

Footcandles to lux		Lux to footcandles	
½	5.4	10 000	930
1	11	5 000	465
2	22	4 000	372
3	32	3 000	280
4	43	2 500	232
5	54	2 000	186
6	65	1 000	93
7	75	500	46
8	86	250	32
9	97	200	19
10	107	100	9
50	540	50	5
100	10 764	10	1

1 fc = 10.764 lux   1 lux = 0.0929 fc

Typical light intensities

		Flood (lux/fc)	Spot (lux/fc)	Measured at distance m (ft)
Fresnel spotlight (tungsten-halogen)	200 W	100/10	700/65	5 (16)
	300 W	125/12	800/75	5 (16)
	1 kW	1 280/119	8 000/743	3 (10)
	2 kW	1 900/176	11 900/1106	5 (16)
	5 kW	875/81	4 600/427	10 (33)
	10 kW	2 200/204	8 600/800	10 (33)
Fresnel spotlight (HMI)	575 W	8 000/740		5 (16)
	1200 W	4 000/370		10 (33)
	2500 W	10 000/930		10 (33)
	4000 W	17 000/1580		10 (33)
Lensless spot	250 W at 90°	350/32 at 45°	1500/140	3 (10)
	600 W at 90°	800/74 at 45°	2400/223	3 (10)
	800 W	400/37	1400/130	5 (16)
	2000 W at 75°	1000/93 at 35°	4000/370	5 (16)
Focusing scoops	1 kW	450/42	1130/105	6 (20)
	2 kW	590/55	2400/220	6 (20)
Soft light	1250 W	800/74		3 (10)
	2.5 kW	600/56		5 (16)
	5 kW	1000/93		5 (16)
Six-light PAR bank	6 × 650 W	1500/140	800/74	10 (33)
Cyclorama units				
Two-light	2 × 1250 W	1100–1500 lux/80–120 fc		3 (10)
Four-light	4 × 1500 W	3770 lux/350 fc		3 (10)
Floor trough	625 W	426–1700 lux/40–160 fc		1.2 (4)

Standard light units

	Technical term	Symbol	Unit	Abbreviation
Light source strength	Luminous intensity (power of light source) (candle power)	I	Candela; candle (USA) (based on luminance of platinum at its melting point)	cd; c
	Luminous flux (light emitted by source) (Pharos)	F, Φ	Lumen (An ideal source of 1 cd radiates a total flux of 4π lumens (12.57 lm) 1 lm falls upon unit area at unit distance, at right angles to the flux) Mean spherical candle power (average F in all directions) Mean horizontal candle power (average F for restricted coverage)	lm m.s.c.p. m.h.c.p.
	Efficiency of source	η	Lumens output per watt supplied	lm/W
Incident light intensity	Illuminance (quantity of light received on unit surface area) (Pharosage)	E	Lux; metre-candle[a] (= lumens/m²) Phot (= lm/cm² = 10 000 lux) Milliphot $\approx$ 1 f.c. Foot candle = lm/ft² = 10.764 lux = 1.076 f.c. milliphot. 1 lux = 0.092903 f.c. or lm/ft² Nox = 10^3 lux	lx; m.c. ph f.c.; lm/ft²
Reflected light intensity, or surface light emission	Luminance (quantity of light emitted from unit surface) ('brightness' is deprecated)	L	Nit (= cd/m²) = 0.292 ft L = 3.14 asb. Foot lambert = $1/\pi$ cd/ft² = 10.76 asb = 3.426 nits (luminance of perfect diffusing surface emitting or reflecting one f.c.) Candelas/m² = 0.092903 cd/ft² = 0.291863 ft L	nt ft L
	Helios	H	Candela per square inch candela per square foot = 3.14 ft L Apostilb = $1/\pi$ nit = $1/\pi$ cd/m² (diffuse luminance of surface emitting or reflecting one lux) Stilb = cd/cm² = 10 000 nits = $1/\pi$ lamberts Lambert[a] = lm/cm² = $1/\pi$ cd/cm² Skot[a] = 10^{-3} asb	cd/in² cd/ft² asb sb
Total light energy	Luminous energy	Q	Lumen-second; lumerg Lux-second	lm-s lx-s
Color	Wavelength	λ	Micrometer = 10^{-6} Millimicron, nanometre = 10^{-6} mm Angstrom = 10^{-7} mm	μm nm, mμ Å

[a] Deprecated term.

Conversion table (linear units)

	Feet	→	Meters
3.28	1		0.304
6.561	2		0.609
9.842	3		0.914
13.12	4		1.219
16.40	5		1.524
19.68	6		1.828
22.96	7		2.133
26.24	8		2.438
29.53	9		2.743
32.81	10		3.040

	Feet	←	Meters

Inches	→	mm
1		025
2		050
3		076
4		101
5		127
6		152
7		177
8		203
9		228
10		254
11		279
12		304

Neutral-density filters

Type	Transmission (%)	Loss (%)	Filter factor	Increase exposure by (stops)
ND 0.1	80	20	1.3	¼
ND 0.2	63	37	1.6	½
ND 0.3	50	50	2.0	1
ND 0.4	40	60	2.5	1¼
ND 0.5	32	68	3.1	1½
ND 0.6	25	75	4.0	2
ND 0.7	20	80	5.0	2¼
ND 0.8	15	85	6.7	2½
ND 0.9	13	87	8	3
ND 1.0	10	90	10	3¼
ND 2.0	1	99	100	6¾
ND 3.0	0.1	99.9	1 000	10
ND 4.0	0.01	99.99	10 000	13¼

Formulae

Current in amps = Voltage ÷ resistance (in ohms)

Resistance = Voltage ÷ current (in amps)

Voltage drop in cable = Current × resistance

= Power ÷ current × current

Power (in watts) = Current (in amps) × voltage

= Current × current × resistance

= voltage × voltage ÷ resistance

Power consumption

Power (watts)	110 V	120 V	220 V	240 V supply
100	0.9	0.8	0.5	0.42
150	1.4	1.3	0.7	0.63
200	1.8	1.7	0.9	0.83
375	3.4	3	1.7	1.6
500	4.5	4.2	2.3	2.1
650	6	5.4	3	2.7
800	7.3	6.7	3.6	3.3
1 000	9	8.3	4.5	4.2
1 250	11.3	10.4	5.7	5.2
1 500	13.6	12.5	6.8	6.3
2 000	18.2	16.7	9	8.3
5 000	45.5	42	22.7	20.8
10 000	91	83	45.5	41.7

Current (amps)	Power (watts)			
	110	120	220	240 V
5	550	600	1 100	1 200
10	1 100	1 200	2 200	2 400
13	1 430	1 560	2 860	3 120
15	1 650	1 800	3 300	3 600
20	2 200	2 400	4 400	4 800
25	2 750	3 000	5 500	6 000
30	3 300	3 600	6 600	7 200
40	4 400	4 800	8 800	9 600
50	5 500	6 000	11 000	12 000

Cables and power connections

Feeder cable

Main feeder cable. Single conductor cable carrying the main power load from company supply bus bar or generator, to a distribution box (*spider box*). Three separate lines are run: positive, negative, neutral. Fitted with *lug* connectors.

Secondary feeder cable. A three- or two-wire cable from distribution box (spider, plugging box) to a number of fused receptacles.

Extension cables. Cables used to distribute power to individual or groups of lighting fixtures, or to further distribution points. A plug (male) at one end, and a receptacle/socket (female) at the other.

Splicer/splitter. A cable with two or three separate receptacles (sockets) at one end, to allow a number of lighting fixtures to be fed from the same supply. (Check the maximum load!)

Adaptors are used to convert one form of plugging system to another.

Cable connections

Internationally, a wide range of connectors is used, of which the following are typical:

Household/domestic plugs (plastic or rubber)
USA 110/120 V maximum load 15 A. No earthing (ground) 2-prong. Maximum load 2 kW.
Heavy-duty household. Maximum load 30 A. Earthed (grounded) 3-prong.
British 220/240 V maximum load 13 A. Fused. Earthed 3-pin. Maximum load 3 kW.

Pin connector plugs. Small rectangular plastic block. Two or three conductor pins on one side. For heavy-duty cables. Location use. Tie cable ends together.

Stage plugs (floor plugs, Kliegl plugs). Rectangular cable plug ½ in/1 in wide has large flat sprung conductor on either edge. Rectangular heavy-duty metal distribution box (with 1–6 connector receptacles). Will accept several plugs side by side. Available in fused or earthed versions. Not color-coded. Supply polarity uncertain. Plug can be dislodged from receptacle, causing poor contact (surface-burning and overheating).

Twist-lock plugs. Plug has three curved metal connectors (one earth) that lock firmly into slots in the socket receptors, with quarter-turn twist.

Single-pin connectors. Separate heavy-duty thick metal pins (color-coded) fit into 3-receptacle sockets.

Technicolor lugs. Large slotted cylindrical copper molding with a securing screw, brazed or crimped onto cables, fixing onto bus bars (wide thick copper strip) of company supplies, distribution point/service panel (head).

Molded rubber plugs. Large molded rubber plugs with three connector pins (one earth), held by retaining skirt in 1–4 way molded rubber receptacle (loads 1–10 kW).

Mareshal decontactor type DS. Special molded weatherproof metal fitting with nylon insert in ratings 30–500 A. Shuttered live contacts for five-pin arrangement.

Flat-blade connector (Tri-Edison). In two forms: central straight key blade (ground) and two 45° angled blades; U-shaped key (ground) and two parallel blades.

Further reading

Arnheim, R., *Art and Visual Perception*, Faber and Faber, London.

Bermingham, A., Talbot-Smith, M., Angold-Stephens, K. and Boyce, E., *The Video Studio*, Focal Press, London/Boston.

Carlson, V. and Carlson, S., *Professional Lighting Handbook*, Focal Press, London/Boston.

CIE, *International Lighting Vocabulatory*, CIE, Paris.

Committee of Colorimetry, *The Science of Color*, Optical Society of America, Thos. Y. Crowell, New York.

Cox, A., *Photographic Optics*, Focal Press, London/Boston.

Detmers, F., *American Cinematographer Manual*, American Society of Cinematographers.

Evans, R. M., *An Introduction to Colour*, Chapman & Hall, London.

Evans, R. M., *Eye, Film, and Camera in Colour Photography*, Wiley, New York.

Evans, R. M., Hanson, W. T. and Brewer, W. L., *Principles of Colour Photography*, Wiley, New York.

Le Grand, Y., *Light, Colour, and Vision*, Chapman & Hall, London.

Graves, M., *Design Judgment Test*, Psychological Corp., New York.

Gregory, R. L., *Eye and Brain*, Weidenfeld & Nicolson, London.

Henderson, S. T. and Marsden, A. M. (eds), *Lamps and Lighting*, Edward Arnold, London.

Hunt, R. W. G., *The Reproduction of Colour*, Fountain Press, London.

Judd, D. B. and Wyszecki, G., *Colour in Business, Science and Industry*, Wiley, New York.

Jule, J. A. C., *Principles of Colour Reproduction*, Wiley, New York.

Langford, M. J., *Advanced Photography*, Focal Press, London/Boston.

Millerson, G., *Technique of Television Production*, Focal Press, London/Boston.

Millerson, G., *TV Scenic Design Handbook*, Focal Press, London/Boston.

Millerson, G., *TV Lighting Methods*, Focal Press, London/Boston.

Millerson, G., *Video Production Handbook*, Focal Press, London/Boston.

Murray, H. D., *Colour in Theory and Practice*, Chapman & Hall, London.

Nurnberg, W., *Lighting for Portraiture*, Focal Press, London/Boston.

Pieron, H., *The Sensations – their Functions, Processes, and Mechanisms*, Muller, London.

Ray, S., *Applied Photographic Optics*, Focal Press, London/Boston.

Ritsko, Alan J., *Lighting for Location Motion Pictures*, Van Nostrand Reinhold, New York.

Rushton, W. A. H., *Visual Problems in Colour*, NPL Symposium, HMSO, London.

Samuelson, D. W., *Motion Picture Camera & Lighting Equipment*, Focal Press, London/Boston.

SMPTE, *Principles of Colour Sensitometry*, Society of Motion Picture & Television Engineers, New York.

Spottiswoode, R., *Focal Encyclopedia of Film and Television Techniques*, Focal Press, London/Boston.

Stiles, W. S. and Wyszecki, G., *Colour Science*, Wiley, New York.

Vernon, M. D., *Visual Perception*, Cambridge University Press, Cambridge.

Woodworth, R. S. and Schlosberg, K., *Experimental Psychology*, Methuen, London.

Wright, W. D., *The Measurement of Colour*, Hilger & Watts, London.

Index

Major references are indicated by **bold** type